World Trends in Tourism and Recreation

American University Studies

Series XXV
Geography

Vol. 3

PETER LANG
New York • Bern • Frankfurt am Main • Paris

Zbigniew Mieczkowski

World Trends in Tourism and Recreation

PETER LANG
New York • Bern • Frankfurt am Main • Paris

Library of Congress Cataloging-in-Publication Data

Mieczkowski, Zbigniew
World trends in tourism and recreation / Zbigniew Mieczkowski.
p. cm. – (American university studies. Series XXV, Geography ; vol. 3)
Includes bibliographical references.
1. Tourist trade. 2. Recreation. I. Title. II. Series.
G155.A1M48 1990 338.4'791–dc20 89-48408
ISBN 0-8204-1197-3 CIP
ISSN 0899-6040

CIP-Titelaufnahme der Deutschen Bibliothek

Mieczkowski, Zbigniew:
World trends in tourism and recreation / Zbigniew Mieczkowski. – New York; Bern; Frankfurt am Main; Paris: Lang, 1990.
(American University Studies: Ser. 25, Geography; Vol. 3)
ISBN 0-8204-1197-3

NE: American University Studies / 25

Printed by Weihert-Druck GmbH, Darmstadt, West Germany

ACKNOWLEDGMENTS

The author is unable to list here all the persons who directly or indirectly contributed to this book. However, several names should be mentioned. Professor John Smallwood of University of Winnipeg read the whole manuscript and Mr. Mike Fay, Jim Johnston (both of Canadian National Park Service) and Gail MacDonald (Lawson Travel) part of it. The author is very thankful for their critical remarks, however, he is the only person responsible for mistakes and omissions. Thanks also are due to the staff in the Department of Geography, University of Manitoba for their excellent work: Trudy Baureiss, Hedy Chambers and Beth Hendricks for typing and Marjorie Halmarson for figures and tables.

CONTENTS

ACKNOWLEDGEMENTS v

CONTENTS vii

PREFACE xiii

Chapter 1 - INTRODUCTION 1
1.1 TOURISM IN THE CONTEMPORARY WORLD 1
1.2 TOURISM RESEARCH 2
1.3 RESEARCH AND CHANGE 5

Chapter 2 - TERMINOLOGY 7
2.1 INTRODUCTION 7
2.2 LEISURE 8
2.2.1 Philosophical concept of leisure 8
2.2.2 Practical contemporary concept of leisure 11
2.3 RECREATION 14
2.3.1 The concept 14
2.3.2 The relationship between recreation and leisure 16
2.3.3 Spatial approach to recreation 18
2.4 TOURISM 20
2.4.1 Introduction 20
2.4.2 Statistical definitions 20
2.4.3 Conceptual definitions 25
2.4.4 The problems of international standardizations of terminology 29
2.4.5 "Tourism" and "travel" 31
2.4.6 "Tourism" and "leisure" 34
2.4.7 "Tourism "and" recreation" 34
2.4.8 Types (categories) of tourism 37

Chapter 3 - HISTORY 43
3.1 INTRODUCTION 43
3.2 PREHISTORY AND ANTIQUITY 45
3.2.1 Recreation 45
3.2.2 Travel 46
3.3 MIDDLE AGES 48
3.3.1 Recreation 48
3.3.2 Travel 49
3.4 RENAISSANCE AND BAROQUE 51
3.4.1 Recreation 51
3.4.2 Travel 51
3.5 ENLIGHTENMENT AND ROMANTICISM 53
3.5.1 Recreation and Travel 53
3.5.2 The Grand Tour 57
3.6 DEVELOPMENT OF MODERN TOURISM AND RECREATION IN 19TH CENTURY EUROPE 58
3.7 HISTORY OF RECREATION AND TOURISM IN NORTH AMERICA UP TO 1900 63
3.8 TOURISM & RECREATION IN EUROPE AND NORTH AMERICA 1900 - 1950 68
3.9 CONCLUSIONS 70
3.9.1 Changing attitudes towards nature as tourism/recreation resource 70
3.9.2 The evolution of attitudes towards leisure. 71
3.9.3 The evolution of travel into modern tourism 72

Chapter 4 - PREREQUISITES FOR MODERN TOURISM 75
4.1 INTRODUCTION 75
4.2 DISCRETIONARY INCOMES 77
4.3 LEISURE TIME 81
4.3.1 Work and leisure 81
4.3.2 The increase of leisure time 83
4.3.3 Future prospects: more leisure time? 87
4.3.4 Leisure time slots 93
4.3.5 Weekend 94
4.3.6 Vacation 97

4.3.7 Daily leisure after work 99
4.4 TRANSPORTATION 99
4.4.1 Introduction 99
4.4.2 The Railroads 102
4.4.3 The Automobile 107
4.4.4 The Autobus 114
4.4.5 The Recreational Vehicle 115
4.4.6 Water Transportation 117
4.4.7 Air Transportation 123
4.4.7.1 Development 123
4.4.7.2 New aircraft technology 126
4.4.7.3 The supersonic transportation 127
4.4.7.4 Economic aspects: air fares 128
4.4.7.5 Problems: infrastructure and safety 133
4.4.7.6 Air passenger transportation network 135
4.4.7.7 Private aircraft 137
4.4.8 Summary: trends in tourist transportation 137
4.4.8.1 Modes of transportation 137
4.4.8.2 Spatial patterns of tourism: long range trends 139
4.4.8.3 The energy crisis scenario 141

Chapter 5 - DEMAND 145
5.1 THE SPATIAL SYSTEM OF TOURISM 145
5.2 DEMAND AND SUPPLY EQUATION 147
5.3 DEMAND AND NEED 149
5.4 DEMAND AND PARTICIPATION 150
5.5 DIMENSIONS OF DEMAND 151
5.6 STAGES LEADING TO EFFECTIVE DEMAND 153
5.7 MARKET SEGMENTATION 154
5.7.1 Introduction 154
5.7.2 Antecedant segmentation 156
5.7.2.1 Demographic segmentation (age, sex) 157
5.7.2.2 Socioeconomic segmentation: (income, education, occupation, racial, ethnic and religious variables) 164

5.7.2.3 Geographic segmentation 166
5.7.2.4 Lifestyle or psychographic segmentation 167
5.7.3 Motivational market segmentation 171
5.7.3.1 Motivations for tourist participation 171
5.7.3.2 Motivations and trends in societal values 175
5.7.3.3 Motivations: quest for variety 176
5.7.3.4 Motivational submarket research 177
5.7.4 Attitudinal segmentation 178
5.7.5 Intervening variables 180
5.7.6 Market segmentations in perspective 181
5.8 ELASTICITY OF DEMAND 182
5.8.1 Introduction 182
5.8.2 Income elasticity 184
5.8.3 Price elasticity 186
5.9 SEASONALITY OF DEMAND 189
5.9.1 Introduction 189
5.9.2 Measuring seasonality 191
5.9.3 Natural and social seasonality 192
5.9.4 Response strategies to seasonality 194
5.9.5 The result of response strategies 198
5.10 SOME TRENDS IN DEMAND 199

Chapter 6 - SUPPLY 203
6.1 INTRODUCTION 203
6.2 NATURAL RESOURCES 206
6.2.1 Introduction 206
6.2.2 Classification of natural resources 208
6.3 NATURAL RESOURCES: SYSTEMATIC APPROACH 211
6.3.1 Space 211
6.3.1.1 Space as a resource 211
6.3.1.2 Space requirements 213
6.3.1.3 Factors moderating space requirements 215
6.3.1.4 Response strategies to land use problems 217
6.3.2 Topography 222
6.3.3. Geology and soils 224
6.3.4 Water 225

6.3.5 Climate 229
6.3.5.1 Macroclimate 229
6.3.5.2 Microclimate 231
6.3.5.3 Climate and human health 232
6.3.6 Bioresources 232
6.3.6.1 Vegetation 233
6.3.6.2 Wildlife 235
6.4 NATURAL RESOURCES: ENVIRONMENTAL SYSTEMS 242
6.4.1 The seashore system 243
6.4.1.1 Beach 243
6.4.1.2 Shoreland 245
6.4.1.3 Hinterland 246
6.4.1.4 Morphological classifications of seashore system 246
6.4.1.5 The climate of the seashore 248
6.4.1.6 Planning and management problems of the coastal ecosystems 248
6.4.2 The mountain system 252
6.4.3 The inland rural system on flatland 256
6.5 WILDERNESS AND PARKS 258
6.5.1 Wilderness 258
6.5.2 Park systems in North America 263
6.5.3 National Parks 266
6.5.3.1 Terrestrial parks 266
6.5.3.2 Marine parks 270
6.5.3.3 The US National Park System 272
6.5.3.4 The Canadian National Park System 275
6.5.3.5 Parks planning and management in North America 279
(The necessity for planning and management; Planning for national parks; Problems of park management)
6.6 CONSERVATION OF ECOSYSTEMS 288
6.7 HUMAN RESOURCES 290
6.7.1 Classification 290
6.7.2 Characteristic features of human resources 292

6.7.3 Human attractions 292
6.7.4 Geographical distribution of human attractions 293
6.7.5 Tourism and the preservation of culture and traditions 297
6.7.6 Tourism industry 298
6.7.7 Accommodations 300
6.7.7.1 General trends 300
6.7.7.2 Hotel management problems 306
6.7.7.3 Development of supplementary accommodation 307
6.7.7.4 Private accommodation 307
6.7.7.5 Camping 310
6.7.8 Tourism settlements 314
6.7.8.1 Classification 314
6.7.8.2 Cities 314
6.7.8.3 Resorts 318
(Characteristic features; Health resorts; Seaside resorts; Mountain resorts; Climatic resorts; Retirement communities; Resort cycles)
6.7.9 Meeting and convention tourism 337
6.7.9.1 Trends for expansion 337
6.7.9.2 Terminology and classification 338
6.7.9.3 Size of the meetings 340
6.7.9.4 Convention facilities 340
6.7.9.5 Convention venues 341
6.7.9.6 Spatial distribution 343
6.7.9.7 Economic impact 344

Chapter 7 - CONCLUSIONS 347

REFERENCES 349

INDEX 361

PREFACE

Our society in its historical development has been concentrated on work as an indispensable activity creating and supplying material goods necessary for our existence. The preoccupation with the economics has permeated all aspects of our civilization starting with "work ethic" and ending with research endeavours. This fact is fully explicable by difficult living conditions for most people, connected with general shortage of material goods and inequalities in distribution of income. In such conditions one of the most important aims of society as a whole, except the thin stratum of elite, has been to produce more, cheaper and better. And above all work, work and work.

The industrial revolution which started in Europe about 200 years ago has created the technical and socio-economic prerequisites to abundance of material goods and more equal distribution of national incomes. This means that the specter of poverty has disappeared for vast majority of population in the Developed Countries (DCs), that with the basic needs met, the society started to devote relatively less attention to production and turn to more pleasant aspects of life, in fact to what appears the aim of life - the pursuit of happiness through leisure. Old attitudes change: sociologists note the gradual reversal of traditional valuation of work above leisure - in fact they are indicating that with all contemporary mechanization, automation and computeriazation many people find less and less pleasure in their jobs and work. Increasingly work becomes only a means to enjoy their leisure time which for them is tantamount to living time. Leisure has become part of the way of life of the modern society, it constitutes an indispensable component of life, and together with work - an important part of the rhythm of human existence. Indeed, with at least one third of the total time devoted to leisure in the DCs, the contemporary homo sapiens (human) has become not only homo faber (worker) but also homo ludens (player) and homo viator (traveller).

Leisure time is spent in various spatial dimensions: most of it is spent at home, watching TV, reading, listening to music etc. Secondly, substantial part of leisure is spent in the vicinity of home, mainly in urban and near-urban environment where there is no necessity to stay overnight outside home. However, an increasing portion of free time is spent farther away where an overnight stay in a hotel, motel, second home or other type of accommodation is required. This phenomenon is called tourism. Tourism essentially occurs during leisure time and is mostly but not exclusively associated with recreation activities. However, for various reasons which will be discussed later, also business and convention travel are regarded as parts of tourism.

The objective of this book is to bring into focus tourism and out-of-home recreation not in isolation but with all its ramifications and conditions of development, and also in its interaction with other aspects of human existence. The emphasis is on trends, especially long range trends, which are either observable and/or can be generalized from the vast research literature. This author travelled to exactly one hundred and one independent countries and while he was unable to do research in all of them his observations, experiences and research generalizations are included in this book. Of course, the author also used the world professional research literature in many languages. However, readers interested in the review of research in the field of tourism have to be referred to other sources. Such a review was beyond the objectives of this book. The objective was to generalize the research results and observations of the author and other researchers in order to identify the trends in world tourism. Generalizations which have become common and widely accepted knowledge by having been repeated many times in various sources are not followed by footnotes. The secondary focus of the book are trends in out-of home recreation, mainly, but not exclusively, in their overlap with tourism. The geographical focus of the book is on the DCs (Developed Countries). Tourism trends in the LDCs (Less Developed Countries) differ, at least in part, from those in the DCs. Therefore, any generalizations pertaining to the LDC are explicitly indicated in the text. But even here one should never forget that the main international and domestic tourism markets and destinations are, and will be still for a long time, in the DCs.

Besides trends there is another aspect of tourism which receives only tangential treatment in this book, solely as far as it overlaps with trends. These are the environmental, economic and socio-cultural impacts of tourism. The impacts, in the opinion of this author, deserve a separate attention although, as stated above, there is a significant overlap with trends.

The introduction to the book contains a general evaluation of the scope and importance of tourism and recreation in the modern world. It also discusses tourism as an interdisciplinary field. The chapter on terminology presents the conceptual framework for the phenomena of leisure, recreation and tourism. The chapters on history and prerequisites discuss the dynamics of socio-economic environment and trends in technical parameters of tourism and recreation in the past and at present. Finally, the last two chapters which constitute the core of the book (especially chapter 6 - supply) reflect the trends at both sides of the demand-supply equation.

The book is essentially designed as a reference source rather than for reading cover to cover. Therefore, it contains some unavoidable overlaps of contents.

CHAPTER 1

INTRODUCTION

1.1. TOURISM IN THE CONTEMPORARY WORLD

Very few people are aware of the fact that tourism is the largest business in the world with annual receipts for both domestic and international tourism exceeding US $ 2 trillion which represents about 12 percent of the world's economy (Waters 1988:4). Tourism is clearly in the lead well in front of the world military spending which is estimated at 900-1,000 billion yearly, even ahead of agriculture considering that a large part of it is subsistence agriculture in Less Developed Countries (LDCs) and thus does not enter the GNP. Indeed, the futurologist Herman Kahn ("The next 200 years") was wrong in 1976 predicting: "By the end of the century tourism will be one of the largest industries in the world." In fact, it is the largest right now. In international trade of goods and services international tourism is the second item after oil and there is a distinct possibility that it will surpass it in forseeable future if the world enjoys reasonable peace conditions with only limited local flareups which seem unavoidable nowadays. "Tourism-passport to peace" the logo of the World Tourism Organization expresses succinctly this association of peace with tourism. Actually "peace-passport to tourism" is equally true.

Returning to the economic aspects of tourism one has to emphasize its dynamism and resilience to crisis: as a rule tourism development is always ahead of the world's economic growth year after year and the economic crises affect it surprisingly little. Tourism proved to function as a good instrument of economic diversification for many countries, especially the LDCs. Many of them suffer from low commodity and raw material prices and tourism brings a much desired element of stability as compared to boom and bust cycles of the primary sector of economy which is relatively shrinking in the DCs as compared to the total economy in terms of employment and its contribution to GNP. The secondary sector in most DCs is decreasing relatively and in some of them even absolutely. Tourism, belongs to the dynamic tertiary sector (services) which nowadays is the most important and vigorous component of economy in virtually all DCs.

However, despite the undeniable advantages one should never forget that unrestrained and unregulated development of tourism results in negative economic, socio-cultural and environmental impacts. In order to maximize the benefits and minimize the disadvantages the society has to undertake a number of measures. An indispensable prerequisite of it is the realization of the importance of tourism in the "post-industrial" society. One has to understand that our society still underestimates the significance of this phenomenon in its failure to meet the pressing needs and demands, in its failure to optimally allocate the limited resources, to provide adequate conditions and facilities and finally to evaluate the economic, social and environmental impacts.

1.2. TOURISM RESEARCH

In the post-World War II years some progress have been achieved in the DCs in the appreciation of tourism's significance and necessary action has been taken to create appropriate mechanisms to cope with problems of tourism development. One important mechanism is tourism research - a discipline which unfortunately still does not enjoy a high status of academic respectability. It is often being associated with fun, play, pleasure reflecting rather sentimental and emotional attitudes, which, because of their subjectivity, allegedly render any analytical study impossible. Many scholars, therefore, regard this field as trivial and not worthwhile as serious study. For this reason our present knowledge of the underlying phenomena and implications of tourism has been limited although the situation in this respect is recently improving.

The need for tourism research is undeniable. The trends in supply and demand, market research, planning, problems and difficulties, the social, economic and environmental impacts of tourism - all necessitate a serious research effort, a respectable scientific inquiry. They require the development of theory and the translation of the theory into practice. The scholarly community, the government institutions and the public at large must become increasingly aware of the need of research. The problems that society faces are dictated by the scarcity of resources to be allocated in view of the desire to achieve increased social and economic efficiency and the improvement of the quality of our environment. The grave environmental problems, the sky-rocketing demand and need for leisure time opportunities also contribute to the urgency of strong, effective and, above all, concerted efforts in the field. Thus it is imperative that in coming decades the human activities conducted in the "free time" will constitute one of the most important subjects of the social sciences, not just for cognitive reasons but also to find practical solutions. The recognition of the fact that old laissez-faire policies are unattainable in the era of limited resources and interdependence of economic sectors necessitates the governments to develop policies aimed at problem-solving. The results of research conducted by scholars has proved to be of high practical importance in this respect providing the decision-makers with important feed-back. Thus research in leisure sciences constitutes an important basis for development of policy, planning, implementation and evaluation of human leisure time behaviour including tourism.

The scholarly field of tourism deals with an exceptional variety of materials, approaches and problems. This consequently necessitates cooperation of many social and natural sciences. In fact, it is difficult to find any discipline which would not have at least some relevance to tourism. Moreover, in conformity with modern trends there are no rigid limits creating isolated subfield within the discipline of tourism: interdisciplinary approach prevails where various social and natural sciences become interwoven, thus contributing to the field as a whole. However, despite this modern interdisciplinary approach to tourism, various disciplines focus on specific aspects, treat them from specific points of view thus contributing to the development of the field as a whole.

In fact, the interdisciplinary approach to tourism has a relatively short history. Until 1950s tourism was the almost exclusive domain of economics due to its contribution to economic development. Today although economics has lost its exclusivity it remains as the main discipline dealing with tourism (macro- and micro-economics and also applied economics like e.g. management). Economics has been mainly focussing on positive contributions of tourism to development, profits, employment etc. The contribution of geography, sociology and anthropology is more recent and has become significant since 1960s when the environmental, social and cultural impacts of tourism on the destinations and host societies became evident. Such topics like environmental impact, social stratification of demand and need, social behaviour and group interaction, social impact, cultural impact on folk cultures, cultural and ethnic tourism etc. have received great attention. Geography, sociology and anthropology in contrast to economics has revealed substantial negative externalities of tourism making the point that growth of tourism cannot continue unchecked and unregulated. Also some applied aspects of sociology have been explored like using tourism and recreation to fight criminality, drug abuse and alcoholism. Educationalists claim successes with low achievement students. The contribution of psychology to tourism consists mainly in market research (more concretely market segmentation, psychographics, motivational research etc.) which is indispensable for marketing. Less significant is the research in psychological carrying capacity. Important is the contribution of history which helps to establish historical trends in tourism development. Law contributes to tourism by investigating the freedom of movements on all spatial levels. On the international level this freedom has been promoted by the 1948 Universal Declaration of Human Rights. Regulations controlling these movements are also subject to legal interpretation. Law is needed to organize and supervise the tourism industry, to protect the rights and security of tourists. Political science overlaps with law in many aspects of tourism - the main emphasis is on political climate in which tourism takes place and political impacts of tourism.

The most important contribution of natural sciences to tourism is the research in its environmental impact. As examples one could mention the contribution made by botany with respect to natural carrying capacity or the contribution of climatology in researching tourists' climatic comfort. Medicine and physical education investigate the health impact of tourism and recreation on participants (balneology, hydrotherapy, medical climatology, medical treatment of many ailments, including psychological disorders, struggle against addiction (drugs, alchol, tobaccco). There is hardly any discipline which would have no implications for tourism: take e.g. such fields as forestry, architecture, landscape architecture, urban studies, religion. As a result tourism offers employment opportunities to specialists in almost all fields.

Geography deals with spatial and environmental systems i.e. with the interaction of phenomena in space and the interaction between society and environment (e.g. environmental impact of tourism). Geography occupies a special place in tourism studies because it overlaps natural and social sciences. Therefore, geographers having a unique background in both, are best qualified as integrationists and synthesizers in the scholarly field of tourism which also straddles natural and social sciences. Geographers, because of their training have the best potential to appreciate and coordinate various points of view. This generalizing-synoptic approach of geography represents a particularly high value for the interdisciplinary field of tourism.

1.3 RESEARCH AND CHANGE

This book is about changing patterns or trends in tourism. Indeed, the crucial task of research, including research in tourism is to investigate new trends which reflect the changing patterns of the societal development. Change has been always one of the most characteristic features of the world. Change is inevitable. In fact, it seems that change is the only constant permanent thing in life. We have to learn to accept change in order to accept future, in order to cope with the inherent inertia of human society to resist change. We have to accept change in order to lose the fear of unknown. Marcus Aurelius, the Roman Emperor and stoic philosopher said: "Observe always that everything is the result of change. Get used to thinking that there is nothing nature loves so well as to change existing forms." Of course, the pace of change is not always and everywhere equal. It is relatively slow for rocks, quicker for living things. For humans the pace was slow in prehistoric time and is dazzlingly rapid in modern society entering the post-industrial age. Indeed, the speed of change at present is so fast that the unprepared suffer from "future shock." Alvin Toffler explains that "future shock" is the dizzying disorientation brought on by the premature arrival of the future. It may be well the most important disease of tomorrow.

As in other aspects of human life change has an significant impact also on tourism and recreation, the most important being that leisure time utilization which moved as a result of change in societal values from marginal importance to a central phenomenon in human life. Thus spending on leisure time, costs of leisure time have lost their luxury imprint and have become an essential part of existence at least in the DCs. Participation in recreational activities, enjoyment of vacations are not a privilege anymore but a right for everyone in the DCs. Changes occur at quick pace and identifying trends is not only a matter of intellectual curiosity or remedy against "future shock", fear of unknown, but also of practical expediency because changes carry important pragmatic implications especially for planners, government officials, private entrepreneurs active in the field. All these leisure time professionals have to adapt to change using imagination, creativity and innovation. Therefore, a substantial part of research endeavours is devoted to identifying new trends, both transitory (short range) or secular (long range), learning as much as possible about them and developing a strategy to cope with them. Indeed, facing the future means facing the challenge of change. The rapidity of change in modern time makes the prediction of future trends more difficult but not impossible. Therefore, the important task of research in tourism and recreation is to recognize the new and the receding models in the field. The task is not easy because of the specific nature of developments: relatively small changes which only later acquire snowballing properties, gradual quantitative changes leading to significant qualitative shifts. Take e.g. the prerequisites for modern tourism and recreation: more money, more leisure time, improved transportation. These changes in the material world are having an impact on psychology, attitudes, values, tastes, motivations which translate in transformation of people's behavioural patterns, in other words quantitative and qualitative transformation of demand for tourism and recreational services.

The identification of trends helps the society to cope with an ever-changing situation, to develop new strategies and tactics in order to minimize the disadvantages and maximize the benefits. This could be compared with a hunter who aims at a target where is going to be, not where it is. This is particularly true with respect to tourism which belongs to the most dynamic but at the same time most complicated phenomena of the modern world.

CHAPTER 2

TERMINOLOGY OF TOURISM AND RECREATION

2.1. INTRODUCTION

The new social science of tourism despite undeniable achievements has encountered problems resulting from very rapid development. These problems are connected to a certain extent with the fact that the task of standardizing terminology has been unfortunately neglected. There is a fair amount of confusion with respect to basic notions and definitions. We are plagued by ambiguous definitions or terms introduced without an effort to define them. In the literature many terms such as leisure, recreation, outdoor recreation, tourism, tourist, visitor, recreationalist, vacationist etc. are widely used in various meanings, leading as a consequence to numerous misunderstandings.

The attitude of people professionally dealing with tourism and recreation towards discussion of terminology is to a certain extent negative: some regard such discussion as dull and uninteresting. They suggest to abandon this hairsplitting and dogmatic activity. Others express the opinion that we should not be overly concerned about determining the exact meaning of terms, they plead for flexibility with respect to terminology rejecting the discussion about "merely semantic problems." Some even flatly refuse to deal with terminology arguing that this is an impossible task never to be accomplished: therefore, it is better to redefine all terms used in a particular study at the outset of each publication.

The arguments against these negative positions are threefold: First of all, any scholarly field has to develop strict and generally accepted criteria and notions to establish proper communication between all interested. Development of terminology establishes a sort of "lingua franca," a language understood as widely as possible to avoid misunderstandings and communication breakdowns. Secondly, establishing acceptable standards valid for extended periods of time is absolutely essential for longitudinal research. Constant and cohesive standards and measures make the statistics and research results comparable both chronologically and geographically. In this way the continuity in research is guaranteed. Thirdly, a clear terminology is necessary for legal-administrative reasons (e.g. distinguishing between various categories of tourists).

Of course, it is understandable that tourism and recreation are dynamic phenomena changing in time and space. Therefore, periodical adjustments of terminology to changing conditions seems necessary.

2.2. LEISURE

2.2.1. Philosophical Concept of Leisure

Leisure from Latin "Licere" - to be allowed (permitted) came into English through the French word "loisir." The first attempts to define leisure were undertaken in Antiquity and reflected the Greco-Roman philosophical ideas on work and leisure. According to these views work was an arduous necessity, means to the end. The real goal of life for the privileged members of Greek society was leisure. Aristotle writes in "Politica:" "We labor to have leisure" (Sellin T. ed. 1957:42) and "Leisure is preferable to work, it is the aim of all work." (Miller M. Robinson D. ed. 1957:42). Developing his ideas Aristotle went a step further. He discussed the quality of leisure. "The capacity to use leisure rightly is the basis of man's whole life. Nature requires us not only to be able to work well but also to idle well" (Ibid.). In "Nicomachean Ethics" he says: "The free time is not the end of work; but rather work ends the free time which should be devoted to arts, science and above all to philosophy" (Ungarisches Fremdenverkehrsamt: 1967: 258). He extolled leisure also in this way: "and the soul's well-being seems to be found in leisure" (Sellin, T. 1957:105).

The relative importance of leisure as opposed to work found its reflection in the Greek and Latin languages. The word for work in both languages denotes the absence of leisure by placing a denying prefix in front of the word leisure what can be translated as "unleisure" or "non-leisure" into English. The Greek work for leisure is (skhole), the word for work is (askholia). In the Latin language we have a similar situation: "otium" means leisure, "negotium" - work.

The Greek word for leisure "skhole" - hence the English word "school" - meant serious activity without pressure of necessity. Aristotle understood the word "skhole" in its subjective motivational sense as an attitude, state of mind or quality of feeling not just as free time. Leisure in this sense depends upon personality, traditions and education of an individual. What for one is work is leisure for another one. What is today work may be leisure for the same person at another time. Aristotle's views are shared by Sebastian de Grazia "Work is the antonym of free time. But not of leisure. Leisure and free time live in two different worlds. We have got in the habit of thinking them the same." (De Grazia, S., 1962:8).

De Grazia is opposed to the contemporary idea of free time from work as a synonym of leisure (this opposition he shared also with Aldous Huxley). His attitude may be illustrated by following quotations: "The word leisure has turned into the phrase free time, and the two are now almost interchangeable. It was through the efforts of the philosophers that leisure found its identity. Today the benefit of their thinking is largely lost to us." (De Grazia, S., 1962:87). "The word leisure has always referred to something personal, a state of mind or a quality of feeling. It seemed that in changing from the term leisure to the term free time we had gone from a qualitative to a quantitative concept." (De Grazia, S., 1962:65) (Quantitative is understood here in terms of minutes, hours etc.). De Grazia points out that "... contemporary phrase "leisure-time" is a contradiction in terms. Leisure has no adjectival relation to time. Leisure is a state of being free of everyday necessity, and the activities of leisure are those one would engage in for their own sake." (De Grazia, S., 1962:327). He says "anybody can have free time" but "not everybody can have leisure.": (De Grazia, S.,1962: 8). "Free time refers to a special way of calculating a special kind of time. Leisure refers to a state of being, a condition of man, which few desire and fewer achieve." (De Grazia, S., 1962:8).

Miller and Robinson support the philosophical understanding of the word "leisure" as a purely qualitative concept. They introduce the notions of "free time" (including discretionary time) and "leisure time" as "part of free time devoted to leisure pursuits." (Miller,N., Robinson,D.: 1963:144). Here are their definitions: "Free Time: time available to the individual after necessary work and other survival activities and duties are accomplished, to be spent at the discretion of the individual." Leisure time is "part of free time devoted to activities undertaken in pursuit of leisure, which may, through recreative processes and playful activities, or may not, be attained." (Miller, Robinson 1963: 5). In other words we could define free time as "empty time," leisure as time filled with specific kind of activities.

To sum up the discussion on leisure in its philosophical meaning let us quote two synthetic definitions trying to bring the various aspects of the notion together. In "Leisure in America: blessing or curse?", we find following characteristics of leisure:

(1) time which the individual controls without external

compulsion;

(2) pleasure, happiness, enjoyment, play;

(3) noncompensated activity;

(4) spontaneous, satisfying experience.

(Davy, T.J., and Rowe, L.A., 1964: 72). Max Kaplan puts it this way: "In my thinking, the key elements of leisure are:

(1) an antithesis to work as an economic function;

(2) a pleasant expectation and recollection;

(3) a minimum of involuntary social role obligations;

(4) a psychological perception of freedom;

(5) a close relation to values of the culture and

(6) often, but not necessarily, an activity characterized by the elements of play."

(Kaplan, M., 1963, VI:79).

The philosophical concept of leisure focuses our attention on its importance in the life of humanity, its role as an agent of progress, as the tool diffusing and developing civilizations on our planet. Arnold Toynbee called the creative use of leisure "the mainspring of civilization." The high value attributed to leisure leads many authors to sharp criticism of "misuse of leisure" as a symptom of moral disease (Collingwood, R.G., 1938:96). Indeed, according to these views the misuse of leisure may lead to fall of civilizations eg. Collingwood believes that Greco-Roman civilization died of such a moral disease which also characterize the contemporary Western society. Even such speculative theories have some merit helping people to realize the importance of leisure.

Despite the undeniable merits of the philosophical concept of "leisure" for cognitive reasons, it has drawbacks from the point of view of an operational definition of leisure. "Leisure" as philosophers and some other scholars (sociologists) understand it, is difficult to define because of its subjective, attitudinal nature. The problem starts already by distinguishing between work and leisure which may overlap. Thus, if we accept the philosophical notion of leisure the distinction between work and leisure may be blurred. What for some is work, for others may be leisure.

Let us assume, however, that "leisure" is "non-work" and that it is a part of "free time" (from work) spent in a specific way as the philosophers and some sociologist see it (Fig.l). Even then the relationship between the two notions remains fuzzy. There is a broad transition zone between them and not more or less clearly defined line. Such an approach is difficult to accept for practical, operational reasons, especially while dealing with numerical indicators.

Figure 2.1

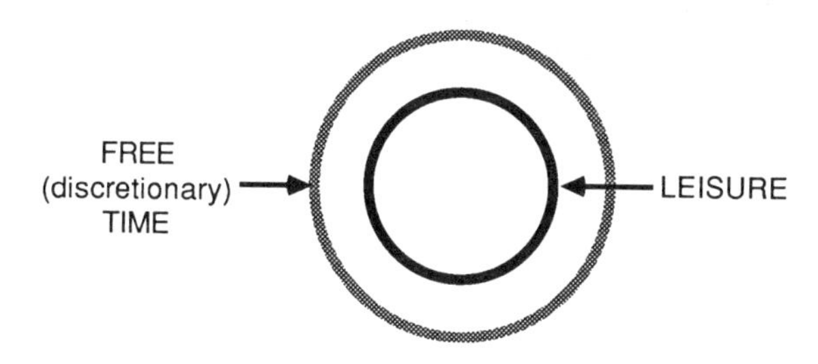

2.2.2 Practical Contemporary Concept of Leisure

Alongside with the philosophical concept of leisure, there is another wider and more pragmatic understanding of the term, identifying it with free (uncommitted, discretionary) time as contrasted to work, work related and subsistance time called by some authors existence time. (Clawson, M., 1964:1).

The authors agree that in many concrete cases the distinction between work and leisure is not clear although to a much lesser extent than in the case of philosophical notion of leisure. M. Clawson (Clawson, M., 1964:1) mentions "some fuzziness around the edges." Indeed, it is the characteristic feature of modern times that some elements of leisure are creeping into work and vice versa. Another overlap occurs between leisure and activities undertaken for the purpose of subsistence e.g. eating which in some cases may be considered as leisure.

There have been attempts to define more exactly on concrete examples of "Time-budgets" what activities or non-activities belong to leisure. The criterion was the degree to which time has been committed as in the "time-budget" classification (Table 2.1):

Another attempt to categorize leisure is on the basis of time slots involved. Generally these time slots are defined in terms of the leisure time that a person has available. Thus the slots can be set out as 1) time available after work, 2) a single day, 3) a weekend usually consisting of two days, 4) a long weekend usually consisting of three days and 5) vacation, usually consisting of four or more days 6) retirement.

The practical notion of leisure, meaning simply free time from work and subsistence, leads to the realization of its importance in human life: leisure constitutes at present well over one third of a working person's time in the DCs. Computed on the weekly basis leisure amounts to 56 hours out of 168 hours. (Chubb, M., Chubb, H., 1981:96). To this vacation must be added which amounts about 1 month in most DCs.

Table 2.1 Time budget

	Fully committed	Partly commited OPTIONAL	
	ESSENTIAL	HIGHLY COMMITTED	LEISURE
Sleeping	Essential sleep		Relaxing
Personal care and exercise	Health and hygiene		Sport and active play
Eating	Eating		Dining and drinking out
Shopping	Essential shopping	Optional shopping	
Work	Primary work	Overtime and secondary work	
Housework	Essential housework and cooking	House repairs and car maintenance	Do-it-yourself, gardening
Education	Schooling	Further education and homework	
Culture and Communication (non-travel)			TV, radio, reading theatre, hobbies and passive play
Social activities		Child-raising, religion and politics	Talking, parties, etc.
Travel	Travel to work/ school		Walking, driving for pleasure

Source: Cosgrove, Jackson 1972:14.

This example illustrates the difficulties in categorizing various activities.

This "fuzziness around the edges" will be probably always with us and should be accepted as an unavoidable.

2.3. RECREATION

2.3.1. The Concept

The term comes from the Latin "recreare", to restore or refresh. It means "refreshment of the strength and spirits after work" (Webster 1976). Recreation in this meaning has the purpose of "re-creating" human physical and mental resources depleted by working activity. Recreation "re-creates" physically, psychologically, spiritually, mentally. It is an experience (an activity or inactivity), aimed at renewal of body, mind and spirit in order to prepare the individual for better performance of work. This teleological character (from Greek: telos - purpose) of recreation was observed by Kaplan: "In its first sense recreation has the purpose of re-creating or revitalizing us so that we may more efficiently go back to activities that are not recreational but fundamentally of a work nature" (Kaplan, M., 1960:20).

The teleological aspect of recreation has prevailed up to the present time as its most important characteristic feature. In fact, in the Soviet block countries the teleological understanding of recreation has completely monopolized the scene. "Restoration of the working capacity" is the main function of recreation in the USSR. (Azar, V.I., 1975:86). "The management of recreational activities is subordinated to the task of large scale reproduction of the labor force and of improvement of production relations" (Azar, V.I., 1975: 87). Recreation in the capitalistic system has another purpose according to Marxist doctrine: it is a tool of increased exploitation of working masses because, although the leisure time has increased, the "re-created" workers face growing intensification of work and this makes them more tired than before (e.g. Azar, V.I., 1975:87).

In the West the teleological approach to recreation has weakened considerably during the last decades. Recreation is being treated as increasingly independent from its preparatory or revitalizing function for work. It contains values "for itself," simply to shape human life in the more pleasurable, stimulating and thus rewarding way. Recreation is nowadays defined as "a non-work activity engaged in for pleasure" (Modern Dictionary of Sociology 1969: 337) enjoyed "during the leisure time" (Dictionary of Sociology 1944: 251). Examples of such non-teleological approach abound in the literature. Chubb regards recreation as "any type of conscious enjoyment (Chubb, M., Chubb, H., 1981:6). Another example of non-teleological definition of recreation:

"All those activities that individuals or groups of individuals choose to do during leisure time, with the object of making life more satisfying and more enjoyable."

(Ontario, Dept. of Education, 1970:65)

The last definition contains also another attribute of recreation: its voluntary character. Recreation is a "voluntary activity indulged in without external compulsion, which results in the revitalization (or recreation) of body and mind (Farina, J., 1961:944). Clawson and Knetsch put it this way: "The distinguishing characteristic of recreation is not the activity itself but the attitude with which it is undertaken. "When there is little or nor feeling of compulsion or "ought to," an activity (or inactivity) is almost surely recreation." (Clawson,M., Knetsch, J.L., 1966:6). Thus recreation is associated with free will of individuals or groups of individuals. The society at large (including recreation planners and other professionals) should avoid moralizing, criticizing, and lamenting about some forms of recreation it does not approve. The public attitudes vary in time and place and one should be rather tolerant unless, of course, some forms of recreation become threats to public welfare and order (e.g. "recreational use of drugs").

These new approaches to recreation do not mean that the traditional teleological interpretation of recreation lost its validity. Indeed the majority of post-World War II authors indicate that recreation is purposeful, constructive and positive. Here are examples of such value-judgmental definitions of recreation: "purposeful activity" (Burton, T.L., 1970:49) "pleasurable and constructive use of leisure time" (Brockman,C.F., 1959:1), "positive use of leisure time" (Patmore, J.A., 1973: 225).

The final remark pertains to the lack of absolute certainty what belongs to recreation. Similarly like in the case of leisure and work there is a continuum between recreation and work observed by Burton (Burton T.L., 1970:216). In addition, what may be recreation for one individual may be work for another one, e.g. in professional sport recreation becomes a job from which vacation has to be taken. Also subjective elements play an important role when distinguishing between work and recreation. The overlapping relationship between work and recreation may be illustrated as follows:

Figure 2.2

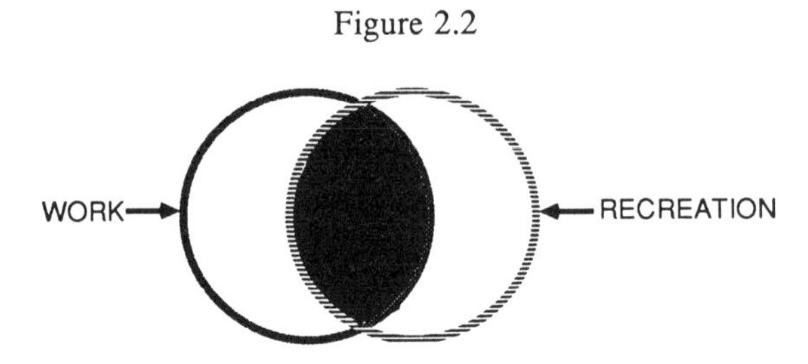

2.3.2. The Relationship Between Recreation and Leisure

Colloquially the terms "recreation" and "leisure" are used synonymously and are almost interchangeable. However, sensu stricto, recreation is content (or more exactly part of the content of leisure (or leisure time). Recreation is "any activity pursued during leisure..." (Dictionary of Sociology, 1944: 251). "If leisure is time available for choosing, recreation is one major activity chosen for such available time," - writes Marion Clawson (Clawson, M., 1964:3). In leisure the emphasis is on time element whereas recreation refers to the content, to the way the leisure time is spent. It is the behavioral pattern which fills this time. This behavioral pattern may be best described as "recreational experience". Most authors, however, express the opinion that recreation consists of "activities" (e.g. "Leisure is time, recreation is activity", Clawson, M., Knetsch, J.L., 1966:27). They try to clarify that this word should not be understood literally and actually includes also "inactivities." Clawson - Knetsch write that "recreation is the activity or activities (including inactivity if freely chosen) engaged in during leisure time" (loc. cit.). Clawson points out that the recreational "activities may range from the most active to the most passive and may take varied forms. The essential element is not what the person does, but the spirit in which he does it and the satisfactions he gets from it. The nature of recreation activities changes with age, family stage, income, health, and many other factors which vary from person to person." (Clawson, M., 1960:7).

After determining that recreation is the content of leisure one has to find out if this content (recreation) fills entirely or only partially the form (leisure). In its broadest sense recreation fills practically completely the leisure time. Neumeyer writes: "broadly speaking recreation includes any activity pursued during leisure, either individual or collective, that is free and pleasureful, having its own immediate appeal, not impelled by a delayed reward beyond itself or by any immediate necessity. Recreation includes play, games, sports, athletics, relaxation, pastimes, certain amusements, art forms, hobbies, and avocations. A recreational activity may be engaged in during any age period of the individual, the particular action being determined by the time element, the condition and attitude of the person, and the environmental situation." (Neumeyer, M.H., 1958:17). This view is rightly opposed by some writers: "Recreation takes place during leisure, but not all leisure is given to recreation." (Clawson, M., Knetsch, J.L., 1966:12). Miller writes: "Among the vast number of activities that occupy this free time are many that are recreational activities" (Miller, N.P., Robinson, D.M., 1963:145). Thus, there are some activities engaged during leisure time which are not associated with recreation e.g. worshipping, studying or visiting the mother-in-law. To this category belong also idleness, overindulgence, crime and other antisocial activities. "Recreational use of drugs" is an obvious misnomer. Thus recreation should be not only pleasurable but at the same time constructive and purposeful (in teleological sense). Here is a good example of such approach:

> "Recreation is conceived of in the generic sense of the word, the "re-making" or "re-creating" of an individual through the use of leisure time in such a fashion as to restore or rebuild what has been depleted or exhausted in his makeup and to add to his knowledge and abilities with the purpose of a fuller, more satisfying life. Mere diversion does not measure up to recreation in this sense." (Little, A.D., 1967:9)

Such a concept of recreation leads to the conclusion that recreation is only part of leisure time. This could be illustrated as follows:

Figure 2.3

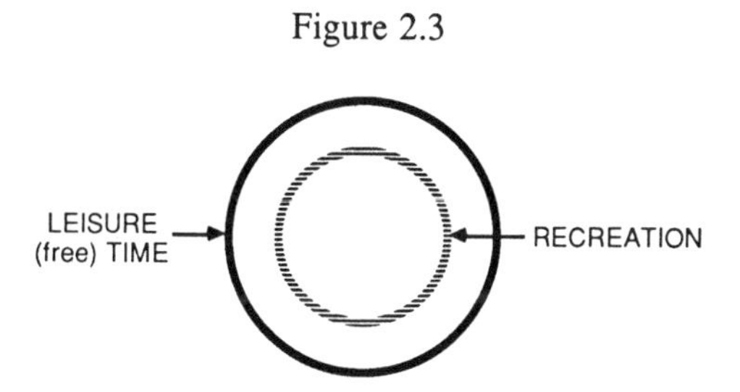

However, if we apply the term "leisure" in its narrower philosophical meaning - then recreation will fill totally the leisure time. This may be illustrated by a quotation from Clawson - Knetsch: "Briefly, if leisure is taken to mean time in which activities (or inactivity) consciously decided upon are undertaken, then the relation of recreation and leisure is very close. On this basis, mere idleness is neither leisure nor recreation." (Clawson, M., Knetsch, J.L. 1966:7).

2.3.3. Spatial Approach to Recreation

The subdivision of recreation according to its location has a great practical importance. In North America a division into indoor and outdoor recreation has been established. This distinction is used in two meanings which causes some confusion:

1. Indoor recreation in a building (mainly home-based) or under a kind of roof cover and outdoor recreation under a free sky. In this sense there is no indication as to the place where the activity takes place in space because some indoor activities may take place in rural areas and outdoor activities in urban setting. Also, this division does not give enough indication as to the nature of activities: there are many examples of the same activities occurring both "indoors" and "outdoors" (e.g. swimming, hockey, tennis, track and field, basketball). The importance of this distinction lies in costs of providing the recreation opportunities: public indoor recreation facilities require not only buildings but also heating and air-conditioning systems. Therefore, their costs are higher than that of outdoor facilities.
2. Outdoor recreation in North America is associated with activities taking place in extra-urban (rural, non-urban, out-of-city) i.e. in Clawson's "intermediate" or "resource-based" areas. The Outdoor Recreation Resources Review Çommission, defines outdoor recreation as "leisure time activity undertaken in a relatively non-urban environment characterized by a natural setting." (Outdoor Recreation Resources Review Commission, 1962:1). This form of "resource-based" recreation depends on natural landscapes and takes mostly extensive forms. It is often called "dispersed recreation" as opposed to concentrated recreation relying on more intensive "user-oriented' forms which normally occur in urban, sub-urban and near-urban setting. Outdoor recreation in the North American understanding is called countryside recreation in Great Britain. The term rural recreation refers mainly to agricultural areas.

The term "outdoor recreation" in its North American understanding (non-urban outdoors) reflects the attitudes of governments and the public prevailing in 1950s and 1960s when the emphasis was on preserving and developing "the great outdoors" to the detriment of indoor and outdoor urban, suburban and near-urban recreation catering to the interests of poor and under-privileged. Only after urban unrest of 1960s in the US, the authorities started to develop both indoor and outdoor recreation opportunities in the cities and near-urban areas. In this connection it became evident that the term "outdoor recreation" should not be monopolized for extra-urban recreation. Although many authors criticized the use of the term (e.g. National Academy of Sciences, 1969:21-22; Chubb, M., 1981:7) and many states dropped the word "outdoor" from the titles of their recreation plans, the term is so deeply ingrained in our vocabulary that a change seems to be difficult.

It seems that much better solution of this terminological problem would be the acceptance of such terms like urban, suburban, near-urban and non-urban (extra-urban) recreation which focus on the location where the recreation activity is taking place. Such terminology suits the purposes of planning and management and pleases especially geographers. Of course, urban geographers deal with urban recreation as an integral component of urban planning. However, most recreational geographers are interested in non-urban, extra-urban or dispersed recreation as requiring space and resources (land areas, water bodies, beaches, forests, etc.) These two aspects: spatial, territorial element and the interaction between resources and men constitute the basis for geographers' claim to participate in the research of the "outdoor recreation." It is certainly a very pragmatic and practical field taking into account that both space and resources are in limited supply. Hence planning based on careful research is imperative.

The categorization of recreation on the basis of location reveals only partially the nature of the activities that are involved. It is the activities themselves and the requirements they have for space and equipment that are indispensable facts for detailed planning. In addition the reason or reasons why a particular individual or group of people decide to participate in an activity at a given time and place must be understood before adequate planning can be undertaken.

There have been numerous attempts to group the recreation activities into such categories like water related or winter recreation. However, there is no widely accepted system and perhaps understandably so, considering the variety of recreational activities..

2.4. TOURISM

2.4.1 Introduction

The linguistic roots of the word tourism derive from Latin words: "tornare" - to turn to round off and "tornus" - wheel, hence circular movement pertaining to change of residence. The modern European languages received the term from the French word "tour" denoting circular tower and circular travel with return to the point of departure which became the basis for the French term "tourisme", Italian "tourismo", German "Tourismus" (substituting for "Fremdenverkehr") English "tourism", Russian "turizm". The word "tourist" was used for the first time in 1800 by Samuel Pegge (1733-1800) in his "Anecdotes of the English language" in the widest possible content: "A traveller is now-a-days called Tour-ist" (Ginier, J., 1969:25). In French the word "tourist" was first applied by Stendhal in 1838 in his book "Memoires d'un touriste." The first definition of tourists appears in the Dictionnare universel du XIXe siecle in 1876. According to it tourists are "persons who travel out of curiosity and idleness ... for pleasure of travel ... (for being able to tell that they travelled" (Sigaux, G., 1965:6).

In this way the term "tourist" and later "tourism" gained a wide approval and use. Although linguistically awkward as connected with the narrow meaning of the word "to tour" it has been widely accepted because of its shortness, usefulness, grammatical flexibility (easy to form the noun and adjective) and similarity in all major languages.

2.4.2. Statistical Definitions

The main practical need for a possibly exact definition of tourism and tourist has arisen from the necessity to establish adequate statistical standards. There have been also legal-administrative considerations in this respect. In fact even before this term has been accepted and used, in 1815 the United Kingdom statistics made a distinction between emigrants and ordinary travellers according to the quality of accommodation on ships. Emigrants were considered to be "all passengers crossing the ocean with the exception of cabin passengers, among whom it was assumed that a certain number were only going temporarily to oversea countries, either for pleasure or business ..." (Ogilvie, F.W., 1933:13).

Thus, the main distinctive quality of a tourist - the temporary character of his change of residence - has been recognized. This characteristics feature of a tourist did not lose its validity up to the present day: considered as tourist for statistical reasons are "travellers whose future and last residence are the same" in contrast to emigrants "whose country of intended future residence differs from that of last residence." (Ogilvie, F.W., 1933:14). Ogilvie suggested one year stay in the foreign country as a dividing criterion between these two categories of travellers.

However, with the development of tourism in the period between the two World Wars the need for more precise statistical definition became urgent. The Committee of Statistical Experts of the League of Nations recommended in October 1936 a definition of "foreign tourist." This definition was adopted by the Tourism Committee of the League of Nations in 1937. According to this definition "foreign tourist" is "any person visiting a country, other than that in which he usually resides for a period of at least 24 hours." This definition was amended by the Economic Committee of the League of Nations in the following form setting the minimum and maximum duration for tourist stay: "Tourist is a person staying in a locality situated outside his place of residence during minimum of 24 hours and maximum of one year." (O.E.C.D. 1963:3). The persons changing their residence for more than one year were according to this definition classified as immigrants.

The 1936-37 definition has been reformulated by the International Union of Official Travel Organizations (present World Tourism Organization) and adopted by the United Nations Conference on International Travel and Tourism held in Rome in 1963. This is at present the official internationally accepted (UN/WTO) definition of international tourism which may be adapted to tourism "at large." It is used by UN Statistical Commission in all its publications.

"For statistical purposes, the term "visitor" describes any person visiting a country other than that in which he has his usual place of residence, for any reason other than following an occupation remunerated from within the country visited.

This definition covers:

1) tourists, i.e. temporary visitors staying at least 24 hours in the country visited and the purpose of whose journey can be classified under one of the following headings:

a) leisure (recreation, holiday, health, study,
religion and sport);

b) business, family, mission, meeting;

2) excursionists, i.e. temporary visitors staying less than 24 hours in the country visited (including travellers on cruises).

Excluded from that definition are:

a) residents (nationals and aliens) returning after a stay abroad not exceeding one year;

b) permanent immigrants (i.e., non-residents intending to remain for a period exceeding one year);

c) temporary immigrants (i.e., non-residents intending to exercise for a period of one year or else an occupation remunerated from within the country); and

d) foreign diplomatic and military personnel stationed in the country.

The statistics should not include travellers who, in the legal sense do not enter the country (air travellers who do not leave an airport's transit area, and similar cases)." (The United Nations Conference on International Travel and Tourism, August - Sept. 1963:14).

In connection with this definition and its subsequent variants following remarks are necessary:

1) Visitors are classified not in terms of their nationality but rather according to their place of residence. Thus nationals of a country residing abroad are regarded as tourists if they visit the country of their nationality. Of course, they should be measured separately.

2) In the inclusive part of the definition visitors are divided into two categories: tourists and excursionists. Tourists stay at least for 24 hours (stay-over visitors). Excursionists (day visitors or day trippers) stay less than 24 hours. In practice this means that they do not stay overnight. A modification of the UN definition was adopted in 1967 when the United Nations Statistical Commission stated that "excursionists" or "day visitors" are persons who do not stay overnight (OECD: 1974:8). Thus the criterion of "overnight stay" has been accepted instead of "24 hours" for classification of visitors into the tourist or excursionist categories. Such an approach seems to be more acceptable considering the economic impact of the overnight stay and the fact that a trip with the overnight stay may last less than 24

hours. This modification has been accepted in 1968 by IUOTO (International Union of Official Travel Organization), today's World Tourism Organization. However, most agencies are still using the criterion of 24 hours (e.g. the UN statistical yearbook) and this seems to be the official standard at present.
3) Tourists can be divided into three categories
a) "pure tourists" who travel in their leisure time, mainly for pleasure
b) business tourists (including congress or convention tourists)
c) other tourists include: students, pilgrims, missionaries (if renumerated from outside the country of residence) persons travelling for health or domestic reasons not connected with pleasure (e.g. travel to funerals).
The inclusion of business and other non-pleasure travel into tourism has constituted the bone of contention for many years even after the 1963 definition was accepted. The 1980 Manila Declaration of the World Tourism Organization includes a reminder in this respect. This declaration "places tourism into the broader context of movements of persons, and any restrictive interpretation of tourism as signifying holidays and recreation alone can only succeed in confusing the issue." (World Travel 1985, 185:63).
4) Following persons figure in the exclusive part of the definition: returning residents, immigrants (staying over one year), migrants (temporary workers staying less than one year), commuters, diplomats and soldiers, transients (including transit passengers). There are some problems in international statistics with the exclusive part of the definition. Some countries have operational difficulties with excluding certain categories of travellers like diplomats and soldiers. However, inclusion of these categories does not distort the statistics significantly. Significant distortions occur if transients are included in statistics (e.g. Bulgaria, Poland) thus inflating the number of visitors.
5) The maximum stay for tourism of one year, as in the League of Nations definition, has been confirmed and is valid today although there were attempts to shorten it to 6 months (UN Interregional Seminar on Tourism Development, 1968:113). Thus a tourist stays between 24 hours and 1 year at destination.
6) There is no minimum distance criterion in this definition.
7) The economic impact of tourism is measured by tourism receipts from visitors (both tourists and excursionists). In practice also some income from the excluded categories like e.g. transients is included.
8) The 1963 definition of tourism refers to international tourism although it has been obvious that it is applicable mutatis mutandis also to domestic tourism. The

1980 Manila Declaration of the World Tourism Organization extends the definition implicitly to all tourism both international and national (domestic).

A comprehensive classification of travellers has been developed by Chadwick (1987:50) indicated in figure 2.4. In connection with this classification one has to indicate the following:

1) This classification reflects the approaches of Statistics Canada which has been known for some time for its keen and thorough interests in problems of terminology.

2) Chadwick's chart visually demonstrates the whole terminology system related to tourism and essentially does not differ from the UN/WTO approach. There is one relatively minor difference: crews are excluded in the chart and regarded as tourists in the UN/WTO definition. While in essence crews are not tourists, however, their economic impacts hardly differs from tourists. Students travelling between home and school are simply commuters and not tourists but students staying and attending school outside their permanent residence are from the economic point of view within the scope of tourism. (Fig., 2.4).

3) Chadwick is still reluctant to call the phenomenon "tourism." Instead he prefers: "travel and tourism." This issue will be discussed in 2.4.5.

Figure 2.4

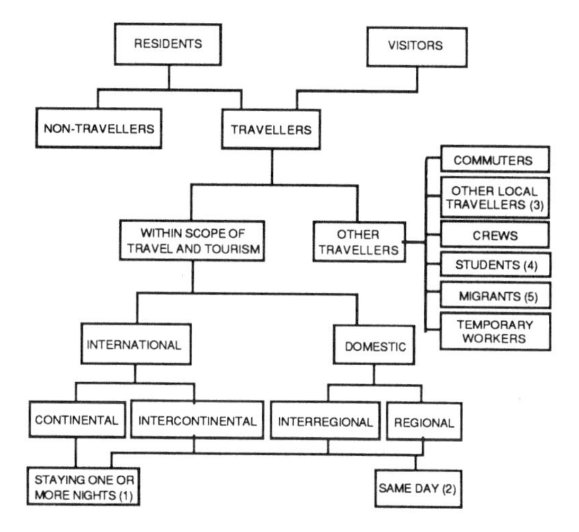

(1) "Tourists" in international technical definitions.
(2) "Excursionists" in international technical definitions.
(3) Travellers whose trips are shorter than those which qualify for travel and tourism, e.g. under 50 miles (80 km) from home.
(4) Students travelling between home and school only -- other travel of students is within scope of travel and tourism.
(5) All persons moving to a new place of residence including all one way travellers such as emigrants, immigrants, refugees, domestic migrants and nomads.

2.4.3. Conceptual Definitions

Alongside with statistical (technical) definitions there are a number of conceptual definitions of tourism. The most important (and controversial) was the 1942 definition (adjusted later) given by two Swiss scholars Hunziger and Krapf: "Tourism is a sum of relations and phenomena resulting from the travel and stay of non-residents, in so far as (travel) does not lead to permanent residence and is not connected with any permanent or temporary earning activity" (Hunziker, W., Krapf,, K., 1942: 21, adjusted in : Hunziker, W., 1959:11).

This definition was adopted by the AIEST (Association Internationale d'Experts Scientifique du Tourisme) and gained quite a wide acceptance. The Hunziker - Krapf - AIEST definition includes the following elements of the phenomenon "tourism":

1) Travel (the dynamic element)
2) Sojourn outside the domicile (static element)
3) temporary character of the sojourn
4) the activity of the visitor must not be gainful

The problems arose in connection with the fourth element which set it apart from the statistical definition. The Swiss scholars regarded the tourist as a pure consumer with all the money spent during the travel earned both directly and indirectly in the place of tourist's permanent residence and not at the destination. The exclusion of business travel from tourism suggested by Hunziger and Krapf was opposed by many authors. This strictly "consumption-oriented" understanding of tourism is not acceptable for several reasons. 1) It clashes with the internationally adopted statistical definition which includes all forms of business travel. 2) Such a differentiation is not feasible in essence, because business or professional travel is usually connected with pleasure tourism and with expenses contributing to the local economy as "pure consumption." Thus conceptually business and other non-pleasure or non-leisure travel can be included into tourism as long as it does not lead to a permanent residence and an occupation renumerated at the destination. 3) "Pure tourists" or pleasure tourists often establish business and professional contacts at the destination which may lead to some economic gain. 4) The enormous impact of business travel on the tourism industry as a whole and urban tourism in particular is so obvious that it is inconceivable to separate them as Hunziger and Krapf want. (Sessa, A., 1971, 1:5-14).

A conceptual definition, in harmony with the statistical, but still based on Hunziger - Krapf - AIEST definition has been proposed by Sessa (Sessa, A., 1971, 1:14) "Tourism is a sum of relations and phenomena resulting from the travel and stay of non-residents as much as this stay does not create a permanent residence." This notion of tourism is linked with temporary change of residence for practically any purpose. This is the prevailing notion of tourism in the literature since 1960s. It may be illustrated by the definition suggested by Medlik: "Tourists - people who are away from their normal place of residence or work, on a temporary, short-term basis, with a view of returning home within a few days, weeks or months." (Medlik, S., 1969: 39). Young puts it still more succinctly referring to a tourist as "one who travels away from home" (Young, G., 1973:29). This definition implies that tourism includes all non-migrant movements of people and this is essentially correct.

An interesting synopsis of definitions has been undertaken by H.P. Schmidhausser who attempts to build bridges between conceptual and statistical definitions of tourism. He examines the definitions from the point of view of nine contentious categories of travel. The definition number 3 has gained most support so far.

Table 2.2

Definitions of Tourism	The definitions of "tourism" quoted in the first column include the following types of inter-communal passenger-traffic and categories of sojourn								
	Business travel	Congress tourism	Commuting	Founding of a business establishment	Short-range recreation in a second home	Day excursions	Visits of rural population or residents of urban agglomerations in nearby cities or regional centers	Sojourn of non-resident students at their place of study or of students at boarding-schools	Sojourn of non-resident patients in hospital
1. **A sum of relations and phenomena resulting** from the sojourn of non-residents as much as this sojourn is not connected with establishing of a residence for a main (principal) permanent or temporary gainful activity (the present definition of AIEST).						x	x	x	x
2. Totality of relations and phenomena, resulting from the travel and the stay (sojourn) of non-residents.	x	x	x	x	x	x	x	x	x
3. Totality of relations and phenomena resulting from the travel and stay of persons in places which constitute neither their principal residency nor place of work. (Definition of St. Gallen).	x	x	x	x	x	x	x	x	x
4. Totality of relations and phenomena resulting from the travel and stay of non-residents, inasmuch their sojourn is not connected with founding of a business establishment.	x	x			2	x	3	x	x
5. Totality of relations and phenomena, resulting from the travel and study of persons, in places constituting neither principal residency nor place of work and no place with central functions for his district of residence (region).	x	x		1	x	x	3	x	x
6. Totality of relations and phenomena resulting from travel, sojourn and over-night-stay of persons for whom the place of over-night-stay is neither principal residency nor place of work.	x	x		1	x			x	x
7. Totality of relations and phenomena resulting from travel and stay of persons for whom the place of sojourn is neither principal residency nor place of work or study.	x	x		1	x	x	x		x
8. Totality of relations and phenomena resulting from travel and stay of persons, for whom the place of sojourn is neither principal residency nor place of work or study and no place of central functions for his district of residence (region).	x	x		1	x	x	3		x

LEGEND

1. For persons, who plan to found a business establishment, the place of sojourn becomes the principal residency and place of work; therefore, according to the definition, they are not considered as tourists.
2. Short-range recreation within the urban agglomeration is not defined as tourism.
3. The attractiveness of cities and regional centers reaches frequently beyond their zone of agglomeration. According to the definitions 4, 5 and 8 one has to regard the shopping spree of a suburban housewife (e.g. suburb of Zurich) in downtown Zurich not as tourism, however, as a similar shopping spree of a housewife from e.g. Schaffhausen, Aarau, Zug or St. Gallen, can be defined as tourism.

(Schmidhauser, H.P., 1971, 2:51-54).

2.4.4. The Problems of International Standardization of Terminology.

It would seem that the existence of an international UN/WTO definition of tourism has created a sound basis for collection of standardized data on tourism a basis for tourism planning and management all over the world. However, unfortunately that is not the case: the UN/WTO definition has not bcen universally accepted. Only the majority of countries but unfortunately not all adhere to this definition and this creates problems. A good example is the United States which does not even have a standard definition valid for the whole country. Such a situation has serious negative practical implications for the tourist statistics: there is complete chaos in the collection of data. National travel surveys conducted by the US Census Bureau, the US Travel Data Center and the US Travel Service deal with "trips taken away from home, overnight or on one-day trip, to a place 100 miles or more from home" (UN Interregional Seminar on Tourism Development, 1968:102). Such a distance criterion greatly underestimates the tourism receipts. The minimum distance was reduced to 50 miles with an overnight stay at the destination (Hudman, L.E., 1980:123) thus interrupting the continuity of data collected and only partially decreasing the omissions which are quite significant. However, according to Chadwick (1987:49) "The National Travel Surveys of the United States Bureau of Census (1979) and the United States Travel Data Center (1984) report on all round trips with one-way route mileage of 100 miles or more. Such an inconsistency does not require any comment. S. Waters is highly critical of the 100 miles criterion indicating that this statistical underestimation" has perpetuated the fiction that the (tourism) industry ranks in second or third place in the nation instead of in number one position (Waters, 1987:21). To add to the statistical problems many US travel surveys omit not only day-trips with their enormous economic impact but also business, convention and essential personal travel (Leisure Systems Inc. 1976:3). By focussing entirely on pleasure travel one takes into account only one part of travel supporting the tourist industry. Besides, the spatial relationships between origin and destination remains obscure as in this definition which defines tourists as "individuals or parties visiting an attraction or area for recreation or touring purposes, who reside outside of an established limit or boundary" (Western Council of Travel Research, 1969: 231). These limits or boundaries are frequently the state boundaries thus relegating the intrastate travel to "non-tourism." This fuzzy concept of distance is also evident in the following definition: "Tourism ... means the travel of persons for pleasure in which one returns to his starting point, usually after travelling a considerable distance involving a time span of several days and is most often interregional. Use of a park by people is visitation - some of which may be by tourists. A simple example of the distinction is: a local man walking his dog in Central Park, New York, is a visitor; a visitor to Central Park from Chicago or Paris, France, is a tourist." (Hart, J., 1966:69).

At the state level reigns total confusion. Definitions of tourism vary from state to state. The common tendency is to regard the out-of-state visitor as tourist excluding intrastate travel almost always and only sometimes (e.g. Nevada) including business and personal travel. In majority of cases these categories of travel are excluded. Alaska subdivides tourists into visitors who visit friends and relatives in Alaska and vacationers who are coming to see Alaska without taking into account that VFR normally also sightsee (Little, A.D., 1967:2). For Arizona tourist is "a nonresident traveller in Arizona while the term "traveller" is used to identify the Arizona resident travelling to places within Arizona. The state of Utah differentiates between travellers and tourists by degree. A tourist will participate in some activity while in the state, while a traveller passes through the state on his or her way to another state (Hudman, L.E., 1980:125).

Thus US definitions are characterized by confusion of terms "tourism," "travel," "visit," "vacation," "outdoor recreation," etc. and no attempt has been made to standardize terminology which is usually "tailored to the purpose of the study in question" (Little, A.D., 1978:7).

The terminological confusion is by no means limited to US. Studying the UN/WTO statistics one has to deal with innumerable footnotes explaining the national variations, significant differences in collection of tourist statistics associated with lack of adherence to internationally accepted terminological standards. Terminology varies not only spatially but also in time e.g. the definition of visitors in Samoa "has changed when different individuals managed the Office of Tourism" (Choy, D., 1984:582). Indeed, one of the important tasks of the WTO is to work systematically to improve the situation in this respect.

2.4.5. "Tourism" and "Travel"

The English word "travel" used in sense of a "journey" comes from the old Norman-French word "travail" (Field E. 1964:406) meaning torment, "trouble", "painful effort or exertion", "toil" and finally in modern French "work" (modern French retains also the former meaning). The French language seems to have inherited the word "travail" from the Latin "trepalium" meaning a three-pronged instrument of torture. Thus the etymology of "travel" indicates its aspect which had very little to do with pleasure. These laborious and troublesome attributes of travel began to decrease with the advent of modern means of transportation although many people think that way about travelling even today.

Thus "travel" means any spatial displacement of people for whatever reasons. It is old as mankind. However, in modern times "travel" in English language is often identified with "tourism" and the two exist side by side. Many English speakers feel that "tourism is a negative word to which some people attach a narrow meaning implying "bargain rates" and the word travel might better cover the scope." (Can. T. Assoc., convention report, April 20-23, Ottawa 1969: 14). E. Eliot writes: "Tourism is a pejorative term the world over. Tourists are sought for their money and despised for their ignorance. It is not a noble role. Travel implies for me something deeper: a sincere attempt to know and understand an alien world, a wish to learn its history and culture and language, to become in some small way a part of it" (Eliot, E., 1974:271).

To avoid the term "tourist," other terms are used: "traveler," "visitor" and "guest," etc. Hawaii has no tourist bureau, rather it has the Hawaii Visitors Bureau.

Travelers become "guests" in hotels, "patrons" in restaurants. Temple Fielding in his guides carefully avoided the term "tourist" substituting instead "pilgrims," "voyagers," or "travellers." The organ of World Tourism Organization is called in English "World Travel" but "Tourisme mondiale" in French. The English version of "Union Internatinale des Organismes Officiels de Tourisme" was: "International Union of Official Travel Organizations." The tourism industry is commonly called travel (or hospitality) industry because of "the negative connotation of "tourism," associated with bargain-rate accommodation and travelling, and the rather unsophisticated behavioural pattern of the participants" (Lundberg, D., 1976:7). S. Waters also advances the argument of negative image against the term "tourism." Additionally, he feels that "tourism" is associated with pleasure travel on vacation. Therefore, according to him, such terms like "visitor" and "visitor industry" would be more appropriate to encompass also business and other non-pleasure travel (Waters, S.,1987:20).

The pejorative meaning of "tourism" and "tourist" may have its roots in the mid-nineteenth century appearance of mass modern tourism at the world scene. The introduction of modern means of mass transportation (steamship and railroad) has initially allowed excluvively the upper and to a lesser extent the middle classes to participate in tourism. Only since relatively recently the vast majority of people in the DCs does take part. Thus the aristocracy has lost its monopoly for participation in tourism and resented the new mass tourism as vulgar. Therefore, the euphemistic substitutes for "tourism" and "tourists" have spread contributing to the terminological confusion. Another reason for widespread use of the term "travel" instead of "tourism" could be suggested: travel may be understood as a wider term encompassing both business and convention tourism as distinguished from "pure" pleasure tourism.

This author pleads for the term "tourism" reserving "travel" for colloquial use as substitute for "tourism" only occasionally especially when the travel element is prominent in the tourism experience. Here is his argumentation: "Travel" is certainly a too wide notion: travel may be undertaken also for reasons of migration, commuting and other movements of people from place to place which are definitely beyond the scope of tourism. Travel has been always with us: the nomads, vagabonds, soldiers, migrants, refugees, explorers, all have been travellers but certainly not tourists. Tourism is a socio-economic phenomenon of modern industrial age and differs from "travel" quantitatively and qualitatively.

Graphically the relationship between "travel" and "tourism" could be depicted as follows:

Figure 2.5

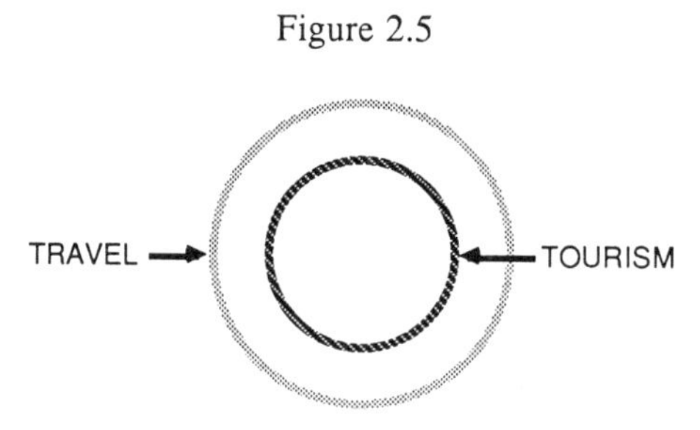

Above we have pointed out that the term "travel" has wider meaning than "tourism." On the other hand, however, travel strictly speaking constitutes only an element of tourism, especially with respect to single-destination tourism when a relatively long stay follows and antecedes a sometimes very short travel, e.g. by jet, to and from the vacation area. One may ask: what is more important "travel" or "stay?" After all "travel" in most cases is only a means to an end: the sojourn. "To learn how to travel is to learn how to stay" - seems a correct statement. These are the reasons why it is difficult to agree with the terminological overemphasis of the dynamic aspect of the tourist experience (travel) over the static aspect (sojourn). Indeed, strictly speaking, tourism includes travel and not vice versa.

Figure 2.6

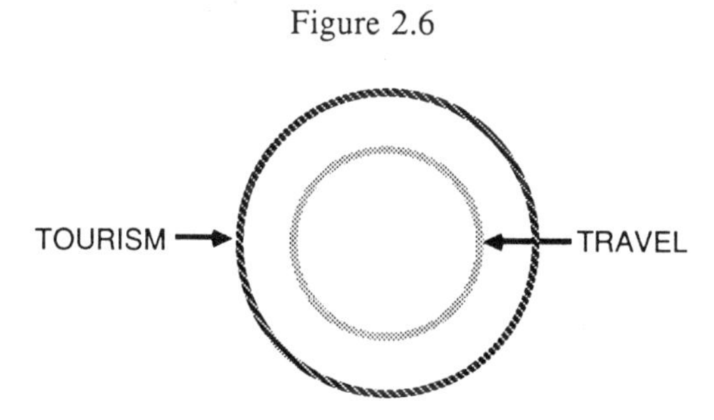

It seems that in more recent time the terms "tourist" and "tourism" gradually gain recognition in English language usage, e.g. the former British Travel Association is now called the British Tourist Authority, The Canadian Tourist Association, named since 1970 The Travel Industry Association of Canada became in 1978 Tourism Industry Association of Canada. Nevertheless, this trend does not suggest that the term "travel" or "traveller" should be abandoned completely as synonymous to "tourism" and "tourist." However, we should be aware that it is less exact.

2.4.6. "Tourism" and "Leisure"

Tourism fills a part of the leisure time. Clearly, much of leisure time (e.g. at home, close to residence) is beyond the scope of tourism. However, a part of tourism is associated with working time (business, meetings, congress, conventions). This relationship could be illustrated as follows (Fig 1.7).

Figure 2.7

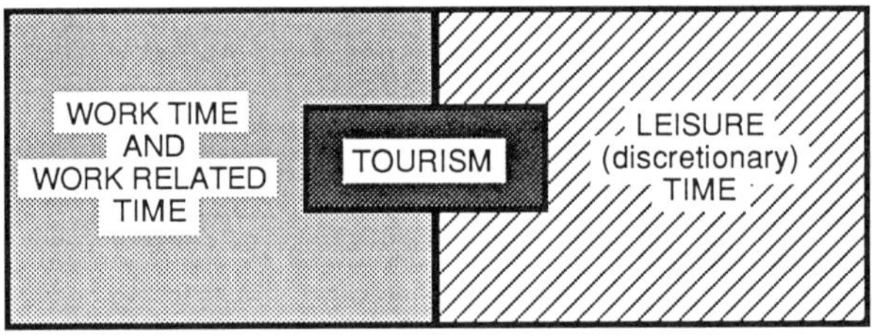

2.4.7 "Tourism" and "Recreation"

The term "tourism" in North America is not only being frequently substituted by "travel" but also often by "recreation." In fact, what is called tourism in Europe means in many cases "outdoor recreation" or even simply "recreation" in North America. (Nelson, J.G., and Scaree, R.C., 1968). This point is explicitly spelled out in an official Canadian government report stating with respect to "tourism" and "outdoor recreation" that these "words are synonymous" ("An analysis of demand trends for tourist accommodation in Canada," 1969: 24). The terms "recreation" and "outdoor recreation" are not only used as substitutes for "tourism" - moreover some authors subordinate "tourism" to "recreation." In this sense tourism (travel, outdoor recreation) are forms of recreation. One frequently refers to "recreation including tourism." Graphically this relationship could be illustrated as follows (Fig. 2.8).

Figure 2.8

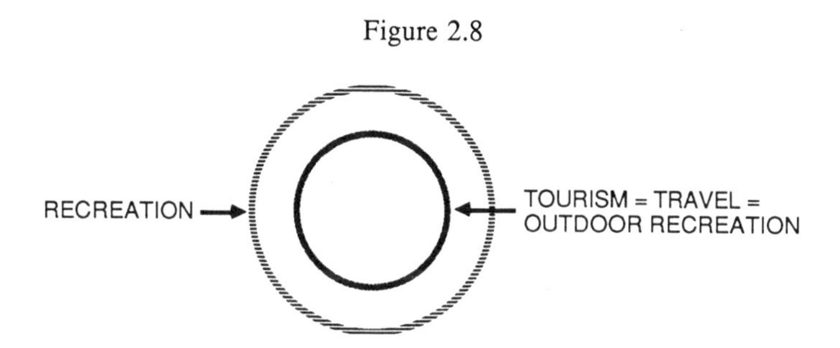

The wide understanding of the term "recreation" has its roots in the acceptance of the premise that recreation is the total content of leisure. Soviet scholars follow this approach defining tourism and suburban recreation as "forms of recreation" (Gerasimov, I.P., 1970: 189-198). In the Soviet Union tourism is determined as "recreational migration or recreation away from home" (e.g. Tverdokhlebov. I, Mironenko N. 1979: 41-48). Thus Soviet researchers focus their endeavors practically exclusively on recreational tourism. Recreation is understood teleologically as means of restoring physical and mental capacities of people for work. Other non-recreational forms of tourism are out of focus and this author is not aware about any publication in this field.

One could argue against these concepts by indicating that the terms "tourism" and "recreation" differ in many significant ways thus making their usage as synonymous, or subordinating "tourism" to "recreation," inadvisable. Here are the arguments:

1. Tourism is associated with not only free (leisure) time but also with the working time (business tourism) whereas recreation occurs totally during the leisure time. Indeed, business, professional, educational, religious (pilgrimages), health tourism (to visit doctors, to undergo operations, to stay in a health resort, etc.), some visits to friends and relatives and other personal

travel (e.g.to a funeral) belong to the class "tourism" but can hardly be associated with recreation. Even some "recreational tourists" rarely engage in specific recreational activities (e.g. camping) - often they just want to change the environment.

2. Tourism in contrast to recreation is always and not sometimes associated with displacement, temporary change of residence of participants. Not all "outdoor recreation" requires change of residence although it occurs in relatively natural, non-urban setting (Clawson's intermediate and resource-based recreation). The participants may travel from home as day-trippers or excursionists. In other words tourism is associated with overnight stays outside the permanent residence whereas recreation (including outdoor recreation) is only partially associated with such stays. In fact, most recreation occurs at home or locally within the community of residence. Indeed, colloquially recreation means use of shorter slots of leisure time mostly locally without temporal change of domicile.
3. "Tourism" and "recreation" not only differ in their meaning but also clash in practice: purchase of recreational equipment may involve financial conflicts in family budgeting with allocation of funds for tourism.
4. Tourism (and the tourism industry) has very definite commercial connotations as part of the economy. Recreation, of course, has also commercial aspects, however, they are weaker and sometimes lacking, especially in cases when recreation is provided free of charge by the government, social organizations, employers etc.
5. Recreation in contrast to tourism centers on activities which increasingly require acquisition of special skills.

The arguments which point out differences between "tourism" and "recreation" should not lead to the conclusion that they have nothing to do with each other. Indeed, there is a significant overlap between them: recreational tourism is the most significant part of tourism. But also there are elements in tourism which cannot be classified as recreation and in recreation elements clearly beyond tourism. This relationship may be illustrated as follows (Fig. 2.9).

Figure 2.9

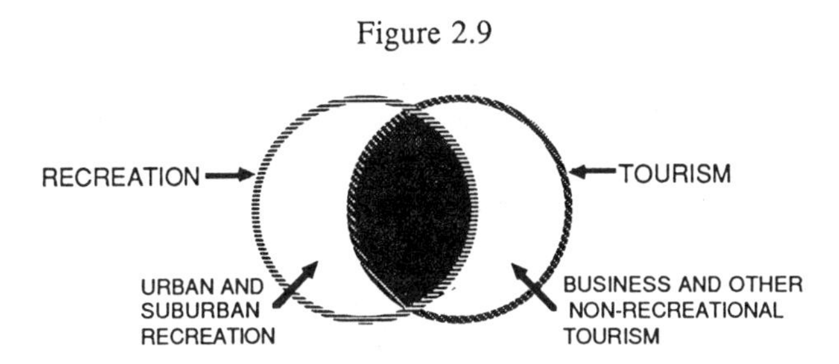

Bringing the relationships between leisure and recreation and tourism and recreation together leads to the diagram.

Figure 2.10

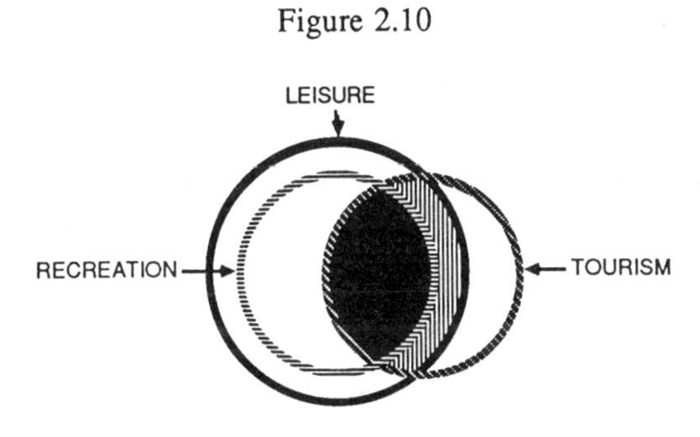

Source for Fig. 2.9 and 2.10: Mieczkowski 1981

2.4.8. Types (categories) of tourism

Tourism can be divided into various categories according to different criteria. The following list has been compiled by the author. Most of the subdivisions have significance for statistics, planning and management.

1. According to the residence of visitors
   a) Resident
   b) Non-resident (out-of-state, out-of-province)
2. According to the residence of visitors
   a) domestic (local, intrastate, interstate, intra- provincial, interprovincial)
   b) international or foreign (subdivided into intra-regional, interregional, intracontinental, inter-continental)
3. According to the age of visitors
   youth tourism, golden age tourism (special rates) etc.
4. According to the duration of the tourist experience:
   excursion - day trips or excursions without overnight stay
   weekend travel - between one and 3 overnight stays
   vacation travel - 4 and more overnight stays
5. According to the season.
   A) Summer - (in Austria, May 1 - Oct. 31)
   B) Winter - a) cold, mountainous regions
   (in Austria, Nov. 1 - April 30)
   b) warm region e.g. Caribbean, Florida
   (summer recreation activities)
6. According to the means of transportation automobile, airplane, railroad, ship, bicycle etc.
7. According to the mode of movement (number of destinations).
   a) sojourn, single stop, single destination, resort tourism, resident tourism (French: tourisme de sejour)
   b) round trip, multiple stop, multiple destination, itinerant tourism, touring. (French: tourisme de passage).
8. According to organizational pattern (arrangement)
   a) individual independent, arranged by the tourist(could be subdivided into: single person, couple, family)
   b) group including "package tourism" (travel package), charter flights, etc.
   The same type of arrangement for the whole group.
9. According to the financial status of tourists:
   a) mass
   b) elite

10. According to the mode of financing:
    a) commercial or free: hotels, airlines etc. involved.
    b) closed or restricted: with certain restrictions in terms of access to health resorts, vacation homes etc.) This type of tourism is fully or partially financed by public or to a lesser extent by private funds and not by the tourist. Such subsidized tourism is often called social tourism. Subsidized tourists in the market economies are mainly but not exclusively people in low income brackets, economically disadvantaged, minority groups, youth etc., in the Soviet block countries - the "nomenklatura" and good workers. Social tourism is the typical, prevailing form of tourism in the Soviet block countries. It is relatively insignificant in North America. In some Western European countries the social services are relatively generous in funding stays in domestic health resorts e.g. in Vichy, France. Another form of social tourism in Western Europe (e.g. France) are vacation of employees totally or partially paid by their companies.
11. According to the type of accommodation:
    a) hotels, motels,
    b) para-hotels or supplementary accommodation: weekend houses (cottages), youth hostels, youth camps, campgrounds, private houses, apartments and rooms.
12. In relation to religion
    a) religious tourism, pilgrimage
    b) secular tourism.
13. According to the type of activity:

A. Recreational - the most important reason for tourism.

Various categories e.g rest and relaxation or related to certain environment e.g: water-related: sunbathing, wading, bathing, swimming, snorkeling scubadiving, boating (including yachting and sailing), canoeing, angling or fishing,water-skiing

Other recreation activities:

a) Summer: camping, picknicking, hiking, prospecting,
b) Winter: skating, skiing, skidooing (snowmobiling)
c) Other: pleasure driving, hunting, photography etc.

B. Non-recreation tourism

a) business
b) professional (incl. conventions, congresses and other professional meetings)
c) Visiting Friends and Relatives (VFR)
d) other personal reasons: shopping, health tourism (including medical treatment, visits to thermal, climatic resorts), religion (pilgrimages), education, spectator sports, entertainment, "tourisme gastronomique."
e) sightseeing for various reasons: cultural,educational, visits to expositions and fairs.

Although listed as non-recreational many of these categories contain strong recreational components, especially points c to e. Indeed, d to e tourists and recreational tourists are frequently referred to as "pure tourists."

14. According to the purpose:

a) pleasure tourism ("pure" tourism)
b) Visits of Friends and Relatives (VFR) which normally has a strong pleasure component
c) tourism for other personal reasons (education, health, weddings, funerals, shopping, etc.)
d) business (including professional travel, conventions)

Business tourism contains mostly some pleasure components. When both business and pleasure are claimed to be equal motivations, the trip should be classified statistically according to the main sources of financing.

15. According to the setting:

a) urban or city tourism
b) resort tourism
c) wilderness tourism "outdoors"

This type of tourism constitutes the major part of American "outdoor recreation" which, however, is a wider notion encompassing recreation in natural setting also outside wilderness areas (Clawson's intermediate and resource based recreation) as long as the "outdoor recreationists." stay in campgrounds, cottages etc. Some "outdoor recreationists" may be urban based (day trippers). However, as indicated above, the notion of "outdoor recreation." although widely established in North America, lacks precision.

16. According to location of the destination

a) urban

b) suburban
c) near-urban
d) exrta-urban

CHAPTER 3

HISTORY OF TOURISM AND RECREATION

3.1 INTRODUCTION

The history of tourism and recreation helps us to improve our understanding of the present situation. It reveals how the changing material basis of human life leads to new concepts, new ideas, blazing the trail to the continuous process of social transformation. Indeed, the past is always the prelude to the future. Analysing the events of the past we come to the conclusion that tourism as modern socio-economic phenomenon pursued mainly in leisure time started only in the 19th century with the development of capitalism associated with the industrial revolution. However, travel (not tourism) is certainly as old as mankind but in the precapitalistic times it differed from modern tourism first of all in quantity (only very few participated), distribution among the social classes (only few could afford it) and finally quality: travel did rather rarely coincide with recreation. In other words travel and recreation were almost totally divorced, travel being undertaken only exceptionally in leisure time and for recreational reasons and if undertaken then as a rule for relatively short distances. The reason was that actually up to the 19th century travel was slow, difficult, dangerous, expensive and almost entirely devoid of the recreational or pleasure elements so characteristic for the modern times. This situation necessitates a separate discussion of travel and recreation in the historical perspective. This approach will be followed in our historical chapter.

Who were these precapitalistic non-recreational travellers? Mainly migrants, explorers, wandering merchants, craftsmen, sailors, pilgrims joined from the 17th century on by a tiny social stratum of rich Englishmen and other eccentrics. However, until about the middle of the 19th century humanity was extremely immobile and travel for whatever reason was an exception. A river, a mountain, heath or swamp, a political boundary or simply a longer distance were enough to create separate human environments with various and unique characteristics. If variety constitutes a characteristic feature of civilisations, human immobility has been a necessary condition for it. Thus a vast majority of people stayed put throughout their lives and rarely ventured more than hundred kilometers away from home. This should not be forgotten while discussing the history of tourism.

In contrast, the development of capitalism in 19th century brought about a movement of people who have been travelling mainly in their leisure time and for recreational reasons. This is tourism in the modern sense of the word. Thus, relationship between modern tourism (travel is a too wide notion) and recreation shows a significant (and increasing) overlap in the area of recreational tourism only since the industrial revolution swept through Europe, North America and other parts of the world.

The transformation of precapitalistic travel and recreation, almost entirely separated from each other, into modern tourism and recreation, overlapping with each other, (fig. 2.9) occurred only when they became available for most people and not only for restricted strata of population. Modern tourism is a mass phenomenon. In this way modern tourism resembles democracy: we tell that there was democracy e.g. in ancient Greece or in medieval and post-medieval Poland but these were certainly not democracies for all people, for the masses, consequently not in modern sense of the word. Certainly, there were some marginal exceptions to this general rule: e.g. Iceland was a democracy in precapitalistic time. The same with tourism: there were some instances when elements of tourism appeared in the precapitalistic era e.g. in ancient Greece and Rome but their scope was minimal and their importance marginal. Therefore, the term "travel" with reference to pre-capitalistic spatial movement of people seems the most appropriate. The term "tourism" is reserved for the modern phenomenon starting in the middle of 19th century.

As far as recreation is concerned - this term may be applied consistently throughout the history although its content and ideas certainly were subject to considerable transformations in the course of time.

This chapter is divided into sections according to the generally accepted historical periodisation. However, to generalize the course of history of tourism and recreation one could distinguish only three eras:

1) Pre-industrial era of travel and recreation with only elements of tourism noticeable.
2) Mass tourism take-off era in the DCs during the industrial revolution in 19th and 20th century endingabout 1945-1950 with the post-World War II reconstruction.
3) Mass tourism era starting about 1950 in most DSs. This era is regarded as contemporary and therefore is treated outside this chapter.

3.2 PREHISTORY AND ANTIQUITY

3.2.1 Recreation

Little is known about leisure and recreation in prehistoric gathering, hunting and fishing civilizations. One thing is sure that there was very little of leisure time (with some notable exceptions) in those difficult periods of human struggle with nature, struggle mainly for bare survival. However, few remains of early artistic endeavors and customs are known today: they reflected human fears with respect to all-mighty nature and have their roots in magic. We also know that in primitive societies there were no sharply drawn lines between work and leisure e.g. hunting meant work and recreation at the same time. Games, fun and art forms were spontaneous and associated with everyday activities.

Only the emergence of agriculture coupled with the permanent settlements could create the surplus of food and goods necessary for sustenance of non-working "leisure elites" - mainly aristocracy and priesthood - which on the basis of their political and economic power were free of manual labor and could concentrate on ruling and administering functions. Thus work and leisure became more clearly delineated. The division of people into two categories Homo faber - man the maker or worker versus homo ludens - man the player, meant that the slaves, peasants and craftsman had to work in order to provide leisure for the elite. This new social order first emerged in the Middle East, India and China at least 6000 years ago and lasted in Europe to about the eclipse of the Roman Empire. The privileged ruling elites constituting a relatively small proportion of the entire population were able to appropriate a major part of the wealth produced by the society ("surplus wealth"). This exploitation created the conditions for the rulers to devote part of the appropriated wealth to promote music, painting, sculpture, philosophy. The "leisure classes" not only actively cultivated arts and education but also patronized them to enhance their status and glory. The oppressed classes at the other hand had to toil hard for their masters without the benefit of leisure as a result of lack of time, energy and wealth. "There is no leisure for slaves" wrote Aristotle (Miller, N.P., Robinson, D.M., 1963:40).

The ancient civilizations culminated in the Greek and Roman culture. In fact, here the foundations under the philosophy of leisure have been laid. Leisure well spent was a Greco-Roman ideal contributing to harmonious balance between physical and mental capacities. This ideal is epitomized in the adage: "Mens sana in corpore sano" (a sound mind in a sound body). Physical fitness per se and for military service was considered a necessity. Olympic Games in ancient Greece were an example of these ideas (alongside with their religious significance).

3.2.2 Travel

Spatial displacement (movement) of people has been one of the oldest aspects of human life. It was motivated for many thousands of years mainly by economic necessities like trading, hunting, grazing, looting, military conquest, settlement etc. rather than by an urge for change or longing for adventure, for displacement as an end for itself. The latter motivation was limited chiefly to vagrants, gypsies and other minority groups outside the mainstream of society.

The first people who travelled voluntarily were the traders, merchants. In the early Hebrew language the word "merchant" was synonymous with "traveller" (Enzensberger, H.M., 1967:186). Long distance trade in the Middle East started as early as 9000 B.C., long before the development of cities (International Herald Tribune 18-12-1986:10). The late Neolitic people developed in their villages a marked division of labor by craft. This specialization necessitated the movement of goods, even at considerable distances. Indeed, prehistoric merchants travelled much earlier than the archeologists thought quite recently. These trade linkages were expanded in the historic period: the great trade routes developed e.g. the Silk Route, the Spice route, the salt routes, the amber routes etc. Indeed, the traders "blazed the trail."

One of the most important reasons for travel in Antiquity was religion (e.g. Epidauros, Delphi, Dodone, Olympia) although in many cases it was used as excuse to do business, to see a theatre spectacle, to consult an oracle, to view or participate in sport competitions (e.g. Olympic games) etc. Much less significant was travel in order to learn more about the world, started by some Greek intellectuals. In this respect interesting travel accounts on Babylon , Egypt and Black Sea region were left in the writings of Herodotos (5th century B.C.) who visited most of the antique world just for the sake of learning about different lands and their inhabitants.

The Greeks were the masters of sea travel, the Romans developed a vast network of about 100,000 km of roads for military and administrative needs of the Empire. These lines of communication "are said to have been more efficient than those in eighteenth century England" (Burkart, A., Medlik, S., 1975:3). The

Roman roads were accompanied by hundreds of inns which provided food and lodging and also served as horse relay stations. Despite the fact that these overland communication lines were aimed at serving the interests of the state, there were factors at work which facilitated also commercial or even some recreational travel. These were the political and economic conditions constituting the basis for a degree of spatial homogeneity and relative security ("Pax Romana"). The "tourist" traffic meant mainly short distance recreational travel of wealthy Roman citizens to their villas at sea or in mountains. They were eager to escape summer heat and/or the unpleasantries and limitations of life in big cities (Rome had 1.5 million inhabitants). The distances involved in recreational travel were rather small (Ostia and Montes Albani from Rome, Baiae and Puteoli from Naples). To some extent these trips to seaside and thermal resorts were linked with health reasons, cure at thermal sources so abundant in Italy but the pleasure element seemed to prevail. The Romans developed also a number of thermal health resorts in their colonies e.g. Aquae Sulis (contemporary Bath, England).

The economically disadvantaged masses of Roman population who could not afford luxurious villas in the suburban resorts stayed in Rome and had to "enjoy" non-creative forms of urban recreation characterized mainly by the extravagant cruelty of Roman arena which was alien to the Greeks.

Short distance recreational travel prevailed in ancient Rome because of the obvious deterrent posed by intrinsic hardships and slowness of long distance travel e.g. Rome-Alexandria by sea 10-21 days, Alexandria-Rome 60 and more days (Casson, L. 1974:22). However, such long distance trips also took place increasingly in the later years of the empire (Sigaux, G., 1965:11-14). Sightseeing and recreation tours involving greater distances, e.g. to Egypt, Greece, Rhodes, Asia Minor were undertaken (Wagner, F., 1970:18-20). Most of the Seven wonders of the (Antique) World were located in this area. Ancient Rome had scheduled ship connections, travel offices, festivals, museums, tourist guides. Indeed, the Romans were the first recreational travellers in history. Friedlaender in his "Sittengeschichte Roms" insists that the travel volume of the Roman Empire was surpassed only in the 19th century (according to Enzensberger, 1967:189). It seems that truly there were some elements of modern tourism in the antique Rome, something like "tourism before tourism" (Enzensberger, 1967). The main difference was that the Roman "tourism" was a minority privilege made possible by the existing socio-economic structure based on slave labor.

3.3 MIDDLE AGES

3.3.1 Recreation

The thousand years after the fall of Roman Empire from 5th to 15th century (conventionally 476 to 1453 A.D.) constitute the medieval period or Middle Ages. It meant a radical deterioration of conditions for recreation and travel mainly as a result of new Christian ideology. Already in the later period of Roman Empire the obsession with some recreational activities, notably, the circus, caused sharp criticism by contemporary writers as inhuman and cruel. Also the Christians criticized the excesses of Roman arenas. The circus games were forbidden after the demise of the Empire. In reaction to the Epicurean pagan ideas the church waged war against Greco-Roman rites and festivals, against the evils of pleasure-loving sinful flesh. This attitude may be illustrated by the quotation: "For the desires of the flesh are against the Spirit and the desires of the Spirit are against the flesh, for these are opposed to each other, to prevent you from doing what you would (Paul, Epistle to the Galatians Ch.5:17 Revised Standard Version Catholic Edition, Nelson, 1965).

In this way the Greco-Roman ideas of harmonious balance of mind and body were assaulted. The immaterial soul was declared superior to the material body. Life concentrated only on seeking spiritual values in complete isolation from "contagion of the matter." The church put an ascetic stamp on work and leisure: the first was extolled, the second cursed. Work was leading to piety, shielding the individual of temptations of flesh. ("Idle hands are the Devil's workshop"). The human life had to follow the pattern established by Genesis: "in the sweat of the face shalt thou eat bread, till thou return to the earth." (Genesis, 3, 19) Recreation of any kind was criticized as St. Paul put it "bodily exercise profiteth little." (First Tim. 4:8). One had to mortify the senses and forget about leisure. Human life was regarded only as a brief transition of ascesy and abstinence - a mere preparation for eternal rewards. These early Christian ideas later found their reflection in Protestant Puritanism and retained some of their influence up to the present time.

However, this does not mean that there was no leisure and recreation in the Middle Ages. The social dichotomy of the Antiquity did not disappear: work for many, leisure for few. The ideas of hard labor and forgoing pleasures have been imposed by the ruling elites partially in order to increase exploitation of the serfs. However, many representatives of the ruling social strata including some members of clergy and monks continued to play the role of the "leisure classes" indulging in luxuries of extravagant living. These hypocritical double moral standards were characteristic for the time. Towards the end of the Middle Ages so many members of the church hierarchy, especially its highest echelons sampled copiously from the pleasures of flesh that this was obviously one of the factors which brought about the Reformation. Especially radical was the reaction of the Puritanism in 16th and 17th century England and New England to the attitudes of the Catholic Church allowing leisure time amusements during holidays. First of all together with the repudiation of saints most of the holidays connected with their veneration have disappeared from the church calendar. The number of idle days in the year declined conspicuously thus increasing the time devoted to labor what meant direct economic gains. Secondly, any leisure time amusements on Sunday have been in most places strictly prohibited by law.

Some of the major recreational activities of the medieval elites were jousting tournaments, theatre, concerts etc. but most important was hunting. Overcropping local game became a fact recognized early in Europe and caused increasingly drastic regulation of hunting activities. In practice this meant the exclusive reservation of hunting rights for the elites, especially for the royalty. Thus, kings, nobles and high clergy established game reserves for hunting. These reserves were probably the first forms of recreational land use. Because they have been protected throughout the ages even in the post-medieval period, many of them constitute today's nature preserves and Natural (or National) Parks of contemporary Europe. Even some urban parks in Europe (e.g. Hyde Park in London, England) are former royal hunting grounds.

3.3.2 Travel

The negative attitude to travel in Middle Ages as useless activity may be illustrated by the following Latin saying: "Coelum non animum mutant qui trans mare currant" - "Who travels through the seas changes skies only but not the soul." Interestingly, this opinion was not shared by the contemporary Islamic world which is evaluated by historians, in many respects as culturally superior to medieval Europe. Avicenna wrote that people do not get wiser by reading more books or living longer but by more travelling.

Indeed, during the Middle Ages the travel pattern changed radically in comparison to the Antiquity. Neither the ascetic ideology of the Church nor the poor economic situation could create a framework for travel expansion. The life became more rural and therefore sedentary then previously. Most of the people became bound to the soil, parochial, immobile. The serfs firmly in the clutches of the feudal system were forbidden to move out (glebae adscripti). The individual mobility decreased, the Roman roads fell in disrepair or disappeared entirely. The security of travellers and the permeability of the space diminished as a result of frequent robberies and feudal barriers. The costs (including taxes, tolls, ransoms, guards etc.) increased. As a result a considerable quantitative shrinking of travel occurred. Moreover, also negative qualitative changes in travel patterns took place. The people travelled only if they had to, they felt more secure at home when isolated from others. The structure of traffic flows changed with the possible exception of merchants who still travelled despite the dangers, costs and feudal barriers, although the widespread poverty certainly kept their numbers relatively low. Of course, travel to health resorts or for sightseeing fell almost into oblivion in this unfavorable situation. The most important category of travellers constituted the pilgrims to the holy Christian places like Rome, Santiago de Compostella or the Holy Land (this in connection with the Crusades between the 11th and 13th century). However, these were not trips undertaken in leisure time for enjoyment - these were spiritual trips to the gates of eternity, frequently engaged in for penance or punishment for various offences. Indeed, the punishment was proportionate to the distance. This reflected the dangers and inconvenience of travel in these times. Most pilgrims had to walk, very few could afford horses, wagons etc. The probability of not returning from a trip was very high.

Another category of travellers during the late Middle Ages were the people undertaking trips in quest of knowledge: scholars and students in connection with the creation of universities in Europe in 14th and 15th century, and in quest of skills: apprentices and craftsmen, members of craft-guilds (market fairs, work in another place).

3.4 RENAISSANCE AND BAROQUE

3.4.1 Recreation

It is interesting to note that although there was a considerable upsurge in travel during 15th-17th century, little occurred in the field of recreation except some modest revival of thermalism, particularly in France. One could speculate for the reasons: wars, epidemics, Protestant puritanism, persistence of negative views with respect to nature could be suggested. The elites continued to hunt on their lands, especially in Great Britain.

3.4.2 Travel

The improvement of the economic situation and cultural development in Western and Central Europe during the late Middle Ages starting from the 11th century brought some increase in travel, especially commercial and religious trips. But the eventual displacement of the Middle Ages by Renaissance in 15th century gradually weakened the feudal isolation. The spatial mobility of people in Europe increased unintermittently although still pertaining only to a tiny minority of population.

The main influence on travel and recreation during the Renaissance period (15th-16th century) was exerted by the new individualistic ideas of Humanism with their earthward and man-centered approaches. The humanists looked for their inspiration in Antiquity. ("Man is the measure of all things" - Protagoras). The expansion of knowledge, the importance of individual experience, the spatial diffusion of new ideas gave a substantial boost to travel. The intellectual value of visiting foreign lands and especially Italy and her antique monuments was widely recognized. These new cultural developments and material progress created more favorable conditions for educational and cultural travel. The Renaissance artists, writers and philosophers undertook trips between various European countries to give and share in the new cultural achievements. A network of inns and guesthouses started to supplement the Medieval accommodation and food facilities in convents and castles. The early 16th century saw the beginnings of regular commercial coach transportation. As a result of the increased traffic on the roads of Europe there was a demand for travel literature: in 1552 the "Guide de Chemins de France" by Charles Estienne was published. In 1589 another guidebook appeared "Voyage de France, dressee pour l'Instruction et la Commodite tant des Francais que des Etrangers." Both books contained useful information not only for professional travellers but also for sightseers (Sigaux, G., 1965:27). In England William Harrisons "Description of England" has been published in 1578.

The 16th century brought also another type of travel literature exceeding mere information and belonging to the achievements of arts. Here especially valuable and interesting were the travel accounts ("Journal de Voyage") by the French writer and humanist philosopher Michel Montaigne (1533-1592). His trip which took him to Switzerland, Germany and Italy in 1580-81 had a definite tourist character. He travelled in the leisure time in order to improve his health in thermal resorts which at his time still played a marginal role. More importantly, his other aim was to obtain personal experience about these countries and their inhabitants. Thus his main motive was intellectual curiosity. Another example of a traveller motivated by intellectual interest was the Dutch humanist and philosopher Erasmus of Rotterdam (1467-1536). The Dutch philosopher and philologist Justus Lipsius (1547-1606) wrote about travel in persue of knowledge: "Of old, and nowadays, great men have always travelled. But the use of any sensible voyage was increased knowledge of manners and customs and constitutions of foreign lands, and a broadening of the mind" (Duchet 1949:75).

Among the interests which prompted Montaigne to travel was his infatuation with arts, very often the only reason for trips among his contemporaries. Indeed, the "tourisme artistique" with the aim to visit and acquire, sometimes by force, the treasures of arts in Italy ("the mother of arts") became characteristic for the upper classes during the Renaissance. Needless to add that these travels contributed greatly to the spread of new ideas, tastes, customs, techniques etc. all over Europe.

The baroque period in the history of Europe which could be roughly placed in 17th century brought rather quantitative than qualitative changes in travel as compared to Renaissance. Further improvements in transportation (e.g. Colbert in France) occurred, educational travel (including the Grand Tour) increased.

The 17th century marked a modest revival, initially limited to France, of, as the Frenchmen put it, "thermal tourism" (tourisme thermal) which used hydrotherapy for cure. It was well developed in ancient Rome but almost non-existent during the Middle Ages and Renaissance (partly for reasons of morality). The royalty and the aristocracy were the protagonists (Sigaux, G., 1965: 44). In 1603 the French King Henry IVth created the positions of "general superintendent and intendant responsible for surveillance of baths and fountains of the kingdom" ("surintendants et d'intendants generaux charges de la surveillance de bains et de fontaines, du royaume") (Duchet, R., 1949:85). The upper classes travelled also to their sumptious summer residences, especially in France (Paris region and the Loire valley).

The upsurge of travel in 17th century Europe found its reflection in the number of "travel reports," an increasingly popular literary field. According to a 1680 estimate of Richelet in his "Dictionaire" there were over 1100 published texts of this kind in existence in France alone (Sigaux, G., 1965:42). With rare exceptions (Chapelle, Bachaumont and La Fontaine) their value was rather informative and documentary then literary (Some of their titles are given by Sigaux, G., 1965:107).

3.5 ENLIGHTMENT AND ROMANTICISM

3.5.1 Recreation and Travel

The socio-economic, philosophical and cultural ideas and trends of this period created the foundations and prerequisites for modern tourism and recreation. First of all they brought a gradual shift in approaches to travel and nature. In the past travel had little to do with pleasure or recreation. It was an ordeal in those times of slow, expensive and uncomfortable transportation. Travelling also was extremely dangerous connected with attacks by pirates, robbers or simply xenophobic local population. Only the 18th and especially 19th centuries brought a significant improvement in security on the world's travel routes.

There were also some psychological and cultural aspects which compounded the discomfort of travel: almost complete disregard for aesthetic, scientific and educational values of seeing different landscapes and visiting different peoples. Voyage has been regarded as exile as unhappy roaming afar from home filled with physical and spiritual sufferings (cf. Homer, Odysee, First song; Shakespeare wrote in his play "As you like it" "Ay, now am I in Arden: the more fool I: when I was at home, I was in a better place: but travellers must be content."). Indeed, pilgrimage was mostly undertaken as penance. The travellers not only failed to comprehend and appreciate the beauties of the landscape but on the contrary disliked and even feared the sea or the most magnificent mountains, which were regarded as disagreeable obstructions to travel and feared as the refuge of bandits, heretics, evil spirits or at best regarded as nuisance. They were crossed not for their aestetic values but for sheer necessity like the Alps on the way between Northern Europe and Italy. These attitudes lingered on well into the 19th century. W.G. Clark reports in 1862 about a priest who was climbing Alpine peaks for penance (Galton F. ed., 1964:246). Similar attitudes prevailed among the early settlers in North America. (Huth, H., 1957:5-7). Thus the attitudes to natural environment throughout the ages were negative: nature being alien, savage and dangerous had to be tamed, dominated, exploited and improved by society. Characteristic were in this respect the 18th century endeavors to mould the raw forested landscape into parks, hedges and fields.

The attitudes to travel and nature changed gradually in 18th and early 19th century. The reasons were technological and economic development evident already in the period antecedent to the industrial revolution which brought significant technological, economic and scientific progress. Secondly, new philosophical, cultural and literary ideas rooted in Renaissance and Humanism spread in Europe: the philosophy of Enlightment (18th century) with its rationalistic scientific elements, and later, in the second part of 18th century Sentimentalism and Romanticism (late 18th and first half of 19th century) coupled with democratic political movements. Under the impact of these new trends and ideas old negative attitudes to travel and nature evolved in positive direction. In fact nature was "discovered." Her beauty was praised first by the sentimentalist J.J. Rousseau (1712-1778) who criticized the corrupting influence of society on mankind and advocated the "return back to nature." He developed "cult of nature" (deism) and is regarded as precursor of romanticism ("I feel therefore I am"). Later F. Schelling (1775-1854) created the system of nature philosophy.

The infatuation with nature contributed alongside with the economic progress to a significant increase of travel in the 18th century. Nevertheless, travel did not acquire mass proportions, still only relatively few individuals e.g. philosophers and writers like Montesquieu, Rousseau, Diderot, Laurence Sterne ("Sentimental journey through France and Italy" 1768) or scientists: geologists, geographers did travel among them Alexander Humbolt (1769-1859) who travelled in the Americas and in Siberia. These people were truly veritable pioneers of modern tourism. They as first climbed the mountain peaks and explored new lands guided by longing for foreign countries or by scientific interest. Their relationship to travel and nature underwent a considerable change under the influence of the contemporary scientific and philosophical approaches. The scientific evidence began to conquer the religious belief that the nature was in chaos and catastrophes were the physical manifestations of God's wrath.

This change was especially pronounced late in the 18th century and in the first half of the 19th century during the Romantic period. The poets, writers, painters undertook long trips in order to "discover" and enjoy the nature according to naturalistic philosophy and the romantic ideals. Their role in the development of modern tourism should not be underestimated. The Romantic literature radically changed the attitude towards nature. Not only the aestetic values of the landscape have been appreciated but it also started to play a role as expression of human feelings. The fear for wilderness has been forgotten, new understanding of the beauty of nature and a romantic drive to the faraway places ("Fernweh") developed. The distant and strange showed a seductive quality. A further logical step forward was the recognition of the necessity to learn more about other regions and countries, increased interest in folklore, history, ethnicity, patriotism, rejection of settled views, accent on individual fulfillment, on uniqueness of human nature.

The most important among these Romantic writers - travellers sensitive to the beauties of nature were:

English poet Shelley (1792-1822) - travelled in Italy.
English poet Byron (1788-1824) - travelled in Switzerland, Italy, Greece.
English poet Keats (1795-1821) - travelled in Italy.
Scottish poet Walter Scott (1771-1832).
English painter W. Turner (1775-1851).
American poet W. Bryant (1794-1878) - author of "A Forest Hymn" created under the influence of Byron
French writer Victor Hugo (1802-1885) - travelled in Switzerland.
French writer Balzac (1799-1850)
French writer Chateaubriand (1768-1848)
German writer and poet Goethe (1749-1832) - influenced by the Schelling's philosophy of nature (nature as an organic whole) travelled in Italy.
Polish poet Mickiewicz (1798-1855)
Russian poet Pushkin (1799-1837)

The Romantic attitude towards nature brought about first, mainly English, tourists coming in the last two decades of the 18th century to Switzerland - the classical land of shepherds depicted in the Romantic literature, mainly avoided by the young noblemen making their "Grand Tour." Of course, this travelling in the pre-railroad era was limited to very few members of privileged classes. Nevertheless, already in the first half of the 19th century Switzerland became the classical tourist country.

Travel started gradually to be regarded as a pleasure, as recreation. In the field of recreation the 18th and early 19th century brought important changes. In this respect instrumental were J.J. Rousseau with his advocacy of harmony between man and nature in the process of education, J. Basedow (1723-1790) promoting physical education as important aspect of education, H. Pestalozzi (1746-1827) establishing the principles of modern education and F. Jahn (1778-1852) - the father of calisthenics. All of them advocated a naturalistic, anthropo-centric approach to educational process. They emphasized the importance of exercise in open air close to nature for the mental and physical development of youth.

The 18th century witnessed also a revival of travel to spas and resorts which slipped into complete oblivion since the Roman Empire and developed only in 17th century France. The rediscovery of human body implied also care of health and this was made possible by the progress of medical science, especially hydrotherapy and balneology. Thus cure and rest were the primary reasons of travel to Bath in England, Spa in Belgium, Karlsbad and Marienbad in Bohemia, Baden-Baden in Germany, Vichy in France, Pyatigorsk in Russia etc. Also the recreational and social aspects became important: the spas and resorts flourished as fashionable meeting places for aristocracy. Taking the waters became a social necessity. Different sorts of amusements e.g. theater and gambling became a vogue. Health cure played to a large extent only an alibi function for socialization and entertainment. In the first half of the 19th century the new class of capitalists took the lead over aristocracy in spa and resort travel.

Whereas the inland spas dominated throughout the 18th century, towards the end of it a new trend developed resulting from the discovery of curative properties of sea water in mid 18th century. Thus in the second half of 18th century first seaside resorts developed - Brighton in England and Nice in France followed by many others, on the coast of Normandy and French Riviera, especially in early 19th century when the trend away from inland resorts towards the seaside was reinforced.

With respect to seaside resorts similar development occurred as with inland health resorts or spas earlier in the 18th century: purely curative aspects of these resorts were gradually eclipsed by social interaction of visitors, by quest for pleasure and amusement. This entertainment aspect included, starting in the middle of the 19th century, gambling casinos (typical example: Monte Carlo in the principality of Monaco). Initially the clientele of these resorts, particularly at the French Riviera, were the English. Later they were joined by elites from all over Europe including Russian nobility. Towards the "fin de siecle" the latecomers appeared at the Riviera: the rich Americans. Under their influence the Riviera changed its character from winter resort area to summer playground because they found the traditional summer resorts in England and Normandy too cool. But this change occurred only in 1920's.

Summing up the developments of 18th century and first half of the 19th century, one may generalize that gradually travel and recreation came closer together, they started to overlap more frequently. However, this was still not modern tourism: the distances involved were as a rule small and the emphasis was rather on the sojourn in one resort. Also, as a result of costs and relatively low technical level and high costs of transportation in this pre-railroad era, only relatively few could participate.

3.5.2 The Grand Tour

This type of voyage developed apart from the mainstream of travel and therefore should be discussed separately. The prerequisites for the Grand Tour were created by the quest of knowledge and diffusion of ideas in Renaissance Europe. Its beginnings may be traced to the late 16th century during the Elizabethan period in England. The Grand tour expanded greatly in 17th century and reached its "Golden Age" in 18th century. The country of origin of the travellers was mainly England and the principal destination was Italy although other European countries, France, Netherlands, Spain, Germany, were also involved both as origins and as destinations of travellers. The travellers were sons of socially and materially privileged families sent for a sojourn of up to three years duration abroad in order to supplement their education and above all gain some important contacts regarded as necessary for their future carriers and also give the young noblemen "a polish" of good manners among their own social stratum, i.e. the aristocratic circles of European capitals (hence the limitations of social milieu they were confined).

The changes in the social structure of late 18th and beginning of the 19th century brought a gradual shift towards more professional - educational travel. Also the social background of the travellers underwent a change: the representative of the bourgoisie substituted for the feudal nobility. Nevertheless the "grand tour" did not develop any behavioral patterns or principles which would act in our contemporary tourism. It simply faded away with the changing socio-economic conditions of the 19th century, although one could see its continuation at the present day in extended many-months journeys undertaken by our young people who take time-off jobs or study to see the world.

3.6 THE DEVELOPMENT OF MODERN TOURISM AND RECREATION IN 19th CENTURY EUROPE

Whatever significant for the development of modern tourism and recreation may have been the new philosophical and cultural trends of the 18th century and early 19th century - still more decisive was the influence of new economic and political upheavals introduced in the second half of the 18th century Great Britain and fully developed in Europe and North America in the 19th century: the industrial revolution and the upsurge of capitalism and democracy laid the foundations for modern tourism. The industrial revolution brought about a significant increase of productivity of labor thus increasing wealth, savings and accumulation of capital stimulating investments and economic growth. The introduction of machines and new sources of fuel and power and the requirements of economies of scale caused concentration of production in large industrial enterprises located at close distances from one another because of their interdependence. As a result large industrial cities developed, industrial societies became urban societies.

One of the results of the industrial revolution - the development of transportation - had a very special impact on tourism because it significantly increased the mobility of travellers. The stagecoach era came to an end. The railroad spread all over Europe and North America in the middle and the second part of the 19th century making travel faster, cheaper and more comfortable. Some researchers regard the construction of railroad networks in Europe as direct basis for modern tourism. (Bernecker, P., 1962:4:18). The railroads made travelling not only cheaper and more comfortable than the horse drawn coaches but first of all much speedier: the coach would make around 15 km per hour whereas a train 50 km in about 1850. The trip duration by train between Paris and Nice was 18 hours in 1889 instead of 13 days by coach. (Hawkins, 1980:215). Also the capacity has been increased substantially making really mass travelling possible. The 19th century development of seaside resorts in England and France, which started in late

18th century, would have been impossible without the mid-19 century expansion of railroad network.

In addition to the steam locomotive another form of adaptation of steam engine to transportation was the steamship. It revolutionized the river transportation and above all the sea transportation. In 1840 Samuel Cunard inaugurated the first regularly scheduled steam ship service between Europe and America. The transatlantic steam ships increased the volume, comfort and speed of traffic. The duration of the trip between LeHavre and New York was 10 to 12 days instead of the sailships 2-4 months. The number of transatlantic passengers increased consistently from 40,000 annually between 1860 and 1870 to about 100,000 in 1880s and 280,000 in 1914.

Great Britain as the pioneer of industrial revolution, first among the European nations gained the material prerequisites for development of mass modern tourism. This cradle of industrial revolution due to her early start in about 1760 was the leading industrial nation throughout the 19th century. Consequently it also lead in tourism. In the late 18th century and the first half of the 19th century the English enjoyed almost monopolistic position as pleasure travellers (Sigaux, G., 1965:60). They were regarded as "the tourists par excellence." Even today in some rural districts of Greece the term "lordoi" means simply tourists (Enzensberger, H.M., 1967:191). Switzerland as a classical tourist land owns its development primarily to the British visitors. They were not only the first tourists in the Alps, starting with the last two decades of 18th century, but also the first alpinists. How far they outpaced the other Europeans may be illustrated by the dates of founding of the Alpine clubs: The British Alpine Club was founded in London in 1857, the French only in 1874. Also the spatial diffusion of tourist flows proceeded gradually from the British center located in South Eastern England: tourists first visited Belgium, Netherlands and Paris, then the Rhine valley and subsequently spread to Switzerland, Italy etc.

The question when did the era of modern tourism exactly start is difficult to answer. Like most of the historical periodisations it could be a conventional date only. This author would like to suggest such a conventional date although he fully understands that in Western Europe modern tourism appeared gradually in the middle of the 19th century in connection with the spread of industrial revolution and the victory of railroads over the coach transportation. Nevertheless, if a fixed symbolic date might be accepted - this was the July 5, 1841 when a group of 570 members of a temperance organization made a 20 miles trip from Leicester,

England, to a park at Loughborough. The trip was organized by Thomas Cook (1808-1892) who chartered a whole train from a railroad company and sold the tickets in retail. Four years later, in 1845, Cook founded the first travel agency in the world which organized a first tour to Scotland in 1846 with 350 participants. A special Tour Guide was published for that occasion. Subsequently Cook organized trips to the world exhibitions in London (1851), Dublin (1853), Paris (1855, 1867, 1878). The number of tourists grew exponentially: Thomas Cook alone brought 75,000 visitors to the Paris Exhibition in 1878. His first "grand circular tour of the continent" including Northern France, Belgium and the Rhine Valley at popular low fares took place in 1856. Few years later in 1863 his tours reached Switzerland, in 1864 Italy and in 1866 the Eastern United States. In 1872 T. Cook led a party around the world in 222 days what inspired Jules Verne's book in 1874 "Around the World in Eighty Days." Thomas Cook started as tour operator and only later expanded his business into retail travel agency. Cook's offices have been opened all over the world and his services expanded to all continents. His merit was in laying foundations for the modern mass tourism and in using for this purpose the modern transportation media especially the railroad. A commemorative plague on Thomas Cooks birth house conveys the message: "He made world travel easier" (Brachon, P., 1968:52-64).

The development of tourism in the second half of the 19th century was significant: let us compare only the figures of visitors to the 1851 world exhibition in London - 6 million with that of in Paris in 1889 - 33 million. In 1871 first railroad line exclusively for tourist purposes, Rigibahn, has been commissioned in Switzerland. Railroad hotels were built under the inspiration and often the ownership of railway companies. These hotels, organized along the lines of the classical prototype of "Der Badische Hof" in Baden-Baden in short time squeezed out of business the old highway inns which had to wait for their revival as motels until the interwar period of the 20th century. The concentration of capital in hotel business lead already in 1880 to the formation of the first hotel corporation - the Ritz-chain. In the US the end of the century witnessed the construction of first mammoth hotels of over 500 rooms.

The railroad era since mid-19 century gave also a big boost to farther development of resorts as more inhabitants of large congested urban centers, plagued by environmental degradation, started to seek recreation not only in suburban areas but also farther away in more peripheral locations. New seaside resorts and mountain resorts were developed. The old spas in mainly inland locations have not expanded significantly beyond their 18th century peak. Some of them have shown only a moderate growth. This was due to the increased doubts about the curative properties of mineral waters. The new seaside and mountain resorts used climate as the main recourse e.g. Davos in Swiss Graubunden founded about 1860 for climatic cure of tuberculosis. The seaside resorts using climate and water for cure expanded greatly in the second half of the 19th century due to the railroad, e.g. Nice had 5,000 visitors in 1861 and 150,000 in 1914 (Hawkins, 1980:215). This could be said about Brighton and Blackpool in England, Ostende in Belgium. New seaside resorts whose creation was made possible by railroad were Biarritz in France, Capri and Sorrento in Italy and Yalta in Russia. All of them until World I catered mainly to the luxurious demand of the economically privileged classes. However, also less luxurious places started to develop for middle and lower middle classes: fishermen or farmers have often rented their premises to families with middle class background, more modest hotels, mountain chalets, alpine huts began to operate.

The lower income brackets urbanites and also other inhabitants of big cities with shorter slots of leisure time had to be content with the increasing provision of open space in urban areas. This trend started in Europe already in late 18th century and was reinforced during the 19th century when the kings and nobility sometimes voluntarily sometimes under pressure relinquished their exclusive parks to public use. As first, in 1789, the year of French Revolution "der englische Garten" in Munich was given up by the Bavarian prince, in 1839 the "royal zoo" in Berlin changed into "People's garden." In London, England the Hyde Park and the Regent's Park have been opened for public use in several stages during the 19th century. Napoleon III of France opened for public use the Bois de Boulogne and Bois de Vincenne in Paris.

Nevertheless, despite all the progress in the field of tourism and recreation during the 19th century, there were still some restrictive trends working against the rapid expansion of leisure time activities. The work ethic was still very strong. In fact, it was strengthened temporarily during the Victorian era. The long working hours of 60-80 weekly did not leave much time for recreation and shortening of working week run into the opposition of not only employers but also religious leaders who feared that more free time may be used for "sinful activities." Free time was practically limited to Sundays but even then there were limitations because of religious reasons. The quick pace of mechanization of industrial work contributed to decrease in heavy manual work but the workers were reluctant to translate these gains into increased physical efforts in the course of recreational activities. As a result the physical fitness of the population was at an all time low in the DCs. Also the participation of women in tourism and many recreational activities was limited as a result of their subordinate position in society.

The travel literature in the 19th century differed substantially from that of the past. The late 18th century (1793) "Guide des Voyageurs" still did not give any information on the natural environment or attractions but concentrated exclusively on the shortest possible routes, prices, schedules, lodging and legal aspects of travel - travel was still a necessary evil. All this changed in the first half of the 19th century at the dawn of the modern tourist era: descriptions of attractions appeared with the publication in London in 1829 of John Murray's "Handbook for Travellers" and his first "Red Book," guide on Netherland, Belgium and Rhineland, where he introduced the star system. In France Didot published in 1837 "Le Guide pittoresque portatif et complet du Voyageur en France" and in 1841 the "Itineraire de la Suisse."

However, the most famous author and particularly publisher of travel guides was Karl Baedecker (1801-1859) in Germany. In 1839 he authored and published his first guidebook "Die Rheinlande." Subsequently the publishing company "Karl Baedecker" published up to the World War I close to 100 titles covering Europe, North Africa, North America (guidebook on Canada was published in 1894 in Leipzig). Initially the language was German only but later also editions in French and English appeared. Baedecker's guides used the star system evaluating the importance of tourist attractions, also the quality of tourist services (e.g. hotels) was assessed in an extremely impartial and honest way. Thus the guides acquired excellent reputation because of their reliability. In fact, tourists used the word "baedecker" as designation of any guidebook of quality.

In the field of novel it is impossible to report here at length about the travel novels of famous 19th century authors, but two only should be mentioned: "Stendhal's "Memoires d'un touriste" (1838) and, as stated above, "Around the World in eighty days" published in 1874 by French writer Jules Verne (1828-1905).

3.7 HISTORY OF RECREATION AND TOURISM IN NORTH AMERICA UP TO 1900

The subsequent remarks are mainly limited to the United States because the material pertaining to Canada is scarce. It may be generally assumed that Canada to a large extent followed the US path. The history of tourism and recreation in North America should be treated separately of that of Europe until the 20th century because of different socio-economic and cultural conditions at work and separate course of history in the period when both continents were relatively isolated from each other. Only in the 20th century the development of tourism and recreation became parallel enough to enable a joint discussion of tourism development and trends in all DCs.

First of all, the impact of Protestant Puritanism in Colonial North America and especially in New England was much stronger than in Europe because the 17th century settlers were mainly Puritans and they managed to gain the overwhelming influence on North America for many decades to come. In addition, the economic situation of the early settlers rendered the ascetic Puritanic attitudes most advisable. They rejected all that was not productive or of immediate practical use, they feared their natural environment, regarded it as hostile and did not appreciate its beauties (Huth, H., 1957:5-7). The austere religious ideas of the Puritans conformed closely with the harsh necessities of the pioneer life in the hostile environment. The 17th century pioneers had to work incessantly in order to survive and later to succeed in taking roots on the new continent. Thanks to the Puritan ideology (Huberman, L., 1961:176-178) Colonial America not only survived the critical period of 17th century but also laid the foundations for its future wealth - an indispensable prerequisite for the development of tourism and recreation. The medieval Catholic church of Europe geared to the feudal economy, encouraged work only for living and not for profit and accumulation of wealth warning that this was the road to hell. The non-work time was to be devoted to prayer and religious rites or celebrations. In contrast, the Puritan doctrine of eternal salvation through hard work elevated labor from means of gaining leisure to an end in itself at least in the framework of human existence. Leisure in such context was unnecessary. Hard work resulted in accumulating wealth and investing it in order to maximize profits for the glory of God. It was certainly beneficial, especially in the pioneer years, on the new continent. Later this attitude served as moral justification for the accumulation of capital in the early stages of capitalism.

The Puritan ethic was production-oriented: success in business was clear sign of God's blessing, poverty was an indication of weak character, personality defects and God's punishment. Thus wealth led to the gates of heaven, poverty to the gates of hell. The industrious self-reliable, energetic and frugal Puritan was supposed not only not to waste money but also not to misspent his leisure time for recreation. Only work was virtuous. Loafing was sin. According to the view all idleness breeds mischief and to seek pleasure is to court the Devil. The short non-work time should be devoted to worship of God.

The ideas of Puritan conduct were strictly enforced. Strictest regulations "in detestation of idleness" have been adopted enforcing work and prohibiting amusements even on Sunday which was to be devoted to compulsory church attendance and pious meditation. Idleness has been branded as evil and industry and frugality as noble virtues (Miller, N.P., Robinson, D.M., 1963: 82 ff; Dulles, F.R., 1965:4-21). The church attendance on Sundays was made compulsory (South Carolina continued to make church attendance compulsory as late as 1885 - Dulles, F.R., 1965:207) in most colonies under the penalty of fine or even imprisonment (Dulles, F.R., 1965:8). In New England and other colonies laws were in effect prohibiting any form of amusement or sport. Dulles quotes a case of a "New England minister who refused to baptize children born on the Sabbath in the belief that they had been conceived on the Lord's Day, only to be confounded when his wife gave birth to Sabbath-day twins" (Dulles, F.R., 1965:8). Theatrical performances were unlawful until 1750, and for 50 years thereafter they were camouflaged under name of "exhibitions" (Williams, W.R., 1958:26).

But "there were always a few rebels who sought their pleasures in and of this world, not content with the promise of them in the next" (Clawson, M., 1960:6). They were punished severely: the best known example was of Thomas Morton of Merry Mount, Mass., who organized a May-pole celebration in 1627. He was imprisoned and returned to England. Dulles mentions a court case involving a couple "for sitting together on Lord's Day, under an apple tree" (Dulles, F.R., 1965:8).

The Puritanic attitudes in America survived well into the 19th century, however already in 18th century the Puritanic regulations of life in the colonies started to lose their allpervading influence not only because of new immigrants were mostly non-Puritans, but first of all under the impact of economic development. Incessant work was simply not necessary anymore. Recreation activities began to expand gradually and loose the sense of guilt which has surrounded them. This process started in the southern and middle colonies earlier than in Puritanic New England and initially pertained mainly the well-to-do classes what was especially conspicuous in the South with its slave labor. Frequently recreation was disguised as something else, preferably a religious celebration. The enforcement of the laws prohibiting recreation became more and more lax even in New England although there the laws were enforced "long after practical justification for such an unrelenting attitude had disappeared" (Dulles, F.R., 1965:5). This was the time when Puritanic austerity started to be harmful to society. Almost complete lack of recreation induced the people to drinking which constituted a major social problem (Dulles, F.R., 1965:16-20). Hypocritically taverns were exempt from Puritan sanction because they meant good business for people in control of politics. Benjamin Franklin influenced directly by European ideas early recognized the social and health values of recreation, but only in the 19th century the social attitudes toward travel and recreation changed decisively and the puritanic restrictions weakened. Yet despite the change of social attitudes towards recreation Puritanism was fighting back not only in the 19th but even in the 20th century. (see Larrabee, E., Meyersohn, R.S., eds. 1958:8 and 353; Dulles, F.R., 1965:203-209; Kaplan, M., 1960:150; Wolfe, R.I., 1952:5). The economic prosperity undermined the ascetic ideas and practice, new European philosophical views and further democratization of life contributed to the change. Still more importantly, the internal laws of quickly developing 19th century capitalism played also against puritanism. In the early years of capitalism the initial accumulation of capital for investments into machinery etc. had an overriding importance. However, developed capitalism needed markets, needed consumption and not thrift and abstinence. This contradiction caused further erosion of puritanic principles in America. The work ethic suffered, the value of leisure time and recreation increased.

Thus the change of attitudes towards leisure combined with economic growth resulted in the development of recreation and recreational tourism: summer vacation in the mountains, at the seashore or lake was common among the well-to-do already in early 19th century. In the first three decades of the 19th century a number of summer resorts and spas like Saratoga Springs, New York, White Sulphur Springs, Virginia and Hot Springs, Arkansas were developed in the eastern and central part of the country close to urban centers. By 1857 the big Saratoga hotels could accommodate twelve hundred guests each (Huth,H., 1957:106). The Atlantic City was open for visitors in 1854. However, the real growth of resorts occurred only later in the second part of the 19th century with the development of railroads, e.g. Niagara area was developed in 1870's. The health resorts in the Western part of the continent were developed frequently by the railroad companies as was the case with Banff developed by the Canadian Pacific Railway. But health resorts in North America never achieved the popularity equal to the spas in Europe. The establishment of steamboat connection between St. Louis and New Orleans contributed to the extension of tourist flows towards this city offering among other attractions the Mardi Gras carnival and gambling. Another form of recreational travel in 19th century America were wilderness (or frontier) excursions. Their outgrowth was the National Park movement. As a whole the idea of vacationing and travelling for pleasure took hold in the minds of the people in the second half of the century: the number of beach resorts increased, (Huth, H. 1957:188), the custom of camping out has been introduced.

The participation in all these recreational leisure time activities was almost entirely limited to the wealthy population strata, including the industrial elite which used the leisure time to display their work, wealth, power and status. A sociological analysis of leisure behavior of this elite was provided by Torstein Verblen in his book "The Theory of the Leisure Classes," published in 1899. Verblen, argued in this book that the rich do not accumulate wealth in order to consume. They consume in order to display accumulation. The rich, according to Verblen, display their wealth by not working. He introduced the terms "conspicuous consumption" and "conspicuous leisure." The wealthy elite travelled in 19th century almost exclusively domestically. Only in 1900-1914 and 1920-1929 they travelled to Europe by ship.

In the first half of the 19th century the initiative in provision of recreation opportunities was almost entirely in private hands. However, the second part of the century marks an increased activity of governments at all levels in the provision of recreation opportunities. It began in big cities where the local governments became involved in provision of urban open space. New York City acquired 340 hectares of land to the north of city limits in 1853 and started to develop it into a city park (Central Park) in 1858 when landscape architects Frederick Olmsted and Calvert Vaux won the design competition. Subsequently Olmsted designed many other city parks in North America including the Golden Gate Park in San Francisco, the Berkeley campus and Mount Royal in Montreal.

Soon after these involvements of municipal governments in the provision of recreational urban space also the federal and state authorities have been drawn into the picture. In this respect a great impact had the writings of George P. Marsh. His book "Man and Nature" published in 1864 (the second enlarged edition was published in 1874 under the title "The Earth as Modified by Human Action") warned against the destruction and devastation of natural resources especially of forests. This was the start of the conservation and park movement in the United States. Marsh's ideas influenced the public opinion in favor of granting the Yosemite Valley to California by the US congress as state park in 1864 and of establishing the first National Park, Yellowstone in 1872, followed by Banff National Park in the Canadian Rockies which was founded in 1885.

All over North America public authorities at all levels started to establish community parks, playgrounds, recreation areas, state, provincial and national parks. One of the most important figures in the park movement was the American poet W.C. Bryant (1794-1878) author of "The Forest Hymn". Also famous painters like F.E. Church (1826-1900) and J.F. Cropsey (1823-1900) discovered and depicted the grandeur of American landscape. Among the conservationists the names of John Muir (1838-1914) and David Thoreau (1817-1862) should be emphasized as "discoverers" of nature and advocates of preservation of natural environments.

In 1891 the U.S. National Forests Service and in 1916 the U.S. National Park Service, the first such service in the world, was established. The act creating the National Park Service stated that the purpose of the areas entrusted to it was to "conserve the scenery, the natural and historic objects and the wildlife therein and to provide for the enjoyment of the same in such manner and by such means as will leave them unimpaired for the enjoyment of future generations." (Collier's Encyclopedia, Vol. 17, 1965:171). National forests although primarily created for the conservation of forest and water resources for multiple use purposes also provided much land for recreational use.

Thus, similarly like in Europe, the 19th century revolutionized tourism and recreation in North America. Changes in attitudes, social values, technological and economic progress, growth of democracy and labor unions opened the doors for mass participation in tourism and recreation.

3.8 TOURISM AND RECREATION IN EUROPE AND NORTH AMERICA

1900-1950

The early years of the 20th century up to the outbreak of the First World War was in both Europe and N. America a period of continued prosperity and political stability and consequently good for tourism. There were already up to 200,000 Americans going every year to Europe in early 1900's (Dulles, F.R., 1965:381). New swift and comfortable ships like "Mauretania" or "Lusitania" made the trip very agreeable. In Europe the resorts were thriving both in the coastal areas and in the mountains. The railway transportation was steadily improving its standards of comfort and convenience and also expanding through construction of new lines. All this development has been interrupted by the outbreak of the war in 1914. The period of economic boom in 1920's allowed tourism to develop quickly until 1929.

In North America the mass produced automobile started to compete successfully in 1920's with the railroad making the tourist flows independent from schedules and more universal. On both sides of the Atlantic various enterprises introduced paid vacations for their workers. The working week shortened. All these happenings significantly increased the demand for tourism and recreation, e.g. the number of Americans crossing the Atlantic for Europe reached the peak of nearly 300,000 in 1929.

Seasonal changes in European tourism involved the concept of summer vacation in the Mediterranean. Since tourism started at the French Riviera in late 18th century it was the place to spend winters in relatively mild climatic conditions. Only in 1923 the owners of a hotel at Cap d'Antibes decided to stay open in summer for the first time. Other large hotels followed and the French Cote d'Azur started to function increasingly as a summer vacation area loosing gradually its traditional importance as a winter refuge for the elite. The reasons were the growing democratization of tourism and possibly the influence of American tourists. The ratio between winter and summer tourism decreased form 10.8 in 1921-22 to 5.0 in 1929/30 (Hawkins, D.E., 1980:219). The luxury winter tourism disappeared form Nice in 1930's. The French Riviera stopped to be heliophobic and turned heliotropic.

The depression years of 1930's meant a significant decline for tourism. Many countries and particularly US, undertook programs of public works to alleviate unemployment. Roads were build in recreation areas, reservoirs created with secondary purposes to provide water based recreation. National and State parks improved and developed.

Despite the general slump of tourism in the depression ridden Europe of 1930's some developments also there augured good for the future of tourism. Particularly important was the spread of paid vacations. Pioneering in this respect was the 1936 legislation passed under the Popular Front Government in France allotting nearly 2 weeks vacation annually for workers with at least 1 year service. This was the French response for vacations in totalitarian countries of Germany and Italy regimented by government organizations Kraft Durch Freude and Dopolavoro. The United Kingdom followed France with the Holidays With Pay Act (1938). The duration of paid holidays has been gradually extended after World War II. Thus the period of "congé payée" in France was extended in stages to achieve 4 weeks in 1978 and 5 weeks in 1980's. Some other DCs initiated the introduction of paid vacations during the interwar period by agreements between individual trade unions and employers. The legislatively regulated paid vacation became common only after World War II e.g. in 1963 in West Germany. In United States paid vacations became universal only in 1960's. Without universal paid vacation mass tourism is impossible and this happened only after the World War II in most DCs.

However, despite some positive legislation of 1930's and the improved economic situation of 1935-1939 tourism did not increase appreciably up to the outbreak of the World War II. The war practically put an end to tourism in Europe and severely limited it in North America.

In the early post World War II years the tourist industry in Europe was in shambles. The Swiss government appointed a special commission composed of the best specialists in the field of tourism to assess the future prospect for the industry. The early 1946 report of the Commission unanimously accepted by its members contained pessimistic conclusions. The commission stated that Swiss tourism industry has no chances for the next 30 years. Therefore, the hotels and other tourist facilities should find some alternative use and the personel should be retrained among others in agricultural professions (Kultura, Paris, 1966, 4:105). Fortunately these predictions were wrong: already in 1950 Switzerland had more tourists than in the best interwar year of 1929. Europe and North America stood at the threshold of unprecedented expansion of tourism. Over hundred years after

Thomas Cook initiated the modern tourism an era of mass tourism encompassing the whole world dawned. What was luxury for few in the past became now a necessity for many, certainly the majority in the DCs and still the minority in the LDCs.

3.9 CONCLUSIONS

3.9.1 Changing attitudes towards nature as tourist/recreation resource

Our historical review of travel, tourism and recreation again confirmed the old adage that "change is the only constant in the world" with respect to society's relationship to natural environment as resource for tourism and recreation. Indeed, until the spread of industrial revolution in 19th century in Europe and North America there was little demand for recreational resources, for space. Most people lived in the countryside and in small towns and did not need to travel to resources. The nature, the fields, forests, mountains were right there, just a few minutes walking time away at the most. However, the industrial revolution has brought urbanization, congestion in the cities and lack of open space for urban recreation. The distances between cities and extra-urban resources have grown significantly. In addition even these non-urban resources became more scarce as a result of environmental degradation, urban sprawl and land use conflicts. Thus resources for tourism and recreation have gone through a transformation from imponderabilia to an important treasure which is in short supply and therefore should be handled with care, should be planned and managed as a common heritage of society.

These were changes with regard to nature caused by shifts in material conditions of life. However, there were also other transformations of attitudinal, ideological nature. Indeed, human feelings, attitudes, evaluations with respect to the natural environment, to the landscape change with the time. In other words our present evaluation of the natural environment for the purposes of spending leisure time should be regarded as a historical category. This topic has been discussed in detail in the historical review. Here only some points are summarized.

Greek and Roman philosophy - rational, inquisitive attitude of nature
Middle Ages - negative attitude
Renaissance, Baroque - positive attitude gradually accepted

Enlightment - rational scientific attitude
Romanticism - romantic emotional attitude

For Rousseau nature was a certain perfect paradise situation: this was an ideal standard to which the mankind should return, because the individual was born by the nature pure and good and has been spoiled only subsequently by the influence of society. The return to "Mother Nature" was an emotional, poetic and religious trend (deism) which laid the basis for the concept of Romanticism in late 18th and early 19th century. The romantic notion of nature as an integral whole established the approach to landscape and to its aesthetic values which prevailed for many decades. Even the British alpinists, the first conquerors of the lofty Swiss peaks attached to their achievements something mystical, some irrational cult of spiritual contact with the "Untouched", an almost religious communion with nature. What a difference with our contemporary unromantic, sober, matter of fact attitudes, when the alpinists climb the mountain tops "because they are there" (Sir Edmund Hillary) treating nature as an athletic challenge, a sort of testing ground of their physical and technical abilities.

3.9.2 The evolution of attitudes towards leisure time

Looking back at the history of leisure and recreation we see that some historical periods were marked by more leisure, recreation and recreational travel (e.g. antique Greece and Rome) some by almost none (e.g. the Middle Ages), but we may generalize that in all of them there was relatively little leisure for the masses. Only the privileged elites could enjoy it because they had the economic means based on political power and consequently they could enjoy their leisure time.

In some preindustrial societies the life was characterized by an integrated pattern, a unity between two segments of existence: work and non-work. There was little formal distinction, no clearly delineated division between work and recreation in gathering, hunting, fishing and even agricultural societies. Non-work activities were based on magic, religion or folklore and participation in them was as mandatory as work itself. The work and non-work activities were inseparably intertwined in a total life style. Thus these societies achieved a sort of "balance" of work and non-work. Although in the course of history the division between work and non-work became gradually more clearly delineated it was not until the industrial revolution that there emerged a clear line distinctly dividing them from each other.

Some of recreational activities in the preindustrial societies had strongly pronounced teleological orientation i.e. had to serve some purpose or aim. Thus they were forms of military training, e.g.Olympic sports in ancient Greece or medieval jousting and archery, rather than for their own sake as enjoyable pastimes. This teleological approach prevailed in 19th century and well into the 20th century when recreation became a distinct differentiated element within the social structure, within the rhythm of life. Indeed, in addition to military considerations, recreation was aimed at increasing the work productivity. In this way recreation was supposed to provide "cannon fodder" and productive labor force. Also other moral and social aspects of recreation have been emphasized like reduction of criminality and drug addiction. The change of values occurred only after the World War II when recreation became more and more appreciated per se, for its own sake, as pure enjoyment, fun. These hedonistic approaches coexist with teleological approaches but seem to gain constantly in relative importance.

Analyzing the past we may also generalize that tourism and recreation as activities engaged in during the leisure time have been loosing, especially in modern time, the negative social stigma attached to them in the past. People no longer need to seek an ulterior motive (magic, religion, military exercise, restoration of mental and physical capacities for work etc.) to have pleasure in engaging in leisure time activities. Homo ludens and homo viator acquired equal rights with homo faber as indispensable elements of human civilization. This change of status may be illustrated by the fame and glory associated with leading entertainers, sport champions etc. Remember that in 19th century to be a professional entertainer carried a disreputable stigma.

3.9.3 The evolution of travel into modern tourism

This author reserves the term "tourism" only to modern socioeconomic phenomenon of temporary mass movement of people mainly, although not exclusively, for recreational (pleasure) reasons. Modern tourism, like democracy, is available for practically all members of economically developed society. However, if we look back at the history of the preindustrial era we find that only a very small percentage of people participated in travel. The vast majority was immobile throughout their lives. Travel in this period meant painful effort, hardships and danger. It was inconvenient, expensive and above all slow.

Probably the most important prerequisite of modern tourism is the use of modern means of transportation which sharply distinguishes it from travel in the preindustrial era. For thousands of years the movement of people was limited by the ultimate speed a horse could attain. Preindustrial travel was unable to overcome spatial and physical barriers that isolated various regions of the world. In this sense for most of human history little progress was achieved. Only the introduction of the steamship and the railroad and later of automobile, motorship and aircraft could in historically short time introduce changes of revolutionary proportion not only with respect to speed of movement but also volume, security and comfort. The two other prerequisites of modern tourism, increased discretionary time and income result from tremendous progress in labor productivity achieved in the course of industrialization and later automation. Indeed, only the advent of industrial revolution with all its economic progress created the prerequisites for transformation of travel into modern mass tourism. The discussion of these prerequisites constitutes the contents of the following chapter.

CHAPTER 4

PREREQUISITES FOR MODERN TOURISM

4.1. INTRODUCTION

The most characteristic feature of modern tourism and recreation in the DCs is that they are really mass phenomena affecting people of all ages of almost all occupations and material status. In the past thousands of years they were limited to a tiny fraction of population who enjoyed a privileged position as a result of concentration of wealth and power in the hands of few. The rest of population had little means and time to enjoy leisure. But now what was the privilege of few is the right of all.

The Industrial Revolution which started in England in the second half 18th century and spread over most of Europe and North America in the 19th century brought a tremendous technological and economic progress to these areas, a progress which first of all substantially increased the productivity of labor. This has enabled the society to save, to invest and to accumulate wealth at a rate never attained before, e.g. the US industrial labor productivity increased more than five times between 1850 and 1950 (Sellin, T., 1957:17). These spectacular achievements in the industrial age created the necessary economic prerequisites for the participation of the entire populace in the leisure activities.

In the past civilizations the low productivity of labor did not allow leisure for masses even if the socio-economic disparities between the wealthy and privileged classes on one side and the poor and exploited on the other side would disappear. The productivity of labor was so low that hardly any savings, investments and accumulation of wealth was possible. Possible was only bare subsistence for the masses and luxury for the tiny minority of powerful and wealthy. Thus slavery and the feudal system gave leisure to few. With the industrial revolution came the advanced technology and greater use of capital. The output of goods grew dramatically thus increasing incomes and savings. In this way the prerequisites for mass tourism and recreation have been set up. Only the machine created the possibility of leisure for all, only on the basis of industrialization could effective mass demand for tourism and recreation services be realized. This was proven by the historical developments of 19th and 20th centuries: tourism and recreation has increased as a consequence of industrialization, with a certain time-lag, of course.

The best example is Great Britain which due to the early industrial start for many decades of the 19th century enjoyed an virtual monopoly in the field of effective tourist demand. Thus the causal relationship between industrialization and modern tourism and recreation can be firmly established, a relationship which created prerequisites for the liberation of masses from long hours of every day toil filling almost exclusively their lives. The whole society and not only the elite, for the first time in its history approached the level of economic development which provides the discretionary time and income for the majority of its members to participate in leisure.

However, there was one factor absent preventing for many decades the people to benefit more fully from the achievements of the industrial age: lack of political and economic democracy. There was an undeniable but relatively slow progress of tourism and recreation in the 19th and the first half of the 20th century in terms of gradually increasing participation. However, a substantial part of population still did not participate fully: they enjoyed their short leisure time on a fairly modest scale either at home or in the immediate vicinity of their residence not having enough discretionary time and money for more extensive and active participation. Real mass tourism and recreation was impossible without democratic changes in the political and socio-economic structure leading to more equal distribution of wealth and legal provisions for the increase of leisure time for all. Although these democratic trends gained momentum in the second part of the 19th century, largely due to the work of social reformers and the development of trade unions, they were not universally accepted until after the second World War when practically the whole society started to benefit from the technical and economic achievements of the modern age.

All this technical and socio-economic progress which has transformed the developed world in the 19th and 20th century has created the necessary prerequisites for modern tourism and recreation. These prerequisites will be discussed in this chapter: increased discretionary incomes, increased discretionary leisure time and greater mobility due to modern transportation. These three factors are working in a multiplicative way rather than additive one which leads to an unprecedented growth of demand for tourism and recreation.

4.2. DISCRETIONARY INCOMES

The first major requirement enabling a person to participate in tourism and recreation is the availability of discretionary income. In this respect the industrial age created the necessary prerequisite in contributing to the unprecedented economic growth and the ensuing increase of wealth. Indeed, the leading industrial nations are at the same time the main tourism generators. The real net national product per capita grew in the US about four times in the period between the decade 1869-78 and the decade 1944-53 (Sellin, T., 1957:17). The United States real income per capita (i.e. in constant dollars) has about doubled in the forty years between the mid-1920s and mid 1960s growing at an average rate of nearly two percent annually (Clawson, M., 1972:444).

The rate of economic growth has accelerated after the Second World War until the 1973 energy crisis. The gross national product in the German Federal Republic grew three times (in constant prices) during 1950-1966 (Tourismus 1980, 1968:19). Between 1950 and 1962 the real wages of industrial workers in West Germany increased 101 percent. (Pfister, B., 1964,3:114). The average real incomes of weekly-paid British workers increased by some 50% between 1955-1967 (Patmore. J.,1970:19). The improvement of the general standard of living may be still better illustrated by the more equitable distribution of incomes. In this respect a definite shift from lower to higher income brackets was visible until about early 1980s. The homogenizing trend in income distribution can be illustrated by the change from a pyramid of incomes to an onion-form where middle incomes constitute the largest group. This phenomenon resulted not only from the general rise in income level but also from changes in professional structure of population: as the level of professional skills grew the share of untrained labor in the labor force decreased drastically and the middle class professions become predominant. These structural changes in the labor force resulted in shifts in the distribution and general rise in income levels. Here are the figures for the United States (Zedek, G., 1968:91). These shifts in income distribution could be regarded as typical for most DCs until the early 1980s.

Annual Family Income(in constant $)	1950	1965
$ 10,000 and over	7 percent	25 percent
7 000 - 10,000	13 "	24 "
5 000 - 7 000	20 "	18 "
3 000 - 5 000	30 "	16 "
3 000 and less	30 "	17 "

The growth of disposable (after-tax or take-home) income was accompanied by important changes in the structure of expenditures. Although the expenditures on so-called necessities of life (food, housing, clothing) increased absolutely with the rising income, the rate of this increase tended to diminish leaving larger share for expenditures on non-essentials. This part of income is referred to as discretionary income and is growing at an accelerated rate in comparison with the total net income in the DCs. In West Germany the discretionary income (excl. expenditures for food, housing and clothing) increased from 8% in 1952 to 32% of the total net income in 1970 (Tourismus 1980:60). In US the discretionary income grew from 8% of total personal or gross income in 1955 to 20% in 1970 (Barnet E. 1971, 4: 148). In Canada personal disposable income adjusted for increases in price level, expanded at an annual average rate of 5% between 1961-1971. Despite significant absolute growth of outlays for food, shelter and clothing (about 40% in 1959-1969), their relative weight in the average family's total annual expenditure decreased from 58% in 1959 to 53% in 1969. Thus the share of discretionary expenditures increased: e.g. for travel and transportation the increase was from 13.3% to 16.5%, for recreation, reading and education from 4% to 6.2%. (Bank of Montreal, Business Review, March 29, 1972:1-2).

With the growing share of discretionary income in the disposable grows the freedom of choice of an individual how to spend money. A growing proportion of discretionary income is being spent on tourism and recreation. According to F. Dulles (1965:391-392) expenditure for recreation amounted in USA about eight per cent of personal income in 1935 and twelve per cent in 1962. The expenditures for tourism in the DCs grew in 1960-1969 at double the rate of other private expenditures. Some countries showed even more accelerated rate e.g. in West Germany the increase of expenditures for vacation grew in 1956-1965 by 410 per cent whereas the disposable income grew only 121 per cent (Der Spiegel, 1970, 28:83).

The tourism expenditures grow usually relatively slowly in the initial stages of the increase of disposable income. Later their growth rate is higher than income's. Only after the most urgent needs in improving the living conditions are satisfied the need for vacation comes to the fore with a time-lag. Tourism gradually loses the character of luxury and acquires the status of necessity for most people in DCs and increases its share in the overall expenditures.

The optimistic picture of vigorous economic growth and the accompanying development of tourism and recreation, which occurred at an accelerated rate as compared with the growth of incomes, pertains to the period 1950-1973. Since the 1973-75 energy crisis the economic performance of the DCs (with some notable exceptions like Japan and West Germany) has been less impressive. The real disposable per capita incomes have grown at a very modest pace in 1970s and 1980s These discouraging trends, hopefully short or medium range, result from various developments such as the structural changes in employment: relative or even absolute (in the US) drop in the number of well-paid unionized manufacturing jobs and the expansion of less rewarding, mainly non-unionized service jobs. For example, "on the average, wages in the US service industries are only 53 percent of those in the manufacturing." (Tourism Itelligence Bulletin 1987 Feb.) To allow the ends to meet multiple-income (typically double-income) families are on the increase, full-time workers work over-time, young and retired people and married women work part-time or even full-time. Moon-lighting flourishes. Thus, if the per capita disposable real incomes grew, incomes per worker have decreased in some DCs like USA (New York Times 30-1-1988). The average employment income of Canadian workers declined in real terms by 3% between 1980 and 1985 (Winnipeg Free Press 14-4-1989). Another discouraging phenomenon is the persistency of poverty in many DCs. The trend to income equalization observed in most DCs up to about 1980 has been reversed in the US, where the gap between rich and poor has been increasing in 1980s (Winnipeg Free Press 26-3-1989). The same trend prevailed in Canada ("The Family in Canada" Statistics Canada 1989) Median family incomes stagnated 1979-85 (ibid.) In the US the percentage of people under the poverty line was 13.6% in 1987. These people, largely single parent families and minorities, have no discretionary income to participate in tourism and have to be totally written off as a potential market.

All these unfavorable developments would mean theoretically the flattening of the development curve for leisure time expenditures or even its decline. In fact, despite some temporary setbacks during deep recessions the growth of leisure expenditures has been greater than the growth of real GNP. This means that people do not regard tourism and recreation as non-essentials, as luxuries. Indeed, they have become necessities of life. Therefore, the old rules do not apply. It is true, however, that the market for recreational durables is rather flat. However, the reason is not so much money but rather the availability of time required to use these items. Thus time not money becomes the constraint.

To provide tourism services to the masses in the late 20th century the societies of the DCs are making a number of adjustments. These adjustments are aimed at people with modest incomes to make tourism affordable to virtually everyone. An example of such measures is social tourism i.e. tourism subsidized by public or private funding. The beneficiaries of such funding are usually children, young or handicapped people. Sometimes companies subsidize vacations of their employees (Western Europe, Japan). In Eastern Europe and the Soviet Union the source of subsidization are special Trade Union funds. Social tourism in the Soviet block countries is the prevailing form of tourism in contrast to market economies where it plays only a marginal role. Social tourism is practiced mainly at national level. This means that it involves mainly domestic tourism. In this sense the Japanese government's generous subsidization of overseas school trips seems to be atypical (Tourism Intelligence Bulletin 1987, Feb.). Sometimes the governments promoting social tourism persue certain policies aimed at fostering patriotism, cementing political unity within the country (e.g. interprovincial youth exchange programs in Canada). On the international scale the International Bureau of Social Tourism (BITS) is attempting to improve understanding among nations of the world but does not enjoy generous funding.

Another set of methods to make tourism affordable to all are the trends to self-catering vacations, budget accommodation, camping, time-sharing, house-swapping, youth-exchange, charters, package tours, intermodality arrangements, travelling shorter distances for vacation etc. These are measures which enable tourism to acquire truly mass dimensions despite individual income limitations and/or economic recessions. When the individual incomes increase and the economy is in good shape people spend more money on tourism, use luxury facilities and travel farther away from home. A good example is the Japanese tourism market which has been developing at a record pace in the 1980s. After decades of hard work and limited leisure time the Japanese tourists, well supplied with strong yens, are "invading" the overseas destinations in millions (Tourism Intelligence Bulletin 1988, October and 1989, February).

To sum up the discussion on the impact of discretionary incomes on tourism in the majority of the DCs one may generalize that the stagnating or even declining incomes per worker and the increasing incomes per capita of population, resulting from the growth of multiple-income families, represent two opposite trends. On balance the availability of discretionary incomes for leisure time expenditures is improving rather slowly in the DCs. In the long-range it is reasonable to expect that in peaceful conditions the relentless rise of real per capita incomes in the DCs will continue and this surely will benefit tourism. Certainly, the average distances between origin and destination will increase and the leisure expenditure will continue to increase its share in the discretionary income. One has also to take into account that tourism spending is growing as a result of another favorable long-range development: leisure tourism is increasing its share in tourism spending at the expense of business tourism and it correlates with the growth of real discretionary per capita incomes as compared to business travel which is limited by the growth of GNP, a less dynamic factor. Another development, favorable to leisure time expenditures is the reduction of the volatility of the busines cycle resulting from the domination of DCs' economies by services. The service industries are known to oscillate less in output and employment than the primary and secondary economic sectors. The rapid increase of productivity of labor in all sectors (including more recently the service sector) accompanied by decreasing rates of growth of labor force as the "baby bust" generation starts working, also augur favorably for future growth of the living standards. These developments are beneficial for the growth of tourism which is closely associated with augmented discretionary incomes.

4.3. LEISURE TIME

4.3.1. Work and Leisure

In addition to discretionary income people need discretionary (leisure) time in order to engage in tourism and recreation. What is "leisure time?" In order to explain it better, let's start with total time which indicates the total number of hours, days etc. available during weeks, months, years or the total of human life. This is the basis for time budget planning: total time in a week is 168 hours, in a year about 8760. Part of total time is "essential time" composed of work, work related time and subsistence (maintenance) time. Work related time includes commuting to work and preparation for work. Subsistence (maintenance) time is filled with activities necessary to sustain life like sleeping, eating, personal hygiene. Thus

Essential time = work time + work related + subsistence time

If one subtracts the "essential time" from "total time" one obtains "leisure time" which means discretionary time, relatively free of commitment. Thus

Total time = essential time + discretionary (leisure) time

As indicated in the chapter on terminology there is no clear-cut division line between leisure and "non-leisure" time (or leisure time and essential time). In order to indicate the existence of such "gray" transition zone between them one may use the term "optional time." Some time slots within it vary according to the degree of commitment: they may be allotted to either the essential or leisure time. Following equations express this

Optional time = time committed to various degrees +
discretionary (leisure) time
Total time = essential time + optional time

These are the notions used for personal time budget planning. Many of them are highly individual. However, if we want to discuss the availability of leisure time for tourism and recreation then we focus on working time in the simplified assumption that

Total time = working time + leisure time

Thus the decrease in working time increases time available for leisure. Work time is normally measured on the weekly basis. Standard working week is legislated or in some way set by the company. Forty hour standard work week is now universal in almost all DCs. It differs from the week actually worked which is for a full-time employee longer as a result of overtime. Another statistical parameter is the average working week which includes all employees, both full-time and part-time. Despite the lengthening impact of overtime work the average working week is shorter than the full time standard work week because of the part-time work which in recent years has a tendency to grow faster than full-time work.

4.3.2. The Increase of Leisure Time

The increased productivity of labor and technological progress characteristic for the industrial age has brought about an enormous quantitative and qualitative development of production, creating an abundance of goods and as a consequence increasing affluence of broad masses of population. This spectacular economic progress gave the consumers not only the financial means to enjoy leisure but also more leisure time as a result of shorter working hours and increasing paid vacation. Many countries introduced legislation regulating the leisure time for the working people. The right to leisure for all people has been stipulated in the Universal Declaration of Human Rights adopted by the United Nations General Assembly on December 10, 1948. Article 24 of this declaration states: "Everyone has the right to rest and leisure, including reasonable limitation of working hours and periodic holidays with pay." ("Everyman's United Nations." 1968:589). These gains of the working people have been called by Knebel "touristic emancipation" (Knebel,H.J., 1960:38) with the assumption that a substantial part of this increased leisure time is being devoted to tourism and recreation.

The struggle for the reduction of work time has been after the wage demands the second most important aspect of the trade union activity. This struggle was especially difficult in the North American society with long traditions of puritanism emphasizing the work as the main purpose of existence, as the greatest good. The resistance against the new forms of recreation has been treated by many authors in some detail (Kaplan, 1960, M., 1960:150-151, Dulles, F.R., 1965:203-210 and 386-387, Larrabee, E., Meyersohn, R., 1958: 8). Even in relatively recent times there were opinions criticizing the shortened working hours. As an example of these pronouncements may serve a 1926 quotation from John E. Edgerton, President of the National Association of Manufacturers: "I regard the five-day week as an unworthy ideal ... More work and better work is a more inspiring and worthier motto than less work and more pay. It is better not to trifle or tamper with God's laws" (Larrabee, Meyersohn, 1958:353). E.H. Gary of US steel declared in 1926 that the five day work week was "impractical, it would imperil the competition with Europe and it violates the commandment: "Six days shalt thou labor" (Kaplan M. 1975:4). Another more recent example may be found in the editorial entitled "Prescription for Happiness" in the "Toronto Globe and Mail" April 19, 1952, in which the writer took it as axiomatic that the worker's leisure was misspent, and that "spare time has for many become a worse bugaboo than the work from which they sought to escape." The suggested cure is as follows: "In contrast to this menacing situation, there is a life which affords great satisfactions to those who have the personal character to appreciate them. Whatever the burden which sixteen hours work a day might seem, for thousands of years the majority of mankind lived and thrived under it. During that period of history, the greatest books were written, and man's deepest thoughts about the universe and his own being were expressed. Life then was not filled with distractions. There was ... no time for them. Men were forced by their way of life to come to terms with themselves as well as with nature. They were happier then, and they could be now, if they accepted the same simple prescription: hard, satisfying, challenging work." (Wolfe, R.I., 1952:5). The advocates of long hours of work voice their opinions even in the 1980's: "Putting obvious economic rewards aside, work contributes to our spiritual betterment. Diligent effort and long hours bring dignity and emotional reward." (New York Times 19-4-1986).

Despite the resistance of puritans and people with vested interests in keeping the working masses laboring for long hours the course of the history could not be arrested: the working time has decreased slowly and gradually. It took over 100 years of gains to cut the work week in half. This process may be illustrated on the following example:

The average weekly working hours in the national economy of North America:

1840 - 84
1850 - 70
1870 - 68
1900 - 60
1910 - 55
1920 - 50
1930 - 48
1940 - 47
1950 - 45.5 in Canada and 41.7 in US
1960 - 41 in Canada, US below 40
1970 - 38 (US)
1975 - 36.1 (US)
1979 - 37.5 (Canada)
1985 - 35.3 (US)

(Compiled from various sources)

Working people nowadays enjoy more leisure time than have to work: work takes only about 20-25% of total time whereas leisure constitutes more than one third. In terms of the entire human life the average person spends nowadays 65,000 hours or 10% of the life, working (Statistics Canada, 1978, 1:1).

However, the statistics showing the decreasing work time in the post World War II era, unfortunately, are not reflecting the reality correctly. The reason is that the average work week has decreased mainly as a result of significant rise in low-paid part-time jobs (to a large extent women and teenagers and retirees employed mainly in services) and to a lesser degree due to other arrangements like job-sharing and work-sharing. In fact the full-time standard work week has remained unchanged in most DCs at the level of about 40 hours. Significant number of full-time workers would prefer part-time jobs if the social benefits (pension plans, health insurance, vacations, job security etc.) were prorated or guaranteed. Changes in legislation pertaining part-time jobs are long overdue. This does not lie in the interests of employers who support part-time work because of better productivity and the fact that they pay less per hour than for a full time job and do not provide social benefits.

The increase of leisure time is correlatd with improvements in labor productivity. However, unfortunately, there is a development which slowed down the improvements in overall productivity of labor between about 1950 and 1980. The productivity in the rapidly expanding tertiary sector (services) did not keep pace with primary and secondary sectors which were shrinking relatively and in some countries absolutely in terms of employment. For years the structural changes in employment manifested in the transfer of labor to the tertiary occupations (including part-time labor which is mainly unwelcomed in secondary sector and widely used in services) served as a kind of safety valve against the unemployment problems connected with the sky-rocketing productivity in primary and especially secondary sector. At the same time, however, this process tended to depress the overall gains in productivity of labor as a result of sluggishness of the tertiary sector in this respect.

Thus, the increased relative and absolute demand for labor in the tertiary sector and the resulting shifts of employment to low productivity services, has offset partially the labor-saving impact of automation and has contributed to increased prices for services. "The more time we save in making goods, the more time we spent for services ... The very nature of the services makes them greedy for time" (Burck, G., 1970: 87-88). Thus low-productivity service employment was eating more and more of total work time. Burck came to the conclusion that "leisure society is a myth because more and more man-hours will be needed to provide ever expanding services" (Burck, G., 1970:89) and concludes with a somber note that "even if the performance of services gets better, the prospects for reducing the hours on life's treadmills very much will keep receding into the future" (Burck, G., 1970:166).

One may agree with the diagnosis that service sector as a low productivity sector tends to decrease the gains in leisure time. However, in 1980s computers and other advances in communication technology invaded this sector leading to substantial productivity improvements in office work. Nevertheless, still the productivity in large part of services is low and is unlikely to increase significantly in the near future. A good example in this respect is the quickly increasing sector of personal services.

Another problem related to work time is the fact that it is not distributed evenly among working population. The unionized employees, typically factory workers, are the winners, whereas many non-unionized employees, typically employed in services, most self-employed and also senior executives and professionals are still working well over 40 hours a week, e.g. US professionals average 52 hours work a week (Tourism Intelligence Bull. 1988, Dec.). All these workers belong to "have nots" as far as the leisure time is concerned. The most successful in shortening the work time have been those performing physically or mentally difficult jobs (usually with strong union backing) like steel workers, miners, airline pilots. Thus one could say: "Leisure for the masses and work for the classes," a reversal of the situation in the preindustrial period. Enzensberger quotes in the same context an ironical German term "Freizeitproletariat" (leisure time proletariat). Fourastié remarks that "the present situation is that the director of a ministry in France works 3,000 to 3,500 hours per year, and his office messenger 2,500" and that "it is felt that a director who doesn't pass 60 hours a week in his office is not doing his duty" (Fourastié, J., 1960:173).

The results of these trends on the labor market are not favorable in terms of availability of leisure time. Indeed, if one studies the UN labor statistics one may find that the recent gains in shortening the work time have been modest, to say the least.

4.3.3 Future prospects: more leisure time?

The significant pace in shortening the work time during the last 100-120 years and the technological achievements of the contemporary period, variously called "the age of mass consumption" (Walter Rostow), "post-industrial age" (Daniel Bell) or "technotronic era" (Zbigniew Brzezinski), prompted the futurologists to extrapolate this trend into the future and predict dramatic gains for leisure time, e.g. Kahn anticipated in "The Year 2000" a drop of work time per year to 1,100 hours (Kahn H. 1967). These optimistic predictions have not and will not come true because alongside factors favoring the extension of leisure time there are also powerful forces working in the opposite direction.

First of all, one has to discuss the people's preferences between increased leisure time and income (leisure time-money trade-offs). In what way do the people prefer to reap the benefits of increased productivity: in form of more leisure time or of more income? Time and money are interchangeable and people are constantly confronted with the choice. Throughout the major part of post World War II period, given a choice, people in majority of cases have been deciding to forego leisure in favor of more income. These desires of majority have been translated into trade unions bargaining positions: the most important priority of North American trade unions is getting pay increases, improving job security, struggle against subcontracting, privatization, part-time jobs, etc. Fringe benefits, like longer vacation time, are of secondary importance at best. Let us take an example of a housewife whose leisure time has increased substantially as a result of rising productivity of housekeeping work (time-saving household appliances, non-iron fabrics, easily cleaned plastic surfaces, refrigeration, convenience foods etc.). How does she use the new gained leisure time? Mostly she goes to work - at least part-time thus substituting income for leisure. Similar tendency is noticeable when the standard working week of her husband is shortened below 40 hours. He works overtime if he can or he takes a second job and thus increases the ranks of multiple jobholders - so called moonlighters. People also work more at home: the high cost of various repairs compels the people to assume a do-it-yourself attitude which to a great extent has nothing to do with "hobby" and constitutes an unpleasant necessity. Much of this additional work does not find the reflection in the statistics. Indeed, people have less leisure time than it appears.

The preferences of many, probably most, workers for more money and less leisure is obvious. Only some young people, employees close to retirement age and married women, many of them part-timers, are more likely to prefer shorter work to additional income. However, married men, especially with children to support, seem to opt rather for more income than leisure.

In economic terms the choice between the leisure time and income is complicated and future developments difficult to forecast: such factors as purchasing power of money, trends in productivity, growth mania at the individual and national level, increased scarcity of time needed for consumption etc. have an important impact in this respect.

The conventional wisdom of rapidly shrinking work time in the modern era is at variance with the conclusions drawn in 1960s by J.K. Galbraith (Galbraith, J.K., 1967:363) that the reduction of work time after the Second World War actually ceased. According to him the "optimists" tend to extrapolate the reductions of work time brought by the Industrial Revolution into the future and do not take into account the modern trends, such as increased rates of participation of women in the labor force, do-it-yourself work, moonlighting and overtime. All these phenomena tend to limit leisure. The expectations of dramatic growth of leisure in the near future (Fourastié, Kahn) sound untrue in this context. There could be also a certain amount of "wishful thinking" in these optimistic predictions.

Discussing the reasons why people prefer to work longer thus increasing their income rather than have more leisure, J.K. Galbraith in "The new industrial state" comes to interesting conclusions. He claims that since the end of World War II the working week actually moderately increased. The reason is that people work more in order to earn more money to satisfy their growing wants. According to J.K. Galbraith these growing wants are being generated and manipulated by the "industrial system" (big corporations) who persuade the consumer by means of news media that goods are more important than leisure time (Galbraith, J.K., 1967:363-367).

Linder concurs with Galbraith in pointing out that the people in the DCs actually have increased their work week since the Second World War: the statistics proving the opposite are misleading because the statistically measured average working week (but not the standard full-time work week) is being reduced by the spreading practice of part-time work among women, teenagers and also workers whose standard work week has been reduced and who take up additional part-time jobs. This statistical reduction of the work week does not really signify any decline in work input, especially as many of those now working part-time were previously not gainfully employed (Linder, S., 1970:136). Linder admits that the overtime is included into the average work week statistics increasing it. However, if overtime takes the form of multiple job-holding, the average work week will be reduced as the second job is usually part-time. Linder comes to the conclusion that on an average, in the USA the weekly hours of the full-time, non-farm work force have remained constant during the postwar period at a level of about 45 hours. If we look at components of the overall data, we find that in this period, the number of people working more than 48 hours a week has doubled or risen from 13 percent of the work force in 1948 to 20 percent in 1965. In l980s some 5 per cent of the work force held more than one job in the US. For married men the figure is substantially higher. There has been no decline in multiple job-holding. As this practice cannot have been widespread until a decline in the work week began, it is reasonable to conclude that this phenomenon has instead been gradually increasing. Some estimates indicate also that multiple job-holding has doubled since 1950 (Linder, M., 1970:136). In the United Kingdom one-third of male workers work overtime, an average of 9 hours per week, another one in ten have some second job (Dower, M., 1970:253). Similar trends have persisted also in 1970s and 1980s. A Harris survey indicates that "the average number of hours Americans work has increased by 20 percent over the last 15 years. Leisure time has decreased by 32 percent". (Tourism Intelligence Bull. 1988, Dec.). In Japan, known for the longest working week among the DCs, a catach-up trend has been developing in the second half of 1980s, a trend to more leisure. "The government has decreed that the average working year, now 2,150 hours should bc cut to 1,800 hours in 1992 (The Economist 25-3-1989). The gradual transition to a five-day work week has already begun and by late 1988 encompassed already about 33% of workers (Time 30-1-1989). The government has set a goal to reduce work time from the current average 261 work days per year to 223 days by 1992 (ibid). Leisure-promotion campaigns are conducted. Tourism is increasing well over 10% annually. However, these trends in work-leisure relationships can be regarded as mere catch-up to other DCs. The Japanese work not only harder but also longer than their counterparts in other DCs: 500 hours longer than the West Germans and French and 200 more than Americans and British (ibid.).

However, since 1970s a new trend in people's preferences between leisure and income (leisure time - money trade-offs) started gradually to appear. The number of people preferring leisure has relatively increased although they still constitute a small minority. This trend is noticeable but its reasons are not entirely clear. One could presume that among them is the growth of antimaterialistic attitudes and the rising interest and concern for environment, both natural and human. There are also indications that the society is slowly moving towards a balanced view on work and leisure. People appreciate work, especially work which they enjoy. At the same time, however, they value highly leisure as they learn to use it. One of the most important aspects in this respect is education. Educated people enjoy leisure time more and continuing learning process will increase its importance during leisure.

As a result of these changing attitudes more and more people nowadays are starting to prefer creative leisure than money. Some take early retirement or even completely drop out of the job market preferring to live close to nature in the countryside. There is an increase of part-time work either voluntarily or as a result of economic necessities connected with work sharing and job sharing arrangements.

In addition to the attitudinal shifts the trends which favor increased leisure have been strengthened since 1970s when the economic recessions have decreased significantly the availability of jobs on one hand and the process of computerization has dramatically increased the productivity of labor, initially in primary and secondary sectors. Its impact started to spread into the tertiary sector (especially office work) in 1980s. Computerization has already released millions of employees in the DCs and its spread into the tertiary sector means that one cannot expect that significant new employment will be created in this sector. Nevertheless, employment in the tertiary sector (to which belong tourism and recreation) still offers relatively best prospects nowadays because of slower productivity gains as compared to the other sectors. In fact, employment in the tertiary sector has surpassed in the leading DCs 60% of the total labor force. However, despite of relative future employment gains of this sector one has to express doubts if any major transfers of labor to tertiary occupations (as it was the case until recently) will be possible in the future. The tertiary sector is rapidly losing its role as safety valve for unemployment in the primary and secondary sectors. Therefore, it becomes clear that in order to combat increasing unemployment, measures must be taken to decrease work time.

The efforts to shorten the full-time work week without loss in pay and benefits is encountering significant obstacles. The main obstacle raised by the business community in Western Europe and North America is international competitiveness, especially against the Japanese and other Asians who are working long hours at relatively low pay. The representatives of business insist that hardly any country can afford such an increase of costs unless the shortening of the work week proceeds more or less parallely in all DCs. And this seems to be happening at a snail's pace. Reductions of standard (legal) work week without loss of pay and benefits are long overdue. The legal standard work week in most DCs is still about 40 hours a week. The North American trade unions presently aim at 35 hours in the long range but it seems that Western Europe will be there earlier because of lower priority associated with this issue in North America.

The main argument in favor of the work time reduction is the above mentioned necessity to decrease unemployment. Indeed, many researchers (among them Nobel prize winner in economics Wassily Leontief - USA) plead for work time reduction as a measure to offset the impact of modern technologies on the labor market. In fact, the number of jobs would increase if the people work shorter time. Additional arguments are as follows:

1. The full-time worker today works too long hours with too much overtime. This increases health risks (Prevention,Sept. 1984:77) and various social problems (drug abuse, family disruption, etc.). Work time reductions are certainly aimed at improving the health standards and labor productivity.
2. More jobs are routine and boring. This increases worker's dissatisfaction.
3. The increase of leisure time stimulates leisure industries creating new jobs in this field belonging to the modern expanding tertiary sector.

The decrease in work time is crucial in the reduction of essential time and thus in the expansion of leisure time. However, it seems appropriate to mention new developments in some parts of North America pertaining to work related time and subsistence time. In the post World War II period the work related time expanded as a result of the increase in the commuting time due to suburbanization. However, commuting time has been shrinking since the 1970s under the impact of energy crises, the spread of home computer work ("telecommuters") and rapid expansion of suburban and adjacent extraurban areas not only as "dormitories" for central cities' workers but also as sites for secondary industries and above all for services. The subsistence time also has decreased in 20th century due to the conspicuous improvement in productivity of housekeeping work and shopping facilities. However, the hours for eating, sleeping and personal hygiene have remained relatively constant. These trends in reducing work related and subsistence time will continue in the forseeable future.

In addition to the anticipated, rather modest, reductions in the work time one has to expect also some significant changes in its structure. This is also important for leisure time planning. Both work and leisure are becoming more flexibly arranged during the week. Unlike the traditional slots of Monday to Friday work and weekend leisure more people will be working on weekends and enjoying leisure on weekdays. There are various reasons for these structural changes: compressed work week requires the introduction of alternating work shifts, considerable capital invested in sophisticated equipment requires the intensive use of facilities even during the weekend to offset high fixed costs and remain competitive. One could imagine various arrangements of work time within the week (including the weekend) or between different weeks e.g. three weeks of 6 days work followed by a free week without reduction of total work time. Anyway, it seems that the traditional distribution of work and leisure during the week will be destroyed. The result will be more even use of general and leisure infrastructure, more flexible choices for workers in terms of more convenient distribution of work and leisure. Other structural changes indicated by pollster Luis Harris (Newsweek 28-3-1988) and Graff require more research. Graff indicates that the number of hours spent watching TV increased in US between 1970 and 1985 by 25% (Graff J. 1987:11). The question is: will the growth in TV watching and video cassette recorders ownership turn us into stay-at-home society? What will be the impact of this development on tourism and out-of-home recreation?

The analysis of the current and probable future trends in relation to leisure time leads to the conclusion that it will be expanding quantitatively, although at a slower pace than anticipated by the futurologists. Especially Japan and the Asian NICs (Newly Industrialized Countries) are expected to increase the leisure time for their working people to catch up with Europe and North America. Also the aging population of the DCs indicates that the retirement slot of leisure time is and will be increasing significantly. From the qualitative point of view there have been also some important changes in the 20th century: "the value" of time has increased from decade to decade by immensely enriching its contents although it remained physically the same. This occurred because of technical and scientific progress. Compare how far one could travel 50 or 100 years ago and today. Technical and scientific achievements made the world accessible in an instant or in very short time. One can achieve much more in an hour today than in weeks in the past. The value of time has gone up, the value of a dollar decreased. Thus time acquires nowadays new dimensions. Income can be theoretically increased indefinitely. Time stays physically the same. Increments in leisure time are possible only to a certain degree. As time becomes more precious, the competition between various leisure time activities increases requiring a constant decision making process in time allocation. Also time-money trade-offs by various intermodality arrangements gain in importance, e.g. vacationers flying to the destination area and then rent a car instead of driving all the way.

However, there is also enough evidence that the increase in leisure time may cause severe problems for individuals or whole segments of society not adequately prepared to face the changes. Indeed, increased leisure time may cause unhappiness for some individuals, like "weekend blues", and social upheavals when some segments of society (e.g. young people) have "too much" free time. In such situations governments and non-governmental organizations (NGOs) should take appropriate measures. Among the many options to translate the quantity of leisure time into quality is continuous adult education, combined with educational and cultural tourism and also development of recreational activities contributing to the health of the people. In this way the high quality leisure time will help the people to move closer to the ideal: "Healthy spirit in a sound body." This will make the individuals not only happier through self-fulfillment but also prepared to face the realities of modern world where flexibility, including flexibility in changing jobs, is the name of the game.

4.3.4. Leisure Time Slots

The leisure time researchers are interested not only in absolute amount of leisure time but also in its distribution i.e. in leisure time slots. The leisure time is available in following slots:

1. Daily leisure during the working or school day of several hours duration. Home and local leisure facilities are being used.
2. Weekends for over 100 days a year, when the leisure time may amount up to 12 hours a day. Leisure facilities used are partially local, partially outside the area of residence.
3. Vacation ranging in DCs from two to six weeks a year. Practically unlimited options for enjoyment of tourism and recreation in any, even most distant destination.
4. Retirement. There is enough time for any kind of activity.

The absolute changes in these four slot categories and the relative shifts between them are of extraordinary importance for leisure time planning because every slot of leisure time creates entirely different demands e.g. daily leisure time commands more home-based and local recreation demand, weekends and vacation more non-urban recreation and tourism. Time slots 2-4 are frequently associated in the geographic displacement of participants. The longer the slots of available leisure time, the more numerous the choices, the wider the opportunities, especially in terms of operational range. Indeed, time slots (alongside with costs) act as selectors between home-based, neighborhood, local, regional, national and international forms of displacement. Because of this researchers have to be aware of trends, changes of attitudes, preferences for one or another slot of leisure time. It seems that generally speaking the people prefer to take their leisure time in larger slots in the form of longer vacations and longer weekends rather than longer daily leisure resulting from shorter working days.

4.3.5. Weekend

One of the specially preferred leisure time slots is weekend. In this respect the last several decades brought more changes than in other slots. Indeed, there has been a strong trend to extend it. The biblical injunction: "Six days shalt thou labor" became obsolete when the two day weekend has been gradually achieved in most DCs after the Second World War. One of notable exceptions is Japan, only now gradually moving towards a two-day weekend. The most recent tendencies foreshadow further extension of the weekend even at the expense of a longer working day. In support of this idea one could argue that the fragmented after-work leisure time during the five-day working week is too short to really enjoy it and that during the working-day it is difficult to accomplish much in one day. Absenteeism is high, especially on Mondays and Fridays, e.g. in US auto industry - 20%. There are problems with dental and medical appointments. Much time is being lost in starting to work and in early preparation to leave. There are problems of traffic congestion, costs and time to commute to and from work.

In order to resolve these difficulties for the benefit of employees as well as employers a new arrangement of the work week started in 1970s: the compressed or four day work week. In North America relatively few workers follow this schedule which is more common in Western Europe. According to this scheme the employees work 4 days a week at 9 or 10 hours per shift or even 3 days at 12 hours weekly. The total amount of work time remains essentially the same or is only slightly decreased. The pay remains proportionate to the number of hours worked. This arrangement leads to 3 or 4 day weekend. In some cases more than one week is involved e.g. nurses in some hospitals are working a week 10 hours a day and then get a week off. Manitoba Hydro clerical employees work four-day week only in summer between 8 May to 11 September. In order to compensate for the shortened summer work week the employees work 30 minutes longer during the rest of the year and 45 minutes longer during the four day summer work week. For our considerations crucial are the increased long weekend leisure time slots these employees are receiving.

According to various sources there were following advantages of the compressed work week for the employees:

1. Longer weekends ("mini-vacations") may be spent for travel to relatively remote areas, more time spent at cottages, with the family and friends.
2. More equal utilization of recreation facilities which up till now are little used during the week.
3. Avoiding rush-hour traffic to work and for weekend, easing traffic congestion, spreading loads on transit and power utilities.
4. Saving time and money commuting to work.
5. Psychological improvements of the workers: they are forgetting the factory or office during the long weekends.
6. The people have more time to develop leisure skills or acquire education.
7. Many people on 4 day working week establish primary residence in the countryside and the second home in the city.

Advantages for the companies:

1. The introduction of additional shifts results in more continuous and extended use of plant and facilities with no additional costs
2. Reduction in the number of costly start-ups
3. Extended service to the clients
4. Increase of labor productivity as a result of improved work morale
5. Decrease of absenteeism and lateness
6. Decrease of overtime

Drawbacks of compressed work week:

1. Impractical by some operations e.g. logging camps in remote areas where there is often little else to do than work. In some cases customers refuse to adapt to the new schedule because of problems in contacting employees. There are also reports of difficulties in terms of inter-staff communications.
2. It is impossible to set a three shift operation on a 10-hour day (but possible to two shifts on a 12-hour day).

3. Some workers complain of fatigue connected with long work day and of the losses connected with cuts in overtime.
4. The increased recreation expenditures burden the family budget.
5. Some people are educationally and psychologically ill-equipped and ill-prepared for more leisure. For families who enjoy things together, especially outdoors or who like reading, travel, etc. - increased leisure time is fine, but for people with interests limited to television shows and spectator sports more leisure means more problems.
6. Disruptions of leisure and social patterns e.g. sports events, TV shows, meeting other family members and friends with differing work schedules.
7. Growth of second jobs (some of them illegal).

As a whole taking into account both the benefits and drawbacks of the compressed work week it seems that the positive aspects prevail and the four day work week will be slowly spreading in the future. Thus the people will be provided with what they wish: a longer weekend. Unfortunately, the four day work week is not at present applied widely in the DCs and five day work remains the norm. To achieve the three day weekend, at least partially, the governments use legislation placing some fixed holidays on Mondays. Such an arrangement exists in Canada (five fixed holidays on Mondays annually). In the US the Congress passed a law which went into effect on January 1, 1971 moving the observance of Washington's Birthday, Memorial Day, Columbus Day and Veterans Day permanently to Mondays. This makes together with Labor Day falling already on a Monday five long weekends (mini-vacations) annually for the Americans. Maybe these five Mondays do represent the nuclei for future development towards Monday as the extra leisure day.

The expanding weekend allows many North Americans to take "mini-vacations" of 3-4 days duration. Those living at the Atlantic seaboard are increasingly using this new leisure time for "weekends in Europe" e.g. the New Yorkers going for a "theatre trip" to London. Reduced winter air fares and charter flights and special all-inclusive "weekend in Europe" package tours (Tourism Intelligence Bulletin 1987, Sept. and Oct.)make such a trip relatively cheap. In North America the trend towards "mini-vacations" during the weekend and long weekend leisure time slots is the most dynamic tendency at the vacation pleasure market. Despite some inconsistencies in the available statititics one may generalize that the "short break pleasure travel" defined as a trip of one to three nights for purposes other than business or a convention" (Tourism Intelligence Bulletin 1987, October) grows quicker than total travel. In US these trips constituted about 34% of all domestic trips. If business and convention trips are added, the total short break travel in the US amounts to 52% of all domestic trips (ibid.) and half of all trips, domestic and international, (Tourism Intelligence Bull. 1987, July). Vast majority of short break pleasure travel takes place on weekends (ibid. 1987, November). The explanation for this trend is that working couples find the traditional US two week family vacation difficult to schedule with conflicting career demands. The developments in Western Europe are similar. There is a significant expansion of weekend and short vacation causing congestion in transportation and facilities, called "almost masochist" (Spiegel 1988, 23).

4.3.6. Vacation

Vacation is the slot of paid leisure time. As discussed in Chapter 3, paid vacation followed the shortening of the working hours with a substantial time-lag: for the white collar salary earners it started in the 1920s, for the European workers in 1930s, but for North America actually only after the Second World War. In 1940 only about one quarter of American workers had the right for paid vacation. Already the figure for 1957 was 90 per cent (Enzensberger 1967:195). Most DCs have the vacation time regulated by law.

The vacations initially very short, have been gradually lengthened. At present, the working prople of Western Europe enjoy the longest vacations among the DCs: The West Germans have the longest annual holiday - an average of 6 weeks (Tourism Canada, 1989B:2), the French have about five weeks and the British 4-5 weeks (Tourism Canada 1988:3) The North Americans enjoy shorter vacation: the Canadians have generally 2 weeks after one year, 3 weeks after 5 years and 4 weeks after 10 years of work. In the US the average is 17 days holiday and vacation time annually (Tourism Intelligence Bull. 1987, Feb.). However, the present emphasis of the North American trade unions is on shorter work week rather than on longer vacations. Understandably, strong unions have been able to secure longer vacations or even sabbaticals for their members.

It is important to note that there is an increasing tendency to split the vacations into two segments: one (usually longer) in summer and one in winter (typically one week). The progress in jet transportation enables the workers of North America and Europe to travel in winter to the Caribbean, West Africa and

Middle East to enjoy the beaches. There is a trend to take even more than two vacations annually. Mini-vacations undertaken outside weekend are very much in fashion. An example of such mini-vacation is the new trend to vacation between Christmas and the New Year's Day. These 5-6 days have been always associated with a high degree of absenteeism and low productivity of labor. As a result many enterprises and institutions either close down completely during this period or maintain only a skeleton staff allowing employees to take vacation. However, so far only Italy has legislated holidays between Christmas and the new year. Since 1977 a part of the 17 official civil and religious holidays has been transferred to this period thus decreasing "the bridge" (il ponte) - the Italian weekend extending device. It is the tendency to bridge the gap between a mid week holiday and weekend by workers taking days off. "The bridge" is, of course harmful to the economy.

Taking vacation at shorter intervals is found more appealing, more stimulating, better from psychological point of view because it releases the work pressure. Many people find long vacations as boring. Also longer family vacations are more difficult to schedule when both husband and wife are working. One of the factors which contributed to the trend of mini-vacation were the energy crises of 1973-1975 and 1979-1982 when vacationers did not need longer holidays because the distances travelled have decreased and they tended to stay at one destination and tour less. Thus the former trend to multi-destination vacations has been weakened. In the US also the migration to the South and West may have had some influence on shorter vacations: more people nowadays are living close to attractions and do not require much time to reach them.

The trend to shorter and more frequent vacations exerts a significant impact on tourism industry. The length of stay decreases resulting in higher tourist turnover. Also the scorge of tourism industry - the seasonality is reduced because tourists more frequently take vacations during the off-season and shoulder months. The trend to winter sport participation also contributes to more equal seasonal demand. Therefore, many mountain resorts became multi-purpose recreational complexes open all the year offering a diversity of summer and winter activities with shoulder periods (spring and fall) offered to the visitors at reduced prices. The decrease in seasonality is also caused by demographic factors - more families without school children independent from school holidays and growing number of retired people. The popularity of cultural tourism in its indoor forms (e.g. theater and museum tours) also contributes to patronage of facilities during off-season.

Off-season tourism is cheaper and this conforms with the trend towards more modest level of expenditure during vacations.

4.3.7 Daily leisure after work

Compared to changes in slots 2-4 (weekend, vacation, retirement)the recent changes with respect to slot 1 (daily leisure after work) are rather small although there is some movement also here. The point is to ensure the employees larger blocks of leisure time within the work day. This system is called Flextime or Flexitime (sliding work schedules or flexible working hours). Introduced in 1960s in West Germany Flextime has spread all over Western Europe but has rather limited application in North America. Under this system the employees are supposed to be at work only during specific core hours of the day e.g. from 9 A.M. to 3 P.M. during the peak hours of the business. The rest of the work can be performed anytime between e.g. 7 A.M. and 7 P.M. i.e. before or after the core hours as it suits them. The total number of hours worked is calculated on the monthly basis with a possibility of transferring a certain amount of "plus hours" or "minus hours" into the next month. The response to this plan is very good, especially by women. The employees easily put up with the time clock. However, this system has limitations and is impossible to introduce in a number of enterprises e.g. on production line or on airlines.

Summing up: the present trends in leisure time distribution indicate that slot 1 will probably remain relatively stable or may even decrease in some cases. However, slots 2-4 are increasing in the long range. This includes the tendency to early retirement as a measure to combat unemployment.

4.4. TRANSPORTATION

4.4.1. Introduction

The third prerequisite for modern mass tourism and recreation is mobility resulting from improvements in transportation. Transportation plays an indispensable role as a link between demand and supply in tourism. This role becomes more crucial as the distances from origin to destination grow. The times when it took a few minutes walk from the city to reach its natural environment are gone and the distances are growing. This is the reason why without relatively inexpensive, comfortable and speedy means of transportation our contemporary tourism would be unthinkable neither in volume nor in distances covered. Without modern transportation travel even at relatively short distances, is an ordeal undertaken only under pressure of necessities, and not a voluntary act with pleasure as an aim. Indeed, transport, both public and private, constitutes an important element in the development of tourism. Progress in transport improves the accessibility of innumerable tourist destinations all over the world.

For thousands of years the masses did not travel. The stage of their lives was limited to a radius of only few kilometers from their village or town. Travel further was a fearsome and exceptional eventuality. The industrial revolution brought a decisive change into this traditional, secluded pattern of life. It raised the degree of mobility. In 1900 an average United States resident travelled less than 800 km per year by all means of transportation. In 1956 the annual average per capita travel increased more than 10 times to over 8000 km (Clawson - Knetsch, 1966:5; Clawson, M., 1960:23). Certainly, the modern people travel in average more during one year than their forefathers did during their entire life.

The repercussions of technical and organizational developments in the sphere of transportation on tourism are certainly very conspicuous: new forms of transportation initiate new forms of tourism and open up new previously inaccessible tourist regions. Improvements in the existing transportation networks increase exploitation of new recreational resources. In other words improved access and mobility creates the urge for travel and promotes usage.

However, transportation not only influences the scope, location and quality of leisure time activities. Also vice versa: the needs of people at leisure in many instances exert their impact on transportation. It is impossible to deny that the quantitative and qualitative evolution of automobile and the airplane, of highway systems etc. has been to a great extent geared to the demands of leisure time market. For some modes of transport one could even speak of determining influence: tourism and recreation creates new forms and new routes of transportation for its almost exclusive use as e.g. scenic parkways or cable cars in the mountains.

Under the impact of technology and economic considerations travel has acquired mass dimensions. Large capacity of transport causes us to call it "mass phenomenon." However, at the same time it is also an "individual phenomenon" due to car ownership and a wide spectrum of choices available.

One of the methods to illustrate the progress in mobility due to technical and organizational progress in transportation is the isochrone method. Isochrones are lines drawn on the map, connecting points of equal travel time from a center. Thus there are isochrones of one, two, three etc. hours travel time. Another method of cartographic representation is the shrinkage of socio-economic space. For this purpose Marion Clawson used the example of the United States (Clawson, M., 1972,:13).

The tourism industry offers products which have peculiar features as compared with products of other branches of economy. Tourism products are immobile (cannot be transported) and with some exceptions cannot be stockpiled or stored. They are perishable. A good example is a hotel room. If not used a night - it's gone and the service cannot be repeated the same night. In tourism the consumer i.e. the tourist has to be transported from the point of origin to the destination where the consumption takes place. In this sense the services of regular transportation are also perishable: an empty airplane seat is lost forever. At the destination not only consumption but also production of tourist product-services takes place. Thus in tourism we have a peculiar situation where the location of production, distribution and consumption are identical. Transport in tourism plays a role within the framework of demand rather than supply. In other economic sectors the costs of transportation constitute an element of supply increasing the price of a product. In tourism the costs of transportation belong to demand and do not influence the price of services e.g. a dinner or hotel room in New York costs exactly the same notwithstanding the distance the tourist has travelled in order to enjoy these services.

With increased distance of travel the costs increase. The greater the distance the greater the share of transportation component in overall tourist expenses. One could imagine various scenarios or better a continuum according to the share of transportation costs in the overall tourists costs. These relationships may be illustrated by a curve (Fig. 4.1) where

S = the sum of equal tourist costs

Sv = costs of transportation (dynamic element of tourism)

Sj = costs of stay (static element of tourism)

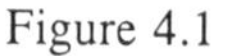
Figure 4.1

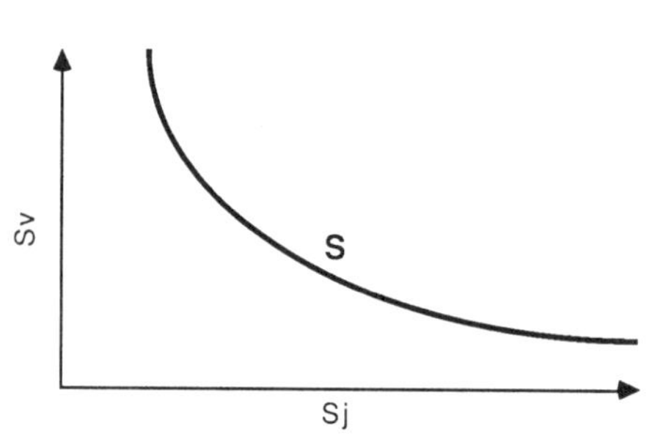

The curve shows various scenarios between a situation when the share of transportation is relatively very high like in touring or at the other extreme relatively low to allow the tourist to spend most of money at the destination (Rutazibwa, G., 1974, 3:96).

4.4.2. The Railroads

The first railroads appeared in Europe and the US in the first half of the 19th century but only during the second half of the 19th century the transportation systems of the DCs have been revolutionized by the development of railroad networks. They almost completely took over the transportation of both goods and passengers and dominated tourist traffic until 1920s in North America and until 1950s in Western Europe. In the vicinity of railroad stations large hotels were constructed to accommodate the passengers. The comfort, speed, relative cost and the ensuing mobility was unequaled by any precedence in the past - the reason why one could point out that actually the railroad created tourism.

Nevertheless, the era of railroad dominance in economically developed countries came to the end yielding to the car and airplane. The train passenger traffic has been declining starting with 1930s and has continued to decrease at accelerated pace after World War II. The elimination of famous express trains like New York - Detroit Wolverine, Spirit of St. Louis in US, Orient-Express (London - Istanbul) are the examples. The number of inter-city trains in the United States has dwindled from 20,000 in 1929 to only 370 in 1970 (New York Times, 20-9-1970). The most severe inroads occurred in North America where the railroad passenger traffic (except commuting in the vicinity of some large urban agglomerations) is nowadays only residual with the abandonment of most long distance routes.

In the 1960s the railroads of Western Europe and in the 1970s the railroads of North America undertook a number of measures aimed at improved services, increasing economy, speed, comfort and safety of their operation. An example of the railroad come-back is the Trans-Europe Express train system established in 1965 and transformed in 1987 into the EuroCity system. It connects more than 200 cities (including all of the capitals) of 13 West European countries operating mostly electric air-conditioned trains with restaurants, telephone connections with all the world, stenographer service, form-hugging foam-rubber seats and large picture windows. These trains at speeds up to 200 km. p.h. save the travelers much of the delays at congested airports and the time-consuming trip to and from airports (many of them have been relocated farther from cities because of environmental and space concerns). Actually as compared with air transportation the passengers are able to save time by trips up to 700 km. Passport and custom inspection in Western Europe are performed aboard moving trains. The trains move on welded rails eliminating the noisy clackety-clack. The West Germany's railroad competing for the tourist patronage instituted early 1960s a special train (Autoreisezug) transporting tourists and their cars to Italy under the slogan "while sleeping win a vacation day." In subsequent years an extensive car-train network has been established in West Germany, France, Great Britain, Benelux, Spain, Italy, Austria and Yugoslavia. The service is long distance, mostly at night. The measures undertaken by Western European railroads to stop the decline of their share in the travel market started to bear fruit towards the end of the 1960s. The fatal trend has been not only stopped but also reversed resulting in absolute increases in passenger traffic.

The West European progress in railroad development has been necessitated by concern about airport and highway congestion, environmental deterioration, high costs of gas, high air fares, lack of parking space in cities many of which are not adapted to car traffic. Relative closeness of urban centers also played a role in the railroad survival. Especially successful have been the French and West Germans who are today together with the Japanese the leaders in the development of newest railroad technology. In September 1981 a major breakthrough was achieved: the world's fastest train. The electric High Speed Train (Train à Grande Vitess - TGV) between Paris and Lyon moving on special track was inaugurated. The cruising speed is 275 to 300 km/h (believed to be close to the limit for rail-riding trains). The average speed is 260 km/h (compared to under 130 km/h for Metroliner between NYC and Washington, D.C.). The airliner passengers pay double as much and need more time city to city than the TGV. Since October, 1983 the Paris-Lyon trip takes 1 hour and 50 minutes and Paris-Marseille 3-1/2 hours. The Paris-Lyon route has been shortened for TGV by 86 km and is almost straight with no level crossings and was relatively cheap to build because steep grades up to 3.5% are used thus making costly tunnels and significant earth moving unnecessary. Fares are remaining the same as for conventional trains and the seats must be reserved. The train takes about 400 passengers in 8 cars between 2 electric locomotives at both ends. The TGV network is spreading all over France and Western Switzerland. In the forseeable future it will extend to Spain (target: the Seville 1992 World Fair), Benelux and England, especially after the completion of the railroad tunnel between France and England due in 1993. There are ambitious plans to extend the high speed trains all over Western Europe within next 25 years (Economist 11-3-1989).

The pioneering country in high speed train development was Japan. Also here the railroads are competing successfully with automobiles and airline jets for passengers. The system called Shinkansen (bullet train) inaugurated in 1964 has spread all over the main island of Honshu with connections to Kyushu in the south and Hokkaido in the north. The comfortable Shinkansen trains travel at cruising speeds of 210 km/h (about 170 km average) covering the 515 km. run between Tokyo and Osaka in about 3 hours. The Japanese can boast world records in railroad tunnel construction: in 1979 they finished the construction of 22.3 km long mountain tunnel on the line Tokyo-Niigata,and in 1988 the 54 km. Seikan tunnel under the Straits of Tsugaru between the islands of Honshu and Hokkaido. The substitution of ferry transportation by under-water tunnels and by bridges stimulates tourism. This circumstance is especially important for Japan. For connecting the island of Honshu with Shikoku the Japanese chose a bridge called Seto Ohashi. Opened in early 1988 this 12.2 km bridge took 10 years to build at the cost of $8.8 billion. The Japanese have also ambitious plans for the future: the development of magnetically levitated (maglev) train with speed of 500 km/h is well advanced and will become operative in 1990s. West Germans are also experimenting with their maglev.

As compared with Western Europe and Japan the attempts in North America to reverse the decline of the passenger traffic on railroads has not achieved the objective. The establishment of government subsidized federal railroad organizations Amtrak in US in 1971 and in 1977 Via Rail in Canada contributed to some streamlining of operations but could not stop further shrinkage of passenger services all over the continent. Even the energy crisis proved to be rather a modest helper for the railroads. The main problem is that the railroads in North America are not well patronized (Rail Road travel accounts for les than 0.5% of intercity travel in US - Winnipeg Free Press 30-3-1989) and the costs of maintenance and operations are high. Therefore, government subsidization is indispensable: Amtrak is subsidized by 30%, Via Rail by 63% (Winnipeg Free Press 3-4-1989. Most railroads of the world are subsidized but this problem is especially acute in North America because of low popularity and long distances. Also the excellent level of other means of transportation adds to the uncompetitiveness of the railroads which are frequently called "the federal money-burning machines". The long-haul lines account for three-quarters of Amtrak's losses (Time 3-6-1985). Less subsidies are required on short-haul lines like Los Angeles-San Diego, Northeast Corridor (Boston-New York City-Washington D.C.) and Windsor-Toronto-Montreal-Quebec City in Canada. On short distances the railroads successfully compete with airlines e.g. the Metroliner between mid-Manhattan (NYC) and Washington D.C. takes only 2 hours 40 minutes. Indeed, more than half of Amtrak's business comes from Northeast corridor where trains have surpassed air travel as the most popular form of mass transit (Time 4-4-1988).

Amtrak has replaced most of the obsolete equipment by modern one. Canada's Via Rail, however, suffers from equipment dating to the early post-World War II period. Both railroads are far from world class in terms of speed, punctuality and general level of services. The logic trend is to further cuts, especially in long distance services used by some diehards, elderly and low-income people. This, however, is not always politically possible. Also, from the economic point of view the elimination of service would result in high costs connected with labor-protection agreements - providing compensation for laid-off workers. Thus the troubled Amtrak and Via Rail are in real quandry. The withdrawal of subsidies would certainly cause their bankruptcy. Total withdrawals seem unlikely at present, however, cuts in subsidies are a distinct reality as exemplified by Canadian subsidy cuts announced in April 1989.

An interesting and successful (Tourism Intelligence Bull. 1987, January) tourist feature of Amtrak is the Auto Train service similar to Europe. It was introduced in 1971 by a private company and taken over by Amtrak in 1984. The Auto-Train ships almost 500 passengers and their cars in separate coaches over the distance of about 1600 km from the outskirts of Washington, D.C. to Florida daily. The overnight trip takes 18 hours and saves on driving, meals and accommodation costs during the trip. It also saves time and enables the passengers to start their vacation rested and with their indispensable car.

The railroads in some other parts of the world are still very important and expanding. The best example is the USSR. The Soviet Union leads the world in the volume of railroad passenger traffic. It is a country where new railroad lines are constructed (e.g. the BAM in Siberia completed in 1984). Freight movement constitutes the primary concern, passenger service is of only secondary importance. The most significant railroad line is the Trans-Siberian, the longest in the world (Moscow-Nakhodka almost 10,000 km). Many international tourists are using this line on their way to the Far East. At the Soviet Pacific port Nakhodka they embark on ships to Yokohoma, Japan. In general, the Soviet trains are slow and uncomfortable with one exception: the Moscow-Leningrad express covering the distance of 600 km. in 4 hours.

In Eastern Europe the railroads are still very important as a result of relatively modest development of car construction industry and rather poor network of highways. There is even some expansion of the network. In Asia the most significant role is played by railroads in China and in India (also for passenger traffic including pilgrims to holy places). In non-Soviet Asia and Africa there is still some new catch up railroad construction whereas in Central and south America and Australia the railroad networks have shrank since the end of World War II. There are plans for a high speed train between Sydney and Melbourne (870 km in 3 hours instead of 12 hours by bus - WFP 28-3-1989). Generally speaking the outlook for passenger railroad traffic in the world seems to be mildly optimistic for the short-haul and in some areas for middle-haul distances.

However, there is a feature in contemporary RR situation which reminds one on cruises in maritime ship passenger transport: part of the famous trains of the nostalgic by-gone era are thriving nowadays to such a degree that one has often to make reservations months ahead. Most of them are luxurious and relatively expensive. Some of the great trains returned to service in late 1940s. By 1960s few survived. The former Orient-Express was revived in 1982 as the Venice Simplon Orient Express (VSOE): it connects London with Venice. Another great train is the Blue Train which connects Pretoria with Cape Town in South Africa. The Indian "Palace-On-Wheels" consists of 13 painstakingly restored cars built for maharajas and British viceroys and takes tourists since 1982 (from October through March only) for a one week or 3 day excursion from New Delhi through Rajastan. Another famous Indian train is the Rajahari Express between New Delhi and Calcutta with about 130 km/h maximum speed. Also the Mombassa-Nairobi Express in Kenya which runs already for almost 90 years reminds one on the old colonial splendor. The Indian Pacific train launched in 1970 between Sydney and Perth (Australia) does not have long tradition but enjoys popularity along with the nostalgic ones. Similarly famous is the express Singapore-Bangkok.

4.4.3. The Automobile

A spectacular increase of mobility has been achieved with the acceptance of the automobile as a mode of mass transportation. This was the "third wave" in the development of travel (the first was the horse and post wagon, the second the railroad the fourth is the airplane). Today, and certainly for many years to come, both domestically and internationally, the private car is the world's most important means of transporting people.

The "automobile age" came earlier to North America than to Western Europe. In 1922 the auto provided half of the total personal transportation in the United States (Clawson, M., 1960:23). The figure for 1941 was already 86 per cent. From that time on this share peaked at 87% in 1956 to decrease slightly in subsequent years. The saturation of the market with cars seems almost complete. The number of registered passenger cars in 1986 was almost 135 million surpassing the ratio of one car per two persons which is regarded as theoretical saturation point. Canada had over 11 million cars in 1987 still more than two persons per car. The North Americans certainly live in the era of "homo automobilensis." Nowhere in the world has the automobile played such a prevailing role in passenger transportation like in North America. The giant car companies did everything in order to destroy their competitors. They bought up urban and intercity transportation systems and then ripped out rail and transmission lines for electric trains and trolley buses leaving the field for the car and the diesel bus. The US automobile industry has also received a multi-billion dollar government subsidy in the form of 1958 Federal Aid Highway Act which authorized large grants for the interstate highway system. Of course, these developments significantly increased the per capita energy consumption making North America more vulnerable to oil shortages. Similar developments have also taken place in other DCs.

The mass use of automobile brought a number of important changes pertaining to the intensity, structure and spatial patterns of tourism and recreation. The personal car increased the intensity of travel, the rate of participation in tourism and recreation by diminishing the friction of distance between origin and destination. It significantly increased speed and above all mobility, enhanced individual flexibility by securing free choice of route, time, duration and interruptions of travel, independence from schedules and regimentation of collective means of transportation. Travel by automobile has broken the barriers between city and countryside providing accessibility to wider areas than the railroad. It had created better possibilities for contact with nature, enabling the travelers to interrupt their trip and enjoy the landscape at practically any point they wish. Driving for most people means also pleasure - so the demand for automobile travel is no longer a demand exclusively for a trip to the place of destination, but also an end in itself.

The individual character, independence, speed, and mobility of the automobile travel tend to shorten the duration of stay in one place increasing the multi-stop (multiple destination) vacations. The motorists interrupt much easier their trip than other travelers, also the degree of their landscape penetration is higher than that of the railroad travelers. They can easier select their destinations and visit less known areas and mountains accessible for a railroad only with difficulty and at a great expense. The relative easiness of interruption dispels fears of some areas and countries that their role may be reduced merely to transit regions bypassed by the tourist boom. Such fears are often being expressed in Benelux countries or West Germany with reference to the tourist flows directed to the Mediterranean area. As a result of automobile flexibility such areas frequently become upgraded from mere transit zones to tourist regions of "intervening opportunity."

The use of the automobile as means of transport for tourism and recreation increases the participation of people with relatively lower incomes. The family car gives them privacy, comfort and convenience. It enables them to camp and use other modest accommodations (their mobility gives them more freedom of choice), to prepare their meals, to minimize further their expenses by visiting bargain regions (e.g. Greece, Yugoslavia, Turkey, Mexico) in order to offset their travel expenses. The maximal utilization of space on family vacation trip helps to keep the per capita travel costs at a minimum. Only few motorists go on vacation alone. Most of them travel in family and other groups because it is cheaper. The total travel expenses, however, are usually greatly underestimated by motorists tempted to count the gasoline costs only. The expenditures connected with the purchase and maintenance of the automobile are generally regarded as work commuting costs. In this way the motorized vacation travel is evaluated not only as the most convenient but also the cheapest family holiday which is not necessarily true. This circumstance gives a significant psychological stimulus to tourist travel by car. However, not only vacationers with below average incomes travelling in more or less self contained manner take their car for holidays: the people in average and above average income brackets, use motels and hotels for accommodation and take food in restaurants. Their share in vacation travel is higher than campers, many of whom have also above average incomes if they use the motor homes. Thus we may generalize that automobile vacationers represent a very wide range of incomes and this has an impact on their behaviour during the trip i.e. on their choice of accommodation and food, on the distance and duration of their travel. However, they have one common characteristic: they travel in groups, mostly family groups and this is their main distinction from vacationers using individually other means of transportation who to a larger extent go alone (of course, excluding various forms of package tours). In addition, among people going by bus or railroad on their holidays there is a higher percentage of elderly or young travellers with below average incomes. In comparison, among the air vacationers there is a higher proportion of people with more than average incomes.

The researchers are interested to determine to what degree car ownership stimulates participation in leisure time activities. Of course, one has to assume that car-ownership itself is not an independent variable, but is strongly correlated with income, occupation and education. The degree of correlation between car-ownership and participation is certainly very high, it pertains not only to rich car-owners but also those with more modest means. However, by analyzing various recreational activities the researchers found strong variations in the extent car-ownership influences them. B. Rodgers (1971:219) grouped the activities into those that show only a slight excess among those with access to a car (under 75% increment in incidence), those showing a moderate excess (75% to 150% increment), and those showing a great excess (over 150% increment). "Low-income activities" like bowling, cycling and team games belong to the first group. Car-ownership increases only slightly the participation rates as compared with people who do not own a car. Moderate effect in encouraging participation has fishing, hiking, skating and outdoor swimming, which have been identified with the second group of activities. The resources and facilities for this group are usually further away from home and thus require quicker and more flexible means of transportation. The third group of activities (camping, mountaineering, hill walking, sea and island sailing) presents obvious access problems often connected with relatively large distances to locations ill-served by public transport. Additional problem is bulky equipment which has to be brought to these remote destinations. In these cases the ownership of a car seems almost indispensable.

B. Rodgers concedes that there are many concealed influences at work in the relationship between possession of a car and recreational performance. However the results of his research at the very least "cast doubt on the naive belief that the car makes physically lazy and recreationally passive." He demonstrates in a most convincing way that the car ownership generates a greatly increased demand on recreational resources of all types.

The most important stimulus for the development of automobile tourism/recreation travel is creating an adequate infrastructure by road construction which understandably is only in exceptional cases undertaken exclusively for tourism purposes. For the tourist traffic important are all-weather-highways, especially freeways (expressways) saving time and shortening the distance through scenically unattractive zones or by-passing urban areas. Also significant are scenic parkways for leisurely drive in the tourist areas. Better utilization of the beauty of the landscape is more and more taken into account by designing new highways. Tourist traffic is certainly benefiting from all technical achievements in the field of highway transportation: computerization, improvements in the quality of the road surface by application of new materials (e.g. waste products, plastics etc.) and new developments in highway designing. All these technical achievements certainly improve the quality of the motorway network. But measures aiming at not only qualitative but also quantitative improvements through construction of new auto routes are of great significance. In this respect several continental motor programs do exist; one of the most imposing is the program for "Europe highways" implemented since 1950. Construction of new freeways, bridges (e.g. Europa bridge south of Innsbruck, Austria or bridges between Danish islands and the continent), tunnels (e.g. Mont Blanc tunnel from Chamonix, France to Courmayeur, Italy) has increased substantially comfort and speed of driving in Europe breaking physical barriers, especially that of the Alps. A special role in tourist travel play the limited-acess expressways prevailing all over Western Europe. West Germany has the best system of expressways

(Autobahnen) of about 8.5 thousand kilometers. The automobiles move on these Autobahnen with an average speed of 110 km but many of them reach 160-180 km. Another example of large scale highway construction programs is the US Interstate Highway system legislated by Congress in 1958. This 75,000 km (42,500 miles) long system has been financed from the Highway Trust Fund fed by federal gasoline tax and other levies. The construction of US Interstate Highway system certainly contributed to the hegemony of the automobile in US transportation. Canada completed the Trans-Canada Highway in 1962 and developed a system of interprovincial highways. A project on the continental scale is the about 30,000 km long Pan American Highway. The Asian Highway connects since 1970 Europe with India with projected extension to Singapore when Burma solves her insurgency problems. The Trans-Sahara road is not ready yet but convoys with goods from Algiers, Algeria to Lagos, Nigeria are already traversing the 3,300 km route in nine days. Although more than half of the distance is over rough desert some tourists dare to cross the Sahara by car sometimes with disastrous results. Of course, the primary purpose for the construction of these highway links is to stimulate non-tourist economic development, but it is beyond any doubt that tourism also benefits. Even the most attractive tourist potential will be never used without adequate transportation routes. Thus the famous "autostrada del sole" linking Northern Italy with Sicily is helping not only to stimulate the economy of the Mezzogiorno but also to bring there millions of tourists, whose expenditures are making the task of the economic development a lot easier.

The impact of automobile upon the landscape is not only limited to highways, bridges, tunnels etc. Still more visible is the new tourism-recreation infrastructure catering to the needs of motorists: gas stations, roadside parks, motels, camping sites, trailer parks, cabins, cottages, private rooms, apartments, holiday villages, drive-in cafeterias etc. The phenomenal expansion of secondary residences in Western Europe and North America would be unthinkable without the automobile. These new tourist infrastructures spread over large areas giving the tourist industry significant locational advantages rendering almost indefinite possibilities for tourist location.

The benefits of the automobile for the humanity in general and for tourism and recreation in particular are undeniable. However, there are also drawbacks: traffic congestion, pollution, heavy death toll cause considerable criticism all over the world. Some critics point out that the private car constitutes an energy inefficient mode of transportation. Contemporary cars are more fuel-efficient than the 1974 cars. However, the problem of energy conservation is still important. The key factor in the public's decision to drive rather than take some form of public transportation is the availability of gasoline. Price is of secondary importance. In fact, gasoline has low price elasticity compared to many other consumer goods. The people in the DCs and especially in North America regard the car as a sort of magic carpet giving them mobility control and freedom. Therefore, they resist anything that threatens the use of the car, like conservation measures which are to a large extent disregarded by the public. The impact of high gas prices can be felt only with respect to long distance travel which has declined in 1970s and 1980s to the benefit of airplane. However, the decrease in gas prices in 1983-1987 increased the long distance car travel. However, most long distance tourists use intermodality arrangements by flying to the destination area and renting a car for local travel. In connection with the fly/drive business the car rentals are booming, growing at about 14% annually in the last years. In the US the leisure travel market increased its share in car rentals from 10% in 1970 to 35% in 1985 (Waters, S., 1986:125). Most of the balance is business rentals. Business rentals prevail on week-days, pleasure rentals on weekends contributing to more equal distribution of demand throughout the week. On the annual basis the seasonality of demand for rentals is pronounced with business rentals decreasing slightly during mid-latitude summer and pleasure rentals skyrocketing. In the US the rental companies operated a fleet of 766,000 cars in 1986. (Waters 1989:135). Another form of intermodality is the car-ferry, widely practiced, especially in Europe. It tends to increase rather than decrease the car travel.

Some critics focus on the motorists' behaviour while travelling. They claim that many motorists are seeking a visual rather than physical contact with the countryside i.e. they seem not to use the possibilities potentially offered by the automobile to increase their contact with the nature and rarely leave their cars during a vacation trip thus reducing their enjoyment of the landscape and their physical fitness. This leads to some bitter comments about the "windshield tourists": "Some people drive their 300-horsepower chesterfields thru the National Parks because we have these great concrete sluceways that take people into one end of our parks and vomit them out at the other end." However, one could see also some beneficial aspects of this situation. The environmental impact of mass recreation is concentrated in relatively small areas in the immediate vicinity of roads and parking zones and decreases rapidly with increasing distance from them. Therefore, regulation of traffic and environmental impact control may be focussing on these high density areas where the overwhelming majority of recreationists tend to stay. Thus the major part of landscape remains relatively untouched for the enjoyment of the few who are making use of scarce natural resources.

The Future

The car as the mode of transport has achieved the dominating position in North America in 1930s and in Western Europe after the World War II. However,at least in North America since the energy crisis 1973-1975, there is a discernible trend of gradual long range drop in the share of the car to the benefit of other modes of transportation, especially of air travel. In the US during the peak period about 20 years ago the use of automobile for all recreational travel was about 90% and below 90% for vacation travel, in 1980 it was already below 80%. In Canada it was barely 50% in 1983 as compared with 73% in 1966 (Canada, Travel Research Notes 1983, 38:1).

As a whole there is no evidence of a trend to significantly decrease the role of automobile in the life of the DCs. The "automobile age" persists and its impact on our lives is not going to decline appreciably in forseeable future. Therefore, some corrective measures are being undertaken by governments: to diminish pollution emission levels are regulated, to improve safety new legislation provides for increased road and vehicle safety, to alleviate congestion new highways are being constructed. This last measure is being criticized on the grounds that more supply results in more demand. Better answer to congestion and energy problems is raising gas taxes and increasing popularity of other means of transportation: airplanes, railroads, buses, bicycles, sailing boats, wilderness hiking, etc.

Future prospects for automobile travel depend closely from the availability of gasoline. With the supply at present assured it is doubtful that the personal car ownership will decline. However, some new patterns in vacation travel are emerging: the family car will be less used for long-distance vacation. Instead the vacationers will make use of intermodality arrangements i.e. they more frequently take fly/drive vacation renting a car at the destination. Also weekend car rentals for recreational purposes are on the increase leaving the weekdays for business rentals. The RV market is thriving with gas available at stable prices but there is a trend to rentals or ownership on time-sharing basis. So-called hybrid campgrounds are also being developed with RV-s parked semipermanently and rented to vacationers at modest prices. Whereas in North America the use of car for tourism/recreation shows a slightly decreasing tendency - this is not true with respect to the other countries, especially in the second and third world. Here the car constitutes still a dynamic element of tourism expansion and its share in recreation travel is increasing. The share of North America in the world car registrations is consistently decreasing (Waters, S., 1987:127). The number of cars in North America is growing only very slightly ahead of population growth whereas this increase is relatively larger in countries which have not achieved the saturation point. S. Waters sees a strong correlation between the increase of world car registrations and the growth of international tourism arrivals (ibid.) and regards this as an indicator and predictor of the role of the autmobile in the development of world tourism.

4.4.4. The Autobus

The bus offers much less independence for the tourist than the car but still plays an important role competing with the railroads and eliminating uneconomical railroad lines on routes of secondary importance and also substituting for the rail traffic in many regions, especially in North America. The sightseeing market is also to a large extent the domain of the bus, particularly in Europe. The bus is more energy-efficient than the car but on long distance routes (over 400-500 km) it has suffered from airplane competition especially after the 1978 US deregulation in North America and high speed railway competition in Western Europe. As a result the buses were left with routes rarely exceeding 500-700 km and middle to lower income, largely elderly or young clientele. They prosper in times of gasoline shortage and lose passengers to private cars in times of gas glut. In the future the market for the bus will remain in the field of intracity, intercity short distance travel, sightseeing tours, charters and vacation tours. According to S. Waters the US "bus charter and tour service grew at a slow but steady pace." (Waters, S., 1987:129).

Bus travel in North America is in bad shape since the last energy crisis 1979-1982. Greyhound carried only 33 million passengers in 1985 as compared with 55 million in 1980 (New York Times 29-8-1986). Airlines, including the new commuter air lines are cutting into the scheduled intercity bus services. Tours and charters are increased their share at the expense of scheduled intercity traffic from 41% in 1976 to 60% in 1984 (Waters, S, 1985:117). The average bus tour in North

America increased from 6.4 days in 1985 to 6.8 days in 1988. Bus tour as travel mode for Americans visiting Canada became more popular: it has increased as percentage of total person-trips of Americans visiting Canada from 6.5% in 1977 to close to 8% in 1987 (Tourism Intelligence Bull. 1988, May). Similar trends to increased use of buses for touris and city sight-seeing can be observed also in Europe.

4.4.5. The Recreational Vehicle

In 1960s a new vehicle gained widespread acceptance in the DCs - the recreational vehicle (RV). There are two types of RVs: on-road and off-road. The on-road RVs appear in several forms: the most popular is the trailer or caravan. It is a separate unit towed by car. Second is a self-propelled unit - the motor home, third is the truck camper: camper unit tops piggybacking a pickup truck. These three types are self-contained i.e. they have provisions for sleeping, food preparation, personal hygiene and temperature control (a veritable home away from home). There are also semi- or non-self contained types of recreational vehicles depending on the degree of provision of these features. Here belong converted vans and fold-down camping trailers. Thus, the on-road recreational vehicle represents an on-road form of more or less self-contained camping. It is economical in terms of family vacation. It allows participants a measure of self-reliance, comfort and family togetherness. Its sales achieved an unprecedented boom of about $2.5 billion annually in US in a five-year period preceding the 1973 oil crisis (New York Times 8-4-1973). The 1969 US production of mobile homes and trailers came to 926,780 including 92,500 truck campers and 23,100 motor homes (New York Times 20-9-1970). The total number of on-road RVs was about 5 million in 1973 (New York Times 11-3-1973). Since 1973 the fortunes of recreational vehicle have been closely tied with the availability and price of gasoline because of its high gasoline consumption. Sales slumped, especially of large motor homes which are gas guzzlers. Light fold-down camper-trailers and light travel trailers were the winners. Again, like with other means of transportation the intermodality is a method which is assisting the industry: renting the RVs at distant location in the framework of fly-drive arrangement has gained some popularity. Another form of intermodality are pontoon boats, usually catamarans, carrying the RVs on lakes for fishing and other water-based recreational trips. Despite setbacks the on-road RVs seem somehow to survive the hard times and even increase in numbers in periods when gasoline is available and prices stabilize. Their sales fluctuate in reverse proportion to gasoline prices (and availability). Also, of course, such factors as interest rates, the level of inflation, consumer confidence, economic growth etc. play a role. Additionally, the decline of US and Canadian dollars on international currency markets and the rise of international terrorism boost the RV demand. To illustrate these fluctuations in sales following figures may be quoted for total RV sales in the US:

1972 - 582,000
1974 - below 300,000
1976 - 541,100
1978 - 526,300
1979 - 307,700
1980 - 195,000
1986 - 360,000

(Waters, S., 1987:128, Green, F., 1978:432; New York Times 20-5-1980, Winnipeg Free Press 19-4-1986). In 1986 30 million Americans owned or rented RV (New York Times 24-5-1987). The number of RVs in US was about 8 million (Waters, S., 1987:128).

There are factors at work which warrant the survival of RV. First of all the structural trends towards increased fuel efficiency and light weight vehicles should be mentioned. Secondly, many users are attached to this form of recreation. The RV people are gregarious members of a very special subculture and they enjoy their lifestyle. Especially strong are these social bonds among retired RV people in the US. Increasing numbers of them roam the country on permanent vacation, spending winter in the South (typically South western deserts) and summer in the North. They have their own magazines, conventions and associations, "Family Motor Coach Association" with more than 100,000 members (NY Times 14-8-1988). Therefore, one may look at the future of RVs with modest optimism.

The off-road vehicles also called All Terrain Vehicles (ATVs) appear in various forms: mini-bikes, trail-bikes, mountain bikes, three-wheelers, four-wheeler dune buggies, snowmobiles. Included into this category could be also probably various permanently installed forms of winter recreational transportation: cable cars (gondolas), chairlifts, T-bars, pomas. However, talking about the ATVs one usually has only the first category in mind.

Car and on-road RV limit the diffusion of tourism and recreation to the road. The ATV allow the users the most thorough penetration of the landscape at any season. Therefore, their environmental impact is much greater than that of the automobiles and on-road recreation vehicles. Their growing popularity has changed radically the recreational patterns of many regions, e.g. snowmobiles put a new life to many areas in North America where winter is a bleak detested period with increased rates of suicides, divorces, murders, etc. The snowmobiles brought pleasure and excitement to the North.

However, despite undeniable recreational and other benefits, the ATVs brought also a number of universal problems like pollution, littering, noise, overhunting, harassment of wildlife, looting of cottages in winter, damage to agriculture, erosion of trails, destruction of fragile ecosystems (e.g. sand dunes) abuse of alcohol, poor safety records (especially the use of motorbikes, three-wheelers by children) etc. Three-wheelers caused 20 deaths a month in US (Time 19-10-1987). The governments have reacted to this by imposing legislative restrictions resulting in decrease of use. An example of such legislation is the outlawing of future sales of three-wheelers by the US Justice Department backed by Consumer Product Safety Commission in January 1988. The Justice Department based its decision on dismal safety records of these ATVs (900 killed in 1983-1987, many of them children). It seems that Canada will not follow this decision in the forseeable future. Even the safety records of the on-road RVs are under attack (CBCs "Market Place" on 19 January 1988) especially these with wooden frame (fragile, flamable). Steel frames, more than one exit and explosion-safe operation of propane stoves were recommended.

The ATVs are mostly enjoyed after work and on weekend, much less on vacations. Therefore, the near-urban areas are most heavily used and here the legislators impose the restrictions. This, of course, prompts the participants to go farther away in search of areas with less use limitations. The future of ATVs lies in places where they are specifically allowed with all other land uses excluded. A good example is the California ban of AVTs in state parks, except in specifically designated areas (Time 13-7-1988).

4.4.6. Water Transportation

The early steam ocean liners appeared in the first half of the 19th century initiating the golden age of ocean passenger transportation. Almost all intercontinental travel up to the World War II was by ocean liners. The travel was luxurious, indeed posh. This term POSH comes from the stamp put on tickets from London to India: portside cabin out and starboard cabin home (port out, starboard home). Such accommodations meant least uncomfortable cabin temperatures during the journey. The most important was the North Atlantic traffic between Europe and North America initiated in 1838. On this route the shipping lines like Cunard maintained fleets of famous luxury liners vying with one another for the "blue ribbon of the Atlantic." The decline came gradually after the second World War when the long distance passenger ship lost in competition with faster and cheaper airplane. In 1946 about 400,000 passengers crossed the North Atlantic both ways by ship and only 100,000 by air, however, already in 1955 the air traffic for the first time exceeded the ocean-born (Dulles, 1965:384). In 1957 the all time maximum of 816,000 ship passengers was reached. In 1969 the figure was only 285,478 ship passengers as compared with almost six million air tourists. The figure for 1978 was about 84,000 and has been decreasing. The airplane is not only faster but also much cheaper. A small jet plane of 100 seat capacity can transport more tourists annually on the transatlantic route than a large liner of 1000 passenger capacity. The era of transatlantic crossings practically ended in 1975 when the Italian government announced the phase-out plan for its subsidized ocean liners. Thus since 1988, after the Soviets and Poles withdrew their ships, only one liner sails the Atlantic regularly: British "Queen Elizabeth 2." She is spending less than half the year crossing Atlantic: the rest of the time, she is cruising. Admittedly, the ocean liners are very expensive to operate, especially when the fuel costs are high. There have been attempts to trim the costs and eliminate waste: cheap labor from developing countries has been employed, self-service cafeterias automated galleys and separately paid food to eliminate waste have been attempted. All these measures, however, could not avert the decline. Another residual of worldwide of port-to-port ship passenger service constitute the freighters supplying a limited amount of cabins for paying passengers. Their services enjoy an excellent reputation but the demand far exceeds the supply which is extremely small. The real joy for freighter tourists involves the time spent at sea (no crowds, no stress) because the sightseeing time at ports has been decreasing consistenly as a result of containerization and modern loading and unloading methods.

In response to the airplane competition the passenger ship traffic has been subject to radical structural changes: since 1960s many of the luxury liners have been directed into profitable cruise business mainly in the Mediterranean,

Caribbean and at the Pacific coast of North America. With the business booming new ships join the present cruise fleet of about 170 ships. Between 1980-1985 the cruise ship tonnage grew at 12% annually (Waters, S., 1986:127). The number of passengers is growing at a rate of over 10% annually (the growth for 1987 was 11% - Time 11-1-1988), reaching almost 2.9 million in 1987 with bookings of about $5 billion in 1987 (ibid.). Croisimer, representing almost all French cruise companies reports 43% increase in revenues between 1985 and 1987 (Le Monde 27-12-1987). The duration of cruise trips vary widely from 24 hours to several months round world tours but most typical are one or two week trips. One week cruises were until recently the most popular. However, as the average duration of cruises has been decreasing there has been a spectacular rise of 3-4 day cruises in 1980s. Indeed, three and four day cruises are the fastest growing segment in the cruise market, increasing at the rate of 35 percent annually (Tourism Intelligence Bull. 1987, March). These cruises are today not only aimed at the first-time market but also most popular among repeaters. Of course, the efforts to tap new markets are still of paramount importance in view of ship overcapacity in the cruise business necessitating intensive marketing, discounting etc. to decrease the gap between supply and demand. The best market is North America (70% of all cruise passengers). Europe comes second. Miami is the world's leading cruise departure port. Other ports are Los Angeles (for Mexican Riviera) and Vancouver B.C. (for Alaska summer cruises). The most important cruising area is the Caribbean followed by the Mediterranean, which has suffered decline in market share from 8.7% to 2.7% after terrorists highjacked "Achille Lauro" in 1985. The 1988 terrorist attack on "City of Paros" was also damaging but the Mediterranean cruise business seems to be on the way to recovery (New York Times 5-1-1989). Also important is North America's Pacific coast, (mainly Alaska cruises in summer and from Los Angeles Mexico cruises in winter) and in 1980s the Atlantic-coast (Northeast summer cruises), between St. Lawrence River and New York City including many inland rivers and canals. New York City lost its former importance as the departure port for winter cruises in the Caribbean to the benefit of mainly Miami, Fort Lauderdale and San Juan, Puerto Rico. Cheap fly/cruise intermodality packages caused this shift of departure ports. The Alaska cruises have their European pendant in Norway (Nord Cap cruises, some of them reaching Spitzbergen). It is interesting that both cold-water destinations enjoy great popularity in recent years. They are more landscape and nature oriented than their southern counterparts. The season for the Caribbean and Pacific coast Mexico is winter, other cruises take place in summer. Cruising on the Black Sea is dominated by Soviet ships. There is also increasing cruising activity in South Pacific (French Polinesia) Southeast Asia and in New Zealand-Australia area. South American cruises center on the coastline between Rio de Janeiro and Buenos Aires. A promising region is the Far East e.g. China cruises including the River Yangtse.

The cruise business is highly competitive, especially in the present overcapacity situation. Therefore, everything is done to cut costs: in many cases the flag of convenience is chosen, frequently even the sanitary regulation transgressed. The most important economies pertain to the labor and fuel costs. In this respect the Soviet Union is a very competitive country and is using this position. Its cruise passenger fleet is one of the largest, about 36 ships (Waters, S., 1987:130). The fares run about 15-20% below most Western ships. The service is spartan but adequate. An added attraction of Soviet cruises: terrorism against Soviet ships is out. Thus operating from Tilbury (England), Rotterdam, and Genoa the Soviets have captured a considerable slice of West European market.

The reasons for the unprecedented boom in the cruise business are many. The cruise is an ultimate vacation. The ship is a floating resort hotel complex which arrives, mostly every morning, at different destinations. The passengers are able to visit various places (land excursions available at extra cost) without the inconvenience of packing up and transfers. On board they enjoy relaxation, refreshment, recreational activities, fitness programs, cultural enrichment, entertainment (including gambling). Various performers and more recently educators are employed. In fact, some of the cruises are specialized (theme cruises) in different on-board activities to cater to specific interests (e.g. education, wildlife, business conventions, fitness programs, country music). All this variety of on-board activities facilitates social contacts between passengers ("Love Boat" connection) and contributes to the increased participation of younger clientele. On most cruises the atmosphere is relaxed and informal (including dress). The cuisine is gourmet which reflects the heightened interest of the public in food quality. The ships are equipped with giroscopically controlled fin stabilizers to reduce pitch and especially roll. Other stimuli contributing to the success of cruises are the intermodality arrangements with airliners which provide for easy air access at reduced fares between the origins of tourists and the ports. Increasingly, the tourists pay a single fare for air/sea (fly/cruise) service (all inclusive package) which is associated with considerable savings on airfare. Also the arrangement called cruise/stay is being offered, combining a cruise with a resort stay. In

October 1982 a new car/cruise service has been introduced for tourists from US Northeast who want to drive their cars during the Florida winter vacation period. The car/cruise/liner connects New York City with Freeport, Grand Bahama. From Freeport there is a feeder service to Miami.

Cruise fares vary not only with the length of the cruise, the individual ship and the type of accommodation, but also with the time of the year. On the Caribbean the season is between 15 December and 15 April with top fares charged during Christmas-New Year and Easter holidays. Summer is the season in the Mediterranean and at the North American Pacific runs. The lowest prices with frequent free air fares to the port are offered during the shoulder weeks when the cruises are still or already running (in the Caribbean: late spring/early summer and fall). Another bargains represent the so called positioning cruises when the ships leave an area for another one with the change of season e.g. the Caribbean fleet changes position to the summer runs at the Pacific coast of North America and in the Mediterranean. Generally, the cruise fares are reasonable. Since 1960 cruises ceased to be the preserves for old and rich. This has opened new markets of younger and less affluent tourists and contributed to the success of cruises. The rich take special luxury cruises e.g. the sailing ships "Sea Cloud" or "Windstar," the 120 passenger ships "Sea Goddess I and II." They can also hire a private yacht.

The future for cruise business looks bright. The business is booming but the pressures of the market (necessity of huge investments, economies of scale etc.) cause a trend to mergers. Thus the companies become larger and larger. The cruise line companies have invested millions in new ships, more fuel efficient and equipped with stabilizers ("The cruise business," 1982, 3:86). The trend is towards the introduction of bigger vessels. Thus the average capacity of cruise ships built before 1980 was 717 passengers and of ships built between 1980 and 1986 was 1099 (Tourism Policy 1987:97). The largest cruise ship is the "Sovereign of the Seas" at 74,000 BRT commissioned in January 1988. It carries 2,284 passengers. The ship "Absolute Dream" nearing completion in Belfast will be still larger with a capacity of over 3,000 passengers. This will be the largest ship ever built (160,000 ton). The cruises offer an increasing variety of activities and expand to new ports and regions, including the LDCs. Many tourists who are apprehensive to travel in these areas (e.g. South America) are willing to participate having a base ship. However, the large increase in new ship capacity (the number of cruise berth will increase from 61,000 in 1988 to 77000 in 1991, Time 11-1-1988) may be a mixed blessing: the industry faces problems of overcapacity. Yet in spite of these short term problems the long range prospects are excellent. There is almost certain that the demand will catch up with capacity because of the large potential market of people who never took a cruise (95% of the US population) and an unusually high number of repeaters.

Besides cruises there is still another form of ship transportation which is expanding. Car ferry constitutes a intermodality between road and water transportation. It is an extremely popular means of transportation especially in Europe with its highly developed coastline. North Sea and English Channel ferries between Great Britain and the continent and Ireland, between Finland and Sweden, Finland and Poland, Sweden and West and East Germany, Spain and Balearic Islands, Southern France and Corsica, Italy and Greece, etc. The car ferries in North America are of local character: e.g. from Vancouver or Seattle to Victoria, B.C., between Nova Scotia and Newfoundland, between New Brunswick and Prince Edward Island. In addition to the car ferry, the intermodality between ship and railroad provides for railroad ferry services e.g. between England and Netherlands. Tourist traffic on water is served also by such "exotic" means of transportation like hydrofoils and hovercrafts. Hovercrafts nowadays can accommodate up to 650 passengers and 75 cars. Hovercrafts service the English channel since late 1960s. Underwater submarine trips are expanding very quickly in such tourist areas like Caribbean and Hawaii. The analysts forsee a $250 million-a-year market with more than 100 submarines worldwide by early 1990s (Time 4-4-1988).

The tourist passenger ship transportation on rivers, canals and lakes plays a significant role. It transports tourists on larger inland water bodies all over the world from Great Lakes and Mississippi to Amazonas and Lake Titicaca, from Swedish and Finnish lakes to Yangtse, Rhine, Danube, Volga and Nile. In recent decades the inland water tourist traffic has been boosted by speedy hydrofoils suitable even in shallow waters and hovercraft which make even marshy areas accessible.

It is interesting that tourism transportation has become an exclusive user of old canals which are today obsolete for transportation of goods. Examples of such exclusive use of old canals by tourist boats is the Canal du Midi in Southern France. It was constructed in 1667-1681 to provide a 240 km connection between the Atlantic and the Mediterranean. Another example is the Rideau canal in Canada linking Ottawa with Kingston (203 km) built in 1826-1832.

The most expanding sector of recreational water transportation in the post-war period pertains to private boats. The use of personal boats is growing very quickly not only on the lakes and rivers of North America and Europe but also on the Pacific and Atlantic seaboards, in the Caribbean area, on the Mediterranean, North and Baltic Seas. Alone in the US there were in 1970 five million outboard boats and ten million in mid-1980s. In addition there were many millions of small boats (like canoes etc.) and over 2 million inboard boats and sailboats in use. At least part of the reasons for these successes lay in the technological developments: plastic materials for the boat's body, drain free outboard motors which completely eliminate spillage of fuel and gas, significant reduction of the outboard noise even with increase of power, rubber engine mounts that stop vibration. The latest achievements in the motor technology is the gas-turbine engine, which although more expensive than diesel, guarantees almost total absence of noise and vibration. The engine is ecologically clean, safe and operates at low cost. Similarly like in case with private planes the accident rates in recreational boating are rather high (6.8 fatalities per 100,000 boaters, 1,063 total fatalities in US in 1984). The fatality rates show a decreasing tendency as a result of safety measures undertaken (e.g. against drunken boating). Judging from several recent well publicized white water rafting accidents with loss of life, the safety records in the fast expanding field of white water rafting, cayaking and canoeing are also not great.

4.4.7. Air Transportation

4.4.7.1. Development

The fourth wave of travel development has been brought by the airplane. Although scheduled air traffic started already in 1920s in Europe and North America the airplane still played a relatively insignificant role in the travel picture until the late 1940s. The world air passenger traffic grew vigorously in the interwar period but achieved only 200 million passenger kilometers in 1930 and 2 billion in 1939. The technical progress during World War II gave impetus to the staggering development of air travel in the post war period: in 1946 there were already almost 20 billion passenger/kilometers and in 1979 the figure surpassed 1 trillion for the first time. The annual rates of growth have been consistently sustained at high incremental levels in all post war period - well over 10% annually up to the 1973 energy crisis. The air traffic dropped by about 5% in 1974 and 1975 to recover to an annual growth level surpassing 10% between 1976 and 1980 and to drop subsequently by less than 10% during the world recession. The annual growth rate in 1983-1986 was 5-7%. There were 1.6 trillion passenger kilometers flown in 1986. The passenger load factor (seats occupied as percentage of seats available) was 67% (Waters, S., 1988:137). The passenger air transportation is heavily concentrated by relatively few carriers representing the leading economic giants of the world. Thus the US carriers prevail with almost 40% of world's passenger kilometers, followed by the USSR (14%) and Japan (5%). This dominance is still more pronounced if one takes domestic traffic only: here the share of the US is 57% and that of the USSR 22% (Waters S. 1988:137). In international services the share of US carriers' is 17% the UK's 10% (ibid).

In the post war period the airplane has acted as the most dynamic element in the world passenger traffic increasing its share significantly to dominate the public transportation in North America: in Canada the airlines control 85% of public transportation as compared to bus companies with 9% and Via Rail with 6% (CBC 28-3-1989). The airplane has no competition in long distance travel. It opens new holiday destinations even in most distant parts of the world. Indeed, the world "is shrinking" as the individual traveller gains access to any place in the world almost within 24 hours and at reasonable price. In intercontinental travel the airplane enjoys virtual monopoly. Its spectacular success in competition with other modes of transportation was the almost complete elimination of the ocean liner from the North Atlantic route where the share of air passenger transportation has grown from 32% in 1949 to 50.2% in 1957 and almost 100% in 1980 (Dacharry, M., 1981:82).

Table 4.1

Passenger Transportation in Seclected Countries in Billions of Passenger - Kilometers

		1938	1950	1960	1970	1980	1986
U.S.A.	railroad	34.9	51.2	34.3	17.3	17.7	17.9
	air	0.9	16.4	62.5	210.3	389.5	530.8
U.S.S.R.	railroad	99.9	88.0	170.8	214	342	390.0
	air	-	1.2	12.1	78.2	160.6	195.9
Canada	railroad	2.9	4.5	3.6	3.7	2.9	2.1
	air	0.02	0.9	4.3	15.4	36.3	-
Great Britain	railroad	32.4	32.1	34.7	35.6	31.7	30.1
	air	0.09	1.3	7.3	19.0	50.2	51.0
France	railroad	22.1	26.4	32.0	41.1	54.5	59.9
	air	0.07	1.1	5.3	13.6	34.1	39.2
Japan	railroad	33.6	39.1	180.9	288	313	333
	air	0.03	0.01	1.5	15.0	51.6	66.6

Source: Rocznik Statystyczny 1986 - 1988

The railroad, private car and the bus has been able to compete with the airplane only up to about 400-700 kilometers losing the battle for passengers at longer distances. Table 4.1 illustrates the history of the competition between the airplane and railroad in selected countries. In Japan and Europe (including the USSR but excluding Great Britain) the railroad is still dominant in passenger traffic. However, in North America it plays only a very marginal role.

One has to emphasize that the airplane is winning in competition with other means of transportation only partially by diversion of traffic: it generates also completely new demand for its services by tapping new markets. Such an untapped market does exist: according to data for 1986 the percentage of US residents who had never flown by commercial airline was 28 (Waters, S., 1987:133).

4.4.7.2. New Aircraft Technology

In the post-World War II period the air transportation received two powerful boosts: the first was the late 1950s introduction of jet aircraft, the second the early 1970s introduction of wide body ("jumbo")jets. The jet aircraft increased considerably the speed from about 400 km/h for propeller aircraft up to about 900 km/h and range (up to 12,000-14,000 km). A good example of longer jet ranges was the introduction of Boeing 747-400 in 1988 making it possible to fly 412 passengers non-stop from New York City to Seoul, Korea. The wide body jets expanded the capacity of airplanes up to 500 passengers. The development of aircraft technology in 1970s and 1980s has been focussed on fuel efficiency, (fuel costs represent an average of 24% of the total operating costs of IATA member airlines - Tourism Intelligence Bull. 1987, March) lighter weight, operating economy, decrease of pollution (including noise pollution), increase of comfort and improved safety. A significant asset is the computerization of ground and flight operations. The development of aircraft technology during the rest of the 20th century will probably focus on increased capacity and range and also improved safety. There won't be any significant gains in speed which will remain at the present level of about 900 km/h. However, efforts to achieve still greater fuel efficiency and safety (engine design, reduction of drag) will continue.

4.4.7.3. The Supersonic Transportation

Many people anticipated the third boost (the first was the introduction of jets, the second the introduction of wide-body jets) for air passenger transportation with the introduction of supersonic transportation (SST). The anticipation did not materialize: the SST is a technical wonder but an economic failure. In 1976 the French-British supersonic Concorde started flights on North Atlantic route with speed of about 2200 km (Mach l=1224 kmh) and stratospheric heights cruising altitude of 15,000 to 18,000 meters. The Concorde cut the duration of North Atlantic flights by about half as compared with subsonic jets (New York-Paris in less than 3 hours and 30 minutes). The fare is about 20% higher than subsonic first class jet.

Concorde run into trouble almost from the beginning. First were the objections about the noise levels. Here the Concorde was successful: it stayed within the legislated airplane noise levels proven by tests despite protests of affected residents. The tests on sonic boom remained inconclusive as to its environmental impact. At present the Concorde is allowed to fly super-sonically only over the ocean. Some scientists raised objections that exhausts from SST (mainly nitrogen oxide) may seriously deplete the stratospheric ozone that protects the life on earth from ultraviolet radiation which could cause skin cancer and eye cataracts. Other scientists expressed fears of stratospheric contamination which would impede solar radiation in reaching earth thus contributing to cooling of the atmosphere with catastrophic results, especially for agriculture. The World Meteorological Organization stated in a 1976 release that if the SST stay under 17-19 kilometers and if there are not too many of them then they would not have a significant effect on the stratospheric ozone (New York Times 8-1-1976).

From the economic point of view the British-French Concorde has been a failure. The research and development costs and the subsidies for the money losing years 1976-1982, paid by both governments will never be recovered. Since 1982-3 the Concorde brings some profits but it has also important drawbacks which will never turn it in a real money-maker as was expected. It costs twice as much as the subsonic jumbo jet Boeing 747 but carries only 100 passengers, significantly less than the Boeing's capacity of 400-500. Concorde is also uneconomical in terms of operating costs because of high fuel consumption and has a relatively short range of 6500 km. As a result of these drawbacks and a disappointing demand there are no new Concordes in construction. The present fleet of about 14 planes is serving only the profitable routes Paris-New York, London-New York, London-Washington, D.C.-Miami. There are also some charters. However, all the other routes of late 1970s have been abandoned.

Recently various press reports indicate that leading airplane manufacturers are working on projects of new supersonic jets with speeds of Mach 3 or even Mach 5. It is, however, doubtful if they will start service in this century.

4.4.7.4. Economic aspects: air fares

The development of air travel depends not only from technological achievements of the aircraft industry but also from other factors. Among them is the fare policy of the airlines. The price elasticity of demand for air services is rather high, especially for pleasure travel - a component which is consistently increasing its share in ticket sales. The potential demand for air travel is huge: there are people who never travelled by airplane and people who would travel if the prices were right. Therefore the airlines have introduced a marketing policy aimed at potential market. Examples of such policy are charter flights and inclusive tours (package tours).

The first charter flights took place in early 1960s. They were offered by special charter airlines. The expansion was initially slow, limited to affinity groups. However, charters have been increasingly recognized as excellent instruments of tapping new markets, improving load factors by creating economies of scale. Thus in 1960s and 1970s the share of non-scheduled or charter services has increased as compared to regular or scheduled services to level off in subsequent years.

The scheduled airlines, especially the trunk carriers and regional airlines, felt initially threatened by booming charter business but soon recognized the correctness of the saying: "If you can't lick them join them." They themselves became increasingly involved in charters. Moreover, sometimes together, sometimes in competition with large tour operators they started to offer package tours which involved charter flights, transfers and block hotel and restaurant bookings. Therefore, the costs of package tours have been low and so have been the prices making even exotic holidays accessible to millions. The package tours (all inclusive tours) or just mere charters are also frequently offered in intermodality combination with car rental or cruise companies as fly-drive or fly-cruise.

Technically the charters are organized not by the airlines but by tour operators (even sometimes in case of charter airlines) who are responsible for marketing. Tour operators rarely sell tickets to the public leaving this task to travel agents.

Some countries do not permit charters in order to protect their own scheduled airlines. Where charters are permitted their use is largely governed by demand: the more popular the destination the more likelihood of charters.

In order to meet the challenge of charters and to penetrate new markets the scheduled airlines have introduced special fares offered either to specific segments of the market (e.g. young people or senior citizens) or the public at large. As the special fares have proliferated the prospective tourist have been facing increasing number of choices. At present the maze of special fares has developed to such a degree that even experienced travel agents have problems to find the cheapest alternative in the multiplicity of ever-changing options.

The most expensive of the special fares is the 14-21 or 22-45 day excursion where the only limitation is the time minimum and maximum. Cheaper is the APEX (Advanced Purchase Excursion) fare. The requirement is, besides a minimum and maximum duration of e.g. 7-60, days the purchase at least 21-60 days before departure. In late 1970s Pan Am developed Budget Fare which allows the airline and not the passenger to select the day of departure within any one week. This fare, at substantial discount, is especially attractive to students, housewives, visiting friends and relatives, teachers and senior citizens. Similar to budget fare is the Standby (Youth) fare which provides for not confirmed reservation.

Among the charters the most important are: Super APEX which requires the purchase 7 days before departure and the ITC (Inclusive Tour Charters) which are sold right up to the time of departure and include hotels and transfers.

The process of air fare reductions in real terms has been throughout the post World War II period limited by strict regulation of airlines: the fares were kept at relatively high level by International Air Transport Association (IATA), an international cartel fixing their levels and thus limiting the competition. IATA has also limited the competition by regulating the allotment of international routes to airlines. Most countries have various government civil aviation authorities which determine the allocation of routes and the entry rights of new airlines. Such an organization was the Civil Aeronautic Board in the US.

In 1978 a new chapter in the history of passenger air transport has started with airline deregulation in the US. This development has given a significant boost to tourism as a whole and to air tourism in particular. Between 1978 and 1987 the number of air passengers have nearly doubled (Waters, S. 1978:139) and the fares dropped to a level nearly 40% cheaper than they would have been without deregulation (Time 6-7-1987:39). Out of 1987 trips 52% were trips for pleasure and personal reasons and 48% for business (Waters, S. 1988:138), presumably including convention travel. The deregulation contributed not only to the spectacular increase in the traffic but also to improved passenger load factors. Freedom of entry and exit with respect to air routes, flight schedules and deregulation of air fares has opened the skies to competition. The airlines compete mainly in fares but also other marketing tools are used with apparently excellent results like e.g. frequent traveller's plans.

The 1978 US deregulation has provided for creation of new airlines, like People's Express (1981-1987). To increase their competitiveness the "no-frill" airlines have undertaken a number of cost-cutting measures like:

- increased daily aircraft operating times up to 10 hours a day
- introduction of late night service (also using cargo planes for passenger transportation)
- cancelling unprofitable routes and even individual flights with unprofitable load factor (an illegal procedure)
- reduction or abolition of ground services (bookings etc.)
- gross overbooking
- abolition of interlining (transfer to other airlines)
- increased number of seats on planes at the expense of passenger comfort
- cuts or abolition of in-flight services which are charged separately e.g. food, check-in baggage
- low wages for nonunionized employees (organized labor is a definite loser)
- employment of part-time employees
- job-sharing arrangements
- labor productivity increases
- requiring employees to purchase company's shares in profit sharing arrangements
- restrictions with respect to bargain rates applicable to certain days of the week and certain hours of the day (outside the rush hours).

Even taking into account that not all the measures have been undertaken by single airlines one can imagine that some of the cuts result in the deterioration of services leading to a multitude of complaints. However, the slashing of the fares has been decisive for the expansion of demand. The gains for the travelling public from the increased competition have been substantial: direct cuts in fares, greater range of low, special-condition fares, cheap services on stand-by basis, increased charter services for all. Not only the public but also the airlines benefited: the decreased costs of flying has increased the highly price elastic demand for air services. Thus the expanded volume of traffic has compensated for lower fares. The old trunk airlines have had to adapt to the new-competitive environment by becoming more cost-conscious and by adopting some of the cost-cutting measures listed above. Fare-slashing wars erupted between airlines for their markets sometimes at suicidally low levels. This led to a number of bankruptcies and airline consolidation (mergers). Even one of the pioneers "People's Express", fell victim of these upheavals.

The issue of airline deregulation is a fairly contentious one. Some of the advantages mentioned above are counterbalanced by drawbacks. Here are some of criticism expressed mainly with respect to the US deregulation:

1) The rapid expansion of air traffic has led to deterioration of safety due to inadequate numbers of well-trained flight personnel and air controllers, congestion of airports and air space, especially during certain periods of the day. The capacity of the existing airports is definitely too small. Also cost-cutting measures may negatively affect maintenance and repair of aircraft. Part of this criticism has proved unfounded. The safety control agencies, The Federal Aviation Administration (F.A.A.) in US and the International Civil Aviation Organization - (ICAO) on international level are going a good job and the number of accidents per passenger/kilometers has actually dropped since deregulation. However, the April 28, 1988 accident of an Aloha Airlines Boeing 737 over Hawaii and a number of similar accidents in 1988 and 1989 have triggered a renewed wave of criticism about the maintenance standards of the aging jet fleet.

2) The amenity of air travel has suffered as a result of deteriorating quality of service, congestion of airports, delays and cancellation of flights etc. Measures have to be undertaken to cope with these problems e.g. the increase of the airport capacity, more functional design of airports, more convenient transportation between city and airport.

3) There were fears that smaller communities services by shorter commuter routes may lose these serviced following deregulation as a result of the "freedom of exit" from unprofitable routes. This happened only to a small extent. These communities may have lost some of the large jets but are serviced,

often more frequently than before, by more economical smaller jets and propeller aircraft owned and operated by new specialized feeder airlines called regionals. From the point of view of tourism these runs are less important than the large hub centers. Tourism concentrates on large airports centrally located near metropolitan areas and major resorts and here the benefits of deregulation are obvious. However, tourists are also well served by regional airlines within the hub-and-spoke systems.

4) The suicidal price wars resulting in bankruptcies and mergers have hurt the industry. This argument was initially valid for the US only but the trend to airline consolidation is spreading in Canada and Europe.

5) The airline industry has become increasingly concentrated as a result of airline mergers. There are fears that it could result in cartel-like oligopolistic conditions, similar to those that the deregulation was intended to supplant. These pessimistic expectations became for the first time evident in late summer 1987 when the 8 main US carriers controlling 95% of the market increased their fares (New York Times 9-9-1987). Since 1987 the trend to increase the airfares has been firmly established. There is also a tendency to decrease the number of bargains by the fewer major (trunk) airlines remaining after bankruptcies and mergers. Non-refundable discount fares have become "a way of life" in US and Europe (Tourism Intelligence Bull. 1987, November). However, a rise in the North American airfares is proven inevitable because the fares were too low to guarantee a long range profitabilty. Nevertheless, despite the fears of price-setting oligopoly the air passenger transportation still remains highly competitive. As for the oligopolistic tendencies among the US airlines, one should note that US antitrust laws could be enforced against them. Nevertheless, the fares are rising, the market share of the top eight US airlines was 94% in 1989 as compared to 80% in 1978 and 74% in 1983 (The Economist 4-2-1989). The new airline strategy calls for maximizing revenues by increasing prices as opposed to the old strategy of increasing market share by discounting prices (Tourism Intelligence Bull. 1988, Oct.).

The on balance positive experience of the US deregulation prompted also other countries to move in the same direction. Canada started to deregulate in 1984 (legislation passed only in 1987) and it seems that this process has been smoother than in US because the airlines have avoided the self-destructing price wars. In 1986 deregulation started in Western Europe leading to reductions of high air fares. This deregulation, locally called "liberalization," is scheduled to be fully effective by 1992 (Tourism Intelligence Bull. 1988, June). The difference between the Western European and North American air traffic is that the charter airlines now carry 60% of all intra-European air traffic (Waters, S. 1988:139), whereas Americans rely more on discounts offered by scheduled airlines. The similarity is that in Western Europe, like in Canada, there is a trend to privatization of government owned airlines.

4.4.7.5. Problems: infrastructure and safety

The growth of air traffic and especially the advent of the high-capacity wide-body jumbo-jets disgorging up to 500 passengers at each landing created a number of bottlenecks in the infrastructure, shortly thereafter designated as the three gaps: surface transport gap, hotel gap and parking lot gap. However, the number of gaps seems to be greater. As a result of runway congestion the waiting time for start or landing permission can be quite long. The departing and arriving passengers face a number of annoying delays and difficulties by check-in, in getting to the airport terminal, receiving their luggage, clearing Customs and Immigration, travelling from the airport to the city, finding hotel accommodation etc. The jet travel from London to Paris takes only half an hour but the trip to and from the airport takes over two hours. Additional delays have been caused by elaborate security measures aimed at preventing highjacking and terrorism.

There are various reasons for these infrastructural inadequacies, the main being that the construction of new airports lags badly behind the expansion of traffic. Very few new airports are being built as a result of problems and costs connected with acquisition of land near large urban centers, pollution (especially noise pollution) controls and also problems associated with the development of adequate transportation between city and airport. The new airports, if constructed at all are located at prohibitably far distances from the city: a good example is the Mirabel airport which was scheduled to serve Montreal but has become a "white elephant" mainly because of the distance from the city. Another factor limiting the capacity of existing airports is the increasingly stringent noise standards which frequently prohibit the utilization of the airport at night.

To cope with these problems a number of measures are being undertaken:
1. To alleviate the congestion on runways: construction of new runways, limitations of runway use by small private aircraft by imposition of high fees or directly by denying them take-off and landing rights, increasing starting and landing frequency (this has, of course, its limits), introduction of all-weather

automatic take-off and landing procedures, lifting of runway use restrictions between midnight and 7 a.m. (the new jets like Boeing 767 are much quieter than the older ones and noise abatement measures are not as essential as in the past).
2. For intercity short distance flights use of helicopters, VTOL (Vertical Takeoff and Landing) and STOL (Short Takeoff and Landing) service. The STOL feeder jets are now at the stage of development. Also the introduction of high-speed trains to distribute the tourist flows from the major air traffic hubs to the regions may eliminate some short-haul air connections.
3. For airport to city service use of helicopters and subways to shorten the time and alleviate the highway congestion.
4. Construction of multistory and underground parking garages at the airports to alleviate the lack of parking space.
5. The introduction of so called "Plane Mates" (elevated mobile lounges), motorized carts of various designs, monorails, moving sidewalks to speed up the transportation of passengers from the airport terminal to the plane.
6. Construction of new grant jetports like e.g. the Dallas-Fort Worth or Atlanta airports. The enormous proportions of these air terminals allow not only for their own rapid transit systems but also for creation of large buffer zones to muffle the noise of jets for people in nearby communities. This seems to be ecologically sound and at the same time the most efficient method to eliminate airport congestion and delays.
7. Streamlining of reservation systems and ticket sales by computerization.
8. Streamlining and speeding up of Customs and Immigration clearance as recommended by the Facilitation Division of International Civil Aviation Organization (ICAO).
9. Streamlining and speeding up of baggage handling at the airports by extensive use of mechanical or even electronic equipment, containers and pallets.
10. Symbiosis of airlines and hotel-chains or even airlines going into hotel business (vertical integration) to increase efficiency, coordination and profits. Thus Pan American owns Intercontinental, TWA Hilton hotels. Many other airlines acquired hotels, especially since 1970s. The airlines offer more and more package tours including hotel accommodation mostly in their own hotels frequently located close to the airport.

One of the most important problems of passenger air transportation is safety. For low flying aircraft there is a danger of collision with birds and loss of lift caused by downdraft winds (wind shear). Some airports are dangerous because of topography, frequent fogs or inadequate technology. The hazards of fire are caused by flammable materials of aircraft interior not to speak about fuel. Growing air space congestion not only around airports but in wider areas (e.g. parts of Western Europe), increases the probability of mid-air collisions. Despite these multiple dangers one has to emphasize that passenger airplane is definitely the safest modern mean of transportation. Safety records of major (trunk) and regional carriers are excellent with local or commuter airlines not far behind.

Nevertheless, new safety features are constantly introduced into the new airplane models and into the ground and air operations. Here are some of them: fuel additives reducing the risk of postcrash explosions, crash-resistant fuel tanks, fire retardant materials for aircraft interiors, reinforced passenger seats and baggage racks, smoke hoods, modernized radar systems facilitating the detection of dangers (other aircraft, birds, downdraft winds etc.), better safety checks (e.g. introduction of high-tech sensors and equipment to spot phony documents used by terrorists and to detect plastic explosives) and maintenance, improved crew training, increased number of air traffic controllers, more sophisticated security measures. Dangerous airports undergo costly operations involving major earth movements close to landing strips, the runways themselves are being extended, antifog devices introduced. The bird habitats in the vicinity of airports are being destroyed and special devices introduced to cope with the bird-airplane collision problems. All these safety measures require funding and this in part explains the trend to higher air fares.

4.4.7.6. Air passenger transportation network

The airway network covers the whole world. One can practically travel between any two, even most distant, points on our globe without stopping for refuelling. Example of such connection is New York City-Tokyo (11,000 km in 12 hours) or Sydney-San Francisco (12,000 km in 13 hours). The need for technical airports which performed the role of refuelling stations like Shannon in Ireland does not exist anymore. An exception is Gander in Newfoundland for flights between Moscow and Havana but this distance is shorter than NYC-Tokyo. The necessity of this refuelling stop results from shorter range of Soviet jets.

Looking at the map of world airway connections one can distinguish obvious patterns in the direction and location of these routes. In other words there are areas where the network of connections is dense, clearly concentrated and linking certain macroregions. The intensity of air traffic in such areas is also higher than in others where the network is less dense. The three leading intercontinental air routes are: Transatlantic, Europe-East Asia and Pacific in the sequence of importance. There are several dozens of connections between Europe and North America over the North Atlantic at various latitudes starting from Iceland and Greenland in the North and ending on Azores and Bermuda in the South. The second most important route is between Europe and East Asia. The air traffic flows follow three basic directions: the route via Middle East, South and South-East Asia developed since the interwar period, the route over the North Pole developed since late 1950s and finally the route over the Soviet Union, the shortest of all (developed since 1970s). The third intercontinental air connection runs over Pacific from North America and Mexico to Japan and other East Asian, South East Asian countries and Australia. The Pacific connections are the most dynamic in their development because of economic boom in the Pacific Rim. However, at least for the time being, the most successful of all air routes has been the North Atlantic one which has grown consistently since the end of World War II, even in periods of economic recessions. It overtook in 1957 the ocean liners and in 1976 reached 13.8, in 1978-18.4 million passengers, 1985-23 million passengers.

The contemporary spatial arrangements of air transportation networks follows the "hub and spoke" pattern: the passengers are transported in wide-body long range jets between major centrally located air hubs (e.g. London, Tokyo, New York, Toronto). From there they may fly by medium or short range planes (jets, turbo-props, propeller) to smaller centers. The introduction of new twin-engined jets like the Boeing 767 and Airbus 310 is slightly modifying this spatial configuration of networks by providing cheaper but lower capacity long distance transportation than the three-and-four-engined jets, also between secondary hubs (e.g. Buenos Aires, Argentina and Harare, Zimbabwe). Thus the number of hubs with international connections is increasing. Some of these hubs, especially secondary ones, are dominated by one or two airlines only. The exigencies of global economy lead to globalization of airlines: in addition to consolidation various kinds of airline linkages and cooperation are evident. As an example of such cooperation may serve the Sabre Computer Reservation System (CRS) shared by major carriers (US and foreign) with regional carriers. This code-sharing arrangement constitutes a sort of marketing merger giving a number of advantages to the members. Some of these advantages are the market intelligence: where the tourists want to go and how much they are willing to pay (Magary, S. 1987:2-3; Economist 4-2-1989).

4.4.7.7. Private aircraft

A new and very quickly expanding phenomenon enters the air travel picture especially in North America: recreational and business travel in private small aircraft. There are well over a million licensed pilots and over 200,000 active corporate and private airplanes registered in US. Choices of where to go are practically limitless: more than 10,000 airports and airstrips. The small aircraft prove to be not only time savers for business but also for weekend and vacation trips. In addition to this travel by private noncommercial plans has inherent undeniable recreational and educational values. To fly such an aircraft is fun, it is interesting and comfortable. Weekend trips of 1000 and more kilometers are common. Propeller aircraft with speed of about 200 km/h prevail in the fleet but there are also small executive jets. The world's best known private jets are Learjets with a fleet of over 1,300. The business-aviation armada of jets and turboprops reached almost 13,000 in 1983 (Newsweek 18-6-1984).

The major problem of small aircraft are poor safety records which in terms of fatal accidents are about 400 times worse than that of commercial airlines (data for US).

An interesting type of recreational flying has gained an importance after World War II. It is the glider which appears in two forms. First is the sailplane which has normally a cockpit for one or two persons and requires a pilot licence. The safety records for the sailplane are good. The second type is the kite-like hangglider which does not have a cockpit and does not require a licence. The safety records are much worse than for the sailplane. In 1970s and 1980s the new form of hanggliders with motors has gained popularity - the ultralight aircraft. These planes fly at a speed of about 70 km/h and have rather poor safety records.

4.4.8. Summary: trends in tourist transportation

4.4.8.1. Modes of transportation

The discussion of various travel modes has given us a broad view on technical progress and changing socio-economic conditions of passenger transportation.

Certainly up to the middle of 19th century the overland transportation has been monopolized by horse-drawn coach. Since about 1840-1860 the railroad began to dominate the picture in overland transportation in the DCs. Maritime transportation was taken over from sail ship in 19th century by steam ship. Later the motor ship substituted for steam ship. After the World War I in 1920s the automobile took over from railroad in North America. In Europe the car started to compete with the railroad already in the interwar period. However, only in 1950s it began to threaten the very existence of the railroads. The railroads responded to this challenge in 1960s by taking a number of corrective measures which have led to their revival allowing them to play a significant role in passenger transportation. After World War II a new competitor - the airplane practically eliminated other travel modes of transportation as competitors at distances over 400-700 km. Such were the changing fortunes of various means of transportation.

At present, we have a situation when diverse travel modes have by and large found their place in the transportation system. They certainly are used for different distances: walking and bicycle for very short range, automobiles, busses and high speed trains are best suited and compete with each other at distances up to 400-700 km, to give way to the jet airplane at longer distances.

The fact that various modes of transportation have found their "niches" in the transportation system does not mean that the relationships between them in terms of their absolute capacity and relative importance remains unchanged. Air passenger transportation which is growing at about 7-10% annually, increases its share in the transportation system quickly. Transatlantic ocean liner and other long distance port-to-port ship transportation disappeared almost completely. The passenger ship transportation is today almost entirely limited to the booming cruise business. Train travel, despite substantial government subsidization, has decreased very significantly in North America but is doing well in many areas and countries among them is Europe, China, India and Japan. Personal car remains the principal mode of tourist transportation in the DCs. However, in North America there is a slight relative decrease in the long range mainly as a result of airline competition with temporary increases when the gasoline is cheap and plentiful. But in most other countries there is a steady absolute and relative increase of automobile passenger transportation. Bus travel has decreased in many DCs. Its fortunes depends in the DCs on fuel availability i.e. it loses to private car when there is a gas glut. In most LDCs bus travel is consistently increasing both absolutely and relatively, in North America its fate is connected with the government railroad policy. If the US and

Canadian governments decide to cut the railroad subsidies bus transportation will expand to fill the gap of shrinking railroad networks, unfairly subsidized competition, according to the bus companies.

The modes of transportation are not only competing but also cooperating with each other in passenger traffic. Indeed, there is a trend to intermodality. Intermodality is a combination of various modes of transportation in one vacation trip. Intermodality tickets are normally sold as packages or all-inclusive tours: fly/drive, fly/cruise, rail/drive, rail/bus, air/bus.

Another trend in tourist transportation is the structural shift in shares of business and "pure tourist" (leisure) travel. Business travel is slowly decreasing its share as a result of newest advances in communication technology: computer hook-ups, long distance direct telephone dialing systems, high-tech phones, teleconferencing, video-teleconferencing via satellites. All of these cut business travel. Their future spread is certain. However, there is another trend which may be working in opposite direction on international level: the increasing globalization of world's economy leads to the expansion of international trade may result in the increase of foreign business travel.

The expenditures for the transportation of tourists from the origin to destination constitutes a large percentage in their overall expenditures. In international tourism they surpass 40% ("International Tourism to 1990": 20). They will probably remain relatively stable in foreseeable future being the subject of two opposing forces:

1) falling air fares in real terms as compared to other tourist expenditures e.g. hotels, food, shopping.

2) the long-range trend to travel longer distances which despite the cheaper costs per kilometer may increase the overall expenditure for transportation.

4.4.8.2. Spatial patterns of tourism: long range trends

Since the beginning of modern tourism in the middle of 19th century the spatial patterns of tourism reflect a consistent long range or secular trend to expand from the core to periphery. This means that the distances travelled by tourists domiciled in core areas (tourist markets, origins) increase as they cover longer and longer distances to the periphery i.e. tourist destination. Thus in 19th century the British tourists gradually expanded their travels from their own country to Northern France with Paris as the main attraction, then to Rheinland, to Switzerland, Italy,

Greece, etc. The range of tourism has increased from domestic to intraregional, interregional and intercontintental. Today it encompasses not only the whole globe but there are prospects of further expansion into outer space. The core area of tourist origin has expanded too. Initially it was Britain. Today it is the developed world of North America, Europe, Japan, Australia and New Zealand which constitute the main markets for international tourism.

There are many reasons for this spatial development of tourism. Of course, the prerequisites are increasing leisure time and incomes. However, without technical and organizational progress in transportation such a development was impossible. Especially important in this respect have been the progress in air transportation. On top of it is the organizational progress in form of charter flights and other special fare arrangements which have improved the capacity utilization rates and thus contributed to the reduction of real costs of travel. Thus tourists who can travel cheaper and quicker venture farther and farther from their origins. There are also some non-economic aspects contributing to the expansion of the range of travel. Tourists "mature". They are becoming more experienced with the time as they participate in changing spatial patterns of tourism. Tourists who stay at home during their vacation engaging in local recreation gradually venture farther from home starting with domestic and then increasing participation in foreign tourism. They usually begin with the neighboring country and then move farther away.

The spatial patterns of tourism evolve frequently under the impact of relative location of attractions at the destination. Indeed, low fares (e.g. charters) and accessibility are instrumental in increasing the tourist flows to the destinations but a cluster of attractions in or close to the destinations constitutes an important asset for the receptor areas. Thus Europe as a tourism destination offers a number of attractive countries located in proximity to each other. Also, the existence of secondary attractions in the vicinity of primary attractions constitutes a clear advantage. Cancun, an important beach resort in Mexico offers excursions to the Yucatan's Mayan ruins by bus or airplane. Chichen-Itza is the most visited of these attractions. Another example of a cluster of attractions located in proximity to each other is Guatemala (Lake Atitlan, native markets of Chichicastenango and Ouezaltenango, the ancient capital of Antigua, Mayan ruins of Tikal). Spain's Costa del Sol offers excursions to exotic Morocco and to a number of nearby inland attractions (Granada, Seville,Cordoba, Ronda).

The transformation of spatial patterns of tourism is of great importance for LDCs. Indeed, the third world is slowly increasing its share in receipts from international tourism originating mainly in the DCs. This share is, at present, relatively small because the international tourism is still concentrated in developed areas: according to various sources, the European and North American destinations receive 80-90% of total expenditures in international tourism. The share of LDCs is increasing slower than expected. There are various reasons for this rather disappointing development, among them the cost of travel which, especially in case of more remote countries, is more than proportionate to the distance because of high air fares, fewer charters and special fare arrangements which nowadays boost tourism in the DCs. The benefits of expanding international tourism are mainly limited to LDCs located relatively close to markets like Mexico, Caribbean, North Africa and parts of Middle East. For these areas the era of mass tourism has arrived in the post World War II period exacerbating the uneven development of tourism within the Third World. Although the LDCs located close to markets benefit from some discount airfares the North-South discrimination in travel costs puts these areas in some competitive disadvantage vis-a-vis the DCs.

Another factor which retards the development of international tourism in the LDCs is political instability in many of them: wars, guerillas, riots, terrorism etc. are putting tourists in physical danger. Also poor economic performance combined with population explosion cause deteriorating socio-economic conditions in many, especially Afriacan and some Latin American LDCs. This in turn causes increase of assaults, robberies, thefts which, equally as political dangers, prompt tourists to stay away.

4.4.8.3. The energy crisis scenario

The long range trend in tourist travel to increase the distances may be temporarily reversed during energy crises. Indeed, the energy crises of 1973-1975 and 1979-1981 had a significant negative, although temporary, impact on tourist transportation. Tourism in terms of activities at destination is not energy-intensive. However, the travel element in tourism does require significant energy inputs and has been vulnerable to energy crisis. Although the energy situation is stabilized at present, the impact of the energy problems will be with us for foreseeable future as the days of cheap energy are unlikely to return. Therefore, a scenario of very expensive energy and even energy shortages cannot be excluded. Such a energy crisis scenario runs counter the long range trend to increase the travel distances but will certainly not cancel it out.

In an energy crisis scenario the problem of accessibility, of distance decay function becomes much more important than before: the mobility of tourists decreases, the patterns of tourism and recreation are becoming less dispersed and more concentrated. The distances travelled shrink. The reduced mobility hurts more the international than domestic tourism, more long-distance destinations than short-distance destinations. Tourists are paying more attention to intervening opportunities i.e. attractions and destinations located at shorter distances between the origin and the more distant destinations which are losing out to tourist localities and regions situated closer to markets. There is more local and regional travel. Urban and near-urban recreation and entertainment increase at the expense of long distance travel. More tourists choose close-to-home vacations. Single-stop vacations gain popularity at the expense of touring (multi-stop vacations). Single-stop destinations are being developed into total resorts that feature a wide range of activities in one spot to cater to various demands within a tourist party (e.g. family). In this connection the average stay increases which counters the long-range trend to decrease the length of stay. As soon as the crisis is over, people return to the old patterns of travel and the distances resume their expansive trend.

An energy crisis causes shifts in demand for modes of tourist transportation: all forms of public transportation increase at the expense of the private car (especially driving for pleasure) and other less energy-efficient transport like RVs. This is especially obvious in long distance travel where commercial carriers like buses, railroads and especially airlines with new-energy efficient wide-body jets are winning clientele. Public transportation has proven to be not only more fuel efficient but also more reliable in terms of fuel availability. Also intermodality arrangements combining air and car travel (fly/drive) and air and sea travel are gaining on importance. In short distance travel the people switch from motorized transportation like private motorboats or snowmobiles to bicycling, hiking, canoeing, boating, cross country skiing and other energy-saving transportation. Also forms of energy-saving recreation like sports, event attractions (festivals, concerts, competitions etc.) are gaining in popularity.

In terms of organizational patterns of tourism in energy crisis scenario there is a trend to less individual travel and more group and package tours. In accommodation there is a tendency to self catering forms such as rented holiday apartments, bungalows, trailers, etc. and less demand for second homes, especially those located far from cities.

The energy-saving shifts in tourism and recreation patterns are increasing some capital investments for facilities. Many of the urban and near-urban activities are user-oriented (according to Marion Clawson's terminology) and require expensive man-made facilities like e.g. swimming pools, artificial ski slopes or fishing ponds. People engage increasingly in concentrated high density and high intensity forms of recreation whereas the extra-urban resource-based forms are less in demand because they require longer travel. Thus, with the distance-decay function gaining importance the dynamic (travel) element in tourism is losing on emphasis in the energy crisis scenario to the benefit of the static element (stay). This fact may have some connection with increased preference of the public for expansion of the weekend leisure time slot rather than vacation and also with the growing substitution of long annual holidays for more frequent short holidays also in winter and during the shoulder periods between season and off-season.

Some of the trends in spatial patterns of tourism discussed above, are of short duration and cease when the energy crisis is over. Many of them, however outlast it and become long range. Some trends were discernible already before the first energy crisis started in October 1973 and the energy crisis merely reinforced them. An example is pleasure driving which ceased to be the most important US recreational activity during the 1973/4 energy crisis and has not regained this position since. Also package tours are loosing their energy-saving aspect in favor of money-saving aspect after the energy crisis is over.

Chapter 5

DEMAND

5.1. THE SPATIAL SYSTEM OF TOURISM

The spatial elements of tourism occur in three zones: origins (or the demand zone which is the focus of this chapter), destinations (the supply zone) and the transit zone:

1) Origin (other terms with similar but not identical meaning: demand area, market, tourism generating area). Here the existing conditions (the stresses related to work, unpleasant climate, overcrowding, environmental deterioration etc.) constitute the push factors causing demand for tourism. The markets may show different degrees of emissiveness (travel intensity) which is measured by number of out-bound tourists in percentage relation to total population. At origin a surprising amount of tourist transactions are taking place because it performs the major marketing functions as seat of tourist promotion, advertising, wholesaling and retailing.

2) Destination (other terms: tourist receiving area, receptor area, supply area, host area, tourist location). This is the area where the tourist product is located. Tourist product is an amalgam of goods, services and intangibles. It is composed of natural and man-made attractions and tourism industry (tourism infrastructure): hotels, restaurants, special tourist transportation, entertainment and other tourism facilities. For the tourism industry to function properly an adequate general infrastructure (transportation and communication systems, water and electricity supply, waste disposal etc.) is essential. The tourism product is immobile. The consumers have to come to the product and not vice versa. Therefore, the product has to exert a certain pull effect on the origin because of its attractiveness. This means it must have some apparent or real qualitative characteristics which are lacking at the origin (pull factors). The prospective tourist learns about these characteristics from ads and promotional literature which constitute the least reliable sources and also from popular and scholarly literature and most importantly from friends who visited the destination (word-of-mouth information).

3) The transit area (routes) where the travel between origin and destination takes place may play various roles in the system of tourism. In one-stop or destination tourism its role is unimportant. It is rather a nuisance, an obstacle which should be

overcome in a minimum of time, discomfort and expense. This situation is especially typical for air travel to beach resorts. However, transit routes as locations of the transportation component in tourism acquire frequently a very important dimension. This is especially the case when certain forms of surface transportation is used (e.g. automobile, sea cruise, bicycling, hiking). Indeed, transit area may become a destination of multi-stop tourism. We talk in such instances about touring. The transformation of the transit area into a destination occurs also in an intervening opportunity scenario. This happens when competitive destinations develop in the area between the market (demand) and destinations or potential destinations (supply). Thus Mexico and the Caribbean constitute an intervening opportunity for the North American tourists who would otherwise visit Central or South America. Florida is an intervening opportunity on the way to the Caribbean and recently developed Cancun on the way to Acapulco. Indeed, Acapulco is no longer the top Mexican tourist destination. It has been outranked by Cancun (Tourism Canada, Sept. 1987: 870928).

Generally, the longer and more expensive the travel between origin and destination the larger the "distance friction" (barrier) between them, the weaker the "gravitational force" between them. This circumstance is of crucial importance for tourism.

The three spatial elements of the tourism system are closely integrated in the experience of a tourist which takes place in all of them. This experience has been discussed by Clawson and Knetsch (1966:33-36). It contains five stages or chronological phases which integrate it both in space and time:

1. Anticipation and planning takes place at the origin. It in some cases extends for many months. It is the time not only devoted to pleasurable consulting of books, guides, maps, experienced friends but also to shopping connected with the future trip. We often forget that a large proportion of the tourist expenditures is being done before the vacation starts.
2. Travel from origin to destination. This phase takes place in the transit area and may be a journey of many thousand kilometers and/or considerable duration. The variation of cost is considerable. The proportion of this phase in terms of time and money spent to the other phases may vary: time-money substitution is frequently practiced. In case of touring it constitutes the main part of the whole experience. However, it is mainly regarded as a necessary nuisance.
3. On-site experiences, take place at the destination. It is usually but not always necessarily the most important phase. The other phases are built around it in terms

of time, satisfaction and expenditures. The sojourn may involve one place (one-stop vacation) or several (multistop vacation) and various clusters of activities including both pleasure and business.
4. Travel back-home takes place again in the transit area. This is probably the least enjoyable of all the phases. The mood has changed, the vacation is almost over, the money spent, the daily routine stands around the corner.
5. Recollection takes place at origin and is usually a very pleasant one. Here the traveller "participates" the third and final time in the trip. Even unpleasant and frightening moments seem to appear in a different light in some cases even as fun. There is certainly much wishful thinking in this phase. The recollection may not only have an individual but also shared character: friends and relatives are told about the trip, shown photos, slides and souvenirs. These talks and discussions may inperceptibly merge with anticipation for another trip based on enriched experience.

5.2. DEMAND AND SUPPLY EQUATION

Two of the elements of the spatial system of tourism, origin and destination, may be approached from the point of view of economics as demand and supply in a non-spatial relationship. They interact with each other in what is called demand and supply equation. This interrelationship is discussed in this section.

Demand in economics means the quantity of goods that buyers are ready to buy at each specific price in a given market at a given time. In the context of tourism, demand means the readiness of tourists to buy certain quantities and qualities of tourist goods and services at specific price levels (taking into account currency exchange rates and prices in international tourism). Thus various price levels will be associated with specific numbers of tourists willing to participate at these levels. In this sense one refers frequently to demand as "a schedule" at which both variables (demand and supply) meet in the market. Indeed both variables in the demand and supply equation are so closely interrelated that it is difficult to discuss them separately. We do so only for analytical purposes. However, we should be aware that demand cannot exist without supply and vice versa. They are inseparable - like two sides of a coin. We must think of them as in constant interactions in both socio-economic and spatial dimensions. "In many ways, the distinction between demand and supply is arbitrary and academic ... for the ultimate concern is the relationships between them" (Patmore, J., 1973:243).

The interaction between demand and supply results in a demand schedule with quantities demanded varying inversely to prices i.e. the lower the price the greater the quantity and vice versa. This is illustrated in Figure 5.1.

Figure 5.l. The demand curves

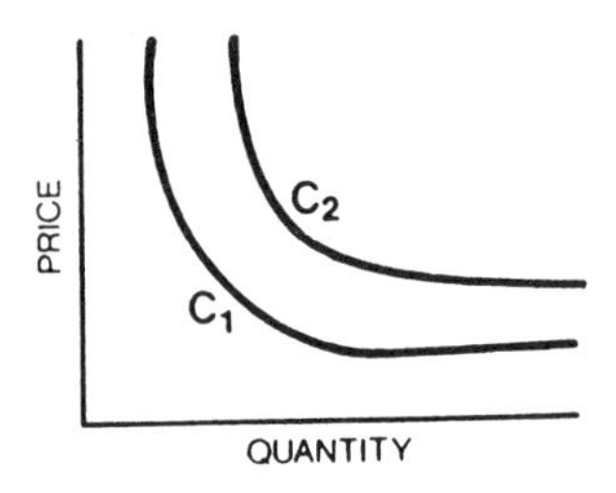

Various products have different curves. In tourism and recreation curves express various demand schedules in diverse situations e.g. for types of roads (direct, gravel, paved) or degrees of attractiveness, availability of leisure time, incomes of total population or population groups etc. The general rule is that changes to the better like improvements of accessibility and/or attractiveness, or increases of leisure time and/or incomes, cause changes in the position of the entire demand curve: it moves to the right further from the intersection of x and y axis. In Figure 5.1. the curve moves from C_1 to C_2 position.

Demand and supply theoretically interact freely at the market place, with demand being more dynamic, more changeable, consequently more difficult to determine, supply more static, burdened with inertia. However, many economists, among them J.K. Galbraith, believe that market in the classical sense does not exist because supply uses its oligopolistic position to dictate and manipulate demand in such a way that it responds to the pressures of marketing. This applied to tourism/recreation would mean that what people do is largely determined by opportunities provided i.e. supply determines the visitor participation and patterns. In reality this scenario is only partially valid. It is true that much of demand is supply-induced but demand is still relatively independent from pressures of supply because the supply in tourism/recreation is less integrated than in other economic sectors and because of high degree of competitiveness. The substitutability seems to be greater in recreation than in tourism, especially cultural tourism (uniqueness of the tourism product). However, tourism demand is price elastic and this increases the competition between various destinations. As a result the demand picture is varied and rapidly changing. Therefore, research in leisure time demand has acquired an extremely important dimension.

5.3. DEMAND AND NEED

While discussing the demand for tourism and recreation people often use the term "need" as a synonym of "demand." This is incorrect. "Demand" and "need" do not mean the same. People may need something but not demand it and demand something which they do not need. Demand is inseparable from supply and prices. Need is in a sense somewhat abstract, detached from concrete possibilities to meet it, detached from opportunities and facilities which could satisfy it. In short: need is detached from supply. Social scientists divide needs into two categories (hierarchy of needs).

1. Basic (primary, low level), mostly physiological (food, sleep, shelter); also, need for survival, safety, defence.
2. Derived or acquired (secondary, higher level) needs which are not necessary for bare survival e.g. education, entertainment, recreation, tourism. These needs indicate that man requires not only to live but to live well.

While it is easy to determine what are the needs on conceptual level, it is difficult to do it on operational level. Thus conceptually many people in DCs regard tourism (especially recreational tourism) and recreation as an essential need, a necessary component of satisfying life for all and not as a luxury. Compelling health, social and economic reasons indicate the necessity of this need. However, because of somewhat abstract nature of the notion "need" there are problems with detailed determination of needs in general and recreational needs in particular. Mercer complains about the transitory character of many needs which "are largely induced by advertising media which deliberately set out to manufacture needs" (Mercer, D., 1973:38). This leads to "faddishness," hurried irrational decisions by consumers who take up and drop activities "in rapid succession."

Despite these problems social scientists are trying to learn more about needs. For this purpose various categories of recreational needs have been established. They will be listed here according to Mercer (1973).

Categories of recreational needs:
1. Normative needs - "more or less precise and objective standards which are set by experts in the field of recreation." e.g. experts tell us arbitrarily that so and so many hectares of open space per 1000 inhabitants should be available for urban recreation. Such a list of standards is almost endless. These standards are used for planning purposes as a sort of "rule of thumb" but their objectiveness is often questioned because of being established by small elite groups.
2. Comparative needs refer to variations in the provision of recreation opportunities. e.g. one area has a swimming pool another has not.
3. Latent needs (identical with latent demand)
These are frustrated recreational needs which for some reason cannot be satisfied.
4. Expressed needs (identical with participation, use)
These are latent needs translated (turned) into action.

5.4. DEMAND AND PARTICIPATION

After stating that "need" and "demand" are two different notions it is necessary to make the distinction between "demand" and "participation." Demand in economic sense refers to the schedule of quantities or volume (visits, user days, etc.) that the people will desire at all possible prices. Recreation demand is "the propensity of a population to participate in a recreational activity at a specific level of recreational opportunity, supply and cost" (Taylor, G., 1969:1). Participation is tantamount to consumption, use or attendance. In other words participation is "demand ex-post," after the demand has met supply at a certain price level and consumption occurred. Demand refers to situations "ex-ante" (before), participation to situations "ex-post," after the interaction (the interface) of demand and supply. The levels of participation not always indicate the level of demand because of lack of opportunities in certain areas may adversely influence participation. Despite these conceptual differences the terms "demand," "participation" or "use" are frequently used alternatively, especially in situations when recreation opportunities are provided free of charge. In such situations changes in consumption primarily reflect the changes in supply of recreation opportunities. Thus consumption becomes the only indicator of demand.

5.5. DIMENSIONS OF DEMAND

Burton, T., 1971:26-27) using the term "demand" in the popular sense as consumption and not as "economic demand" distinguishes five dimensions of demand: existing, potential, induced, diverted and substitute.

The two first dimensions of demand (existing and potential) are the most important and the research attention is focussed on them. The other three dimensions are supply-related.

1. Existing or effective demand is demand which currently exists and is measured in units of use. It is a desire for certain tourism/recreation experience together with the ability to pay for it. It is a market form of demand.

2. Latent or potential demand. This is a non-market form of demand. This demand is not effective for some reason but "would be so in other circumstances. It is a demand which is frustrated by such factors ad the non-existence of facilities" or perceived lack of time, money etc. Potential demand for tourism/recreation means the untapped market which is ready to be opened up, ready to be translated into effective demand. Of course, some environmental, socio-economic and attitudinal prerequisites are absolutely necessary for a demand to be classified as potential. The existence of these prerequisites is characteristic for this latent or "below the surface" demand as opposed to "need" which is not necessarily associated with the existence of such prerequisites. The number of Quebecers visiting National Parks is relatively small especially when compared with the inhabitants of Canadian West. This does not mean that the people of Quebec have a natural aversion to visit National Parks. It simply means that there are only two National Parks in Quebec. The latent demand for outdoor swimming and boating comes suddenly forward to fruition when a reservoir (e.g. Diefenbaker Lake in South Saskatchewan) is constructed in an arid area. The development of ski lift facilities in the Rocky Mountains meant a significant upsurge in Alpine skiing. These increasing opportunities for various recreational activities cause more than proportional increase in participation indicating previous latent demand which could not be met because of lack of facilities located at a reasonable distance from the market. L. Brooks puts it this way: "Pressures simply build up until the explosive force of this demand is revealed when an outlet is provided" ... (Resources for Tomorrow, 1961 Vol, 2,: 962). In some cases the limitations of recreational opportunities may be caused not by lack of facilities but by inherent deficiencies of the resource base e.g.

flatland is not conducive to down-hill skiing. However, even here some corrections are possible in constructing artificial hills such as Blackstrap in Saskatchewan.

Potential demand is extremely important for tourist business. There are people who could but for some reasons have not participated. One has only to tap this market by creating better supply conditions e.g. by special airfare arrangements and by appropriate marketing. Of course, adequate prerequisites must exist: hungry Ethiopians do not constitute potential market for tourism.

To find out the dimensions of the potential demand surveys are conducted and the results are amply published. Figures for percentage of population taking annual vacations, travelling outside the area of residence, participating in certain recreational activities, people who have never flown etc. are important indicators of the volume of potential market which must be targeted by marketing and other actions of tourism industry. Thus, e.g. in the DCs only 50-75% of population take annual holidays away from home and only about 2% of world population participate annually in international tourism. The researchers also supply us with data about competition in the leisure market, competition for time and money between home based recreation where TV, VCR and radio and audio are the strongest competitors and urban, extraurban recreation and tourism. There are indications that tourism industry is increasing the Tourism/Leisure ratio by using various marketing strategies (Mazanec 1981). There is also competition between recreation and tourism e.g. purchase of expensive equipment like motorboats may have an adverse impact on demand for foreign travel.

3) Induced demand. Created as a direct result of the provision of facilities e.g. a new swimming pool in a community creates additional demand to any latent demand which was previously unsatisfied.

4) Diverted demand occurs e.g. when people switch to a new swimming pool in their district from an older one located at greater distance in another area.

5) Substitute demand is similar to diverted but refers to different recreation facility e.g. a new swimming pool may attract people who previously played tennis. Thus they have substituted their swimming demand for tennis demand. The question of substitutability in recreation is very interesting and important. Unfortunately, too little research has been conducted in this field. If the degree of substitutability is high then it is possible to substitute some recreation activities which require scarce resources (e.g. golf) or are regarded as socially and medically undesirable (e.g. boxing) by other more acceptable activities. Substitutability in recreation tourism is higher than in cultural tourism.

Burton recommends a flexible, not a dogmatic, approach to these five dimensions. In some instances they may pertain to the same case, but the point of view is decisive e.g. there is latent demand if there is no swimming pool in a community and both induced and effective demand if a new pool is built.
6) Additionally one could distinguish deferred or pent-up demand if no purchases were done for a longer time e.g. following the end of the World War II there was pent-up demand for tourism/recreation. Similarily, pent-up demand increases during economic recessions or energy crises. It is related to potential demand.

5.6. STAGES LEADING TO EFFECTIVE DEMAND

The demand for any product including tourism and recreation is shaped by many variables before it reaches its final form of effective or existing demand. To introduce a sort of order into this seemingly inpenetrable maze of causations which contribute to these final patterns of specific choices among endless offerings on the supply side of the equation, one could think about certain definite sets of variables which step by step i.e. in more or less well defined stages contribute to the final stage of effective demand. The concept of these stages has been developed in numerous writings by B.L. Driver. The set of variables at the first stage are - to use B. L. Driver's terminology - "the antecedant conditions" e.g. age, sex, education, lifestyle which form a general profile of a consumer.

At the second stage the consumer formed by these antecedent variables develops a number of approaches to tourism and recreation. These approaches are called motivations. Motivations bring the consumer closer to the supply (product) but still on a more general level without linking the consumer with specific products. This link with the products is established only at the third stage when consumer attitudes to specific products are formed. These attitudes range at a scale between very favorable to very unfavorable and are based on certain images of the products. The third attitudinal stage contributes only to the formation of potential or latent demand. It means that the consumer is not yet completely ready to participate. Indeed, motivations and attitudes do not lead directly to participation. The fourth and last stage before consumption (in our case: participation) takes place. This stage is, to use L.B. Driver's terminology, the stage of "intervening variables" i.e. all sorts of obstacles which may frustrate demand like lack of money, time, ill health, family obligations, government restrictions and regulations etc. Only after moving through all these four stages from antecedant conditions to intervening variables the demand finally acquires the final shape of effective or existing demand, ready for the act of consumption of the tourist/recreation product. The stages in formation of effective demand can be summarized as follows: (modified after Driver, B.L., 1970 and his unpublished papers).

I Antecedant conditions (demographic, socioeconomc, geographic, lifestyle variables)
II Motivations (push factors prevailing slightly)
III Attitudes towards the products: favorable or unfavorable images (perceptions) of the product (pull factors prevailing very strongly)
IV Intervening variables
V Effective demand and consumption (participation)

The stages in the formation of effective demand contain sets of interacting variables. The interactions or combinations of these variables account for the variance in consumer behaviour i.e. in practical terms, tourism/recreation choices of specific products (destinations, recreational activities). The analysis of these variables is called market segmentation which takes place at all four stages.

5.7. MARKET SEGMENTATION

5.7.1 Introduction

The development of policies and planning for tourism and recreation and also marketing them is inconceivable without thorough knowledge of the market, without market research which would with reasonable accuracy predict the leisure time behaviour of the people. Until recently the market was understood in a very simplistic way. Promoters envisaged a mass amorphous market which was treated as a monolithic entity without internal differentiation. Averages were regarded as sufficient in determining people's needs and demands. However, with the advent of serious research effort it has soon become clear that there is no average consumer. Therefore, averages pertaining to the whole population are today much less used than in the past because the researchers have realized that averages are derived from innumerable decisions based on personal individual values and preferences which are difficult to generalize for the population as a whole. Thus, there is no one monolithic market but rather a number of submarkets or market segments with relatively homogeneous characteristics. Within a submarket intra-segment differences are minimized and the inter-segment differences maximized. The recognition of this fact has important implications for planning in both the developmental and promotional aspects.

An appreciation of the fact that the market for tourism and recreation consists of an array of segments, was a major step forward moving the industry from a state of primitivism to one of sophistication. While it is still possible to hear the tourist market or the recreation market referred to as if they were really monolithic, a good deal of serious attention is now being given to methodologies that will enable the segments to be identified, described and understood.

The practical importance of market segmentation research is based on the recognition that various market segments behave differently with respect to tourism and recreation: they are characterized by varying leisure time slots, discretionary incomes, propensities to spend time and money, lifestyles etc. Consequently they differ in their attitudes to and choices of activities or clusters of activities (a recreation/tourism experience is mostly composed of a mix or complex of activities). And this is what the planners and marketers want to know. Planners need the information on market segments in order to determine the needs and demands of population and to enable them to plan for more rational allocation and utilization of resources. Marketers want to maximize and optimize their promotional endeavours in targeting various market segments (so called "market strategy") in order to match prospective tourists with tourism products. In practical terms such market strategy boils down to promotion of tourist and recreation products in media which are preferably used by various submarkets. Also the contents of the promotional messages is adjusted accordingly to maximize their impact on the target segment. The necessity of market segments study increases because our modern society becomes less conformist with increasing pluralism of lifestyles. The multiplicity of markets necessitates research and marketing to match the markets with required particular product combinations for the satisfaction of people involved. Thus research differentiates not only the market segments on the demand side but also recognizes the fact that these segments express their demand not only for tourism and recreation as a whole but also, more importantly, for concrete products in this field.

The marketers in their supply-demand matching endeavours can use two basic marketing strategies: they may modify the supply to fit the needs of identified market segments e.g. by offering special youth or senior citizens' fares. Secondly, they may present the same destination to different submarkets using different promotional appeals e.g. emphasizing educational and cultural aspects of Italy as tourist destination for one submarket and Italian beach activities and fashion for another. They may attempt to capture the retirees market or the yuppy (young urban professionals) market etc.

5.7.2. Antecedant segmentation

The market segmentation at the stage of antecedant conditions contains four sets of variables:
1. Demographic: age and stage in the lifecycle, sex, household size, marital status, etc.).
2. Socio-economic (income, education, occupation, race, ethnicity, religion).
3. Geographical (location, regional submarkets, size of community). Sometimes groups 1-3 are called, not very aptly, demographics.
4. Lifestyle or psychographic research uses clusters of variables including "traditional" (the first three groups) but the emphasis is on behavioural variables.

There is a market segment beyond this classification: increased attention is recently payed to disabled travellers. Not only the tourism industry but also non-profit organizations are interested in this market segment (Tourism Intelligence Bull. 1988, Feb.).

Market segmentation is based on assumption that the differentiated submarkets vary in their demand patterns. The exercise of segmentation is futile by identical demand patterns. Also, it is important that the submarkets can be reached by promotional activities via identifiable media which they preferably use. If the submarkets use the same media they are difficult to reach. The same happens if the submarkets are of too small size because this makes marketing endeavours uneconomical.

The major problem connected with market segmentation research is that the variables are frequently interrelated. They overlap. This cross-correlation makes it difficult to isolate the influence of individual variables. One has also to take into account that some behavioral variables may be almost intractable or not measurable. This lack of precision may lead sometimes to questionable results. Recently some of the commercial marketing research firms were taken to court for malpractice when their promises proved to be inaccurate (Time 28-12-1987). Despite these problems the practical utility of market research is beyond doubt.

Nevertheless, this note of caution seems appropriate before entering into discussion of market segments.

5.7.2.1 Demographic Segmentation

Age and stage in lifecycle

One of the most important demographic variables influencing demand is age and the stage in the lifecycle. In fact, the age structure of the population and its changes are of vital interest to tourism and recreation planners. The high birth rates (baby-boom) during the first 20 post World War years (1945-1965) in North America augured an increased share of young people in the total population. However, subsequent drop in birth rates has radically changed the demographic situation. The less numerous generation ("baby-bust generation" of people born after 1965) has already entered the market. In 1980s the cohorts of young people below 20 years of age represent a smaller share in total population than in the past, whereas the "baby boom generation" is now 25-45 years old. This age group has increased its share in the total population (demographic bulge). At the other end of the age spectrum is the 20th century trend of the steadily increasing share of senior citizens in the total population. It will certainly continue in the foreseeable future, reinforced by falling birth rates and by increasing life expectancy. (For more statistical information see Chadee, Mieczkowski 1988).

The age segmentation research indicates that the maximum participation rate in tourism/recreation occurs in the age bracket of about 18-45. Particularly young people (15 to 29 years old)) mostly single and relatively free of professional and family obligations show a higher degree of mobility, a higher propensity for participation, especially in more active risk-taking, adventure forms of leisure time activities than the other age groups. The young are eager to see the world, to experience something new. The urge to escape from the daily routine is especially strong. Vacation ranks high in their system of priorities. The participation patterns emphasize peer socializing, particularly by teenagers. Lack of family responsibilities in most cases removes the constraints. Lack of money may constitute an obstacle although the smaller numbers of young people nowadays should bid up the wages and raise their disposable incomes. The "youth market" constitutes a specific segment of demand because young people tend to participate in cheaper forms of recreation and tourism, especially cheaper transportation and accommodation. They also tend to travel more independently and use less package tours. The tourism industry responds to this market by introducing special youth fares and other price discounts.

The efforts of the tourism industry to promote youth tourism are based purely on economic motivations to increase profits and to expand markets and lack social, cultural, educational and political aspects characteristic for modern tourism. These may be provided only on the government level. In this respect especially interesting are the documents and resolutions of World Tourism Organization (WTO) and in particular the 1980 Manila Declaration (World Travel 1985, 183:57-60; 184:33-34 and 43). The WTO urges the governments of the world to see beyond the purely economic effects of youth tourism which may not be that great but to recognize also the social, cultural, educational and political impacts of youth tourism not only on participants and also on destination areas. Therefore, "the governments should promote youth tourism by:

1) incorporating tourism into educational programmes for the young generation.
2) including the youth tourism into the framework of their national policy of cultural and educational development and national integration.
3) promoting youth exchange projects in the fields of culture, education and sports, summer camps, study tours, internship tours, international building sites.
4) providing financial and fiscal incentives to expand facilities for the young (e.g. simpler and more functional forms of accommodation like youth hostels and schools or providing special transportation fares subsidizing youth travel agencies)".

The role of the WTO is exhortative and it cannot enforce this proposal for expansion of social (subsidized) youth tourism. It is limited to providing the definition of youth tourism (15 to 29 years of age of participants), consulting the interested parties and coordinating efforts aimed at expansion of youth tourism.

The "baby boom" generation (born between 1945 and 1965) constitutes a distinct age segment of the market. Increasingly, these people marry late or do not marry at all and have limited family responsibilities. In addition a certain proportion of them earn relatively high incomes in managerial and professional positions. All this puts relatively high discretionary incomes in their hands. Indeed, some of them belong to the new category of yuppies (young, upwardly mobile urban professionals) whose participation in tourism/recreation is very high. Many of the yuppies are not young anymore but act young. Despite this rather optimistic trend for participation in tourism/recreation one has to point out that the

25-45 age group is the prime buyer of durable goods. This feature is increasing the competition for the discretionary income and certainly has a negative impact on participation. Also depended children "compete" for discretionary income of this age group.

The middle age market (about 40-64 years of age) segment is the most important in terms of demand volume despite their lower participation rates than the younger people. The relative importance of this group will increase in the next 20 years largely at the expense of younger cohorts as the demographic bulge of baby boom generation is already moving into this age bracket. In this age group cultural and educational tourism gains in importance. Their expenditures for tourism/recreation are high because these people have the highest earning power as compared with other age groups. They marry later in life, if at all. The share of singles in this age group has increased in recent years in DCs and this also enhances participation (the singles' travel segment focusses increasing attention of the tourism industry). The middle aged, similarly to young people, have less children than in the past or no children and this fact works also in the same direction. Presence of small children in the family is a factor decreasing participation especially in strenuous recreation activities and long distance tourism. There are some attempts by the tourist industry to cater for families with small children e.g. the Austrian "baby hotels". Nevertheless, in tourism childless couples and singles show higher participation and expenditure rates, particularly in air and long distance travel. Indeed, small children in the household constitute a hindrance in participation which is correlated to the number of children.

In the late middle age period the participation rates tend to diminish, especially with respect to physically demanding forms of tourism/recreation. Certain reluctance to forgo everyday comforts, unwillingness to change the environment and possibly health reasons decrease the participation rates. There is evidence to suggest that the percentage of people taking vacation trips declines after age 50. At the same time some factors work just in the opposite direction: the children grow older and leave the family, the incomes rise to reach a peak between the age of 50 and the retirement.

An increasingly important, but until recently neglected and untapped, market is the age segment of elderly people. By societal and official measures old age starts at 60-65 years of age and the lifecycle stage of retirement begins. Retired people in the 65-75 age category, called "self-compensators" are especially eager participants because only now they have the chance to do things which were out of their reach while working. The numbers of retirees are increasing quickly in DCs as a result of growing longevity and the trend to early retirement. They have the precious commodity at their disposal - time. They are full-time at leisure. The decline in overall participation rates in tourism was in the past caused by lack of money and deteriorating health. However, nowadays, there are factors working in the opposite direction and the retirees market gains on importance not only as a result of quantitative growth. The average health of old people has improved significantly as a result of progress in medicine and medical care. They also have more financial means as pension schemes become more common and savings accumulate. Indeed, in the USA "the real incomes of old people have been rising faster than the real incomes of working people" and "the social security recipients are on average more affluent than the population as a whole" (Newsweek 9-11-1987) and "their poverty rate is lower than that of the under 65 population" (Newsweek 18-4-1988). In the US "adults 50 years of age and older account for half of all discretionary spending power" (Tourism Intelligence Bull. 1987, May). The same age cohort controls, 80% of Canada's wealth (MacLeans 9-1-1989). No wonder that a new term has been coined for these seniors: "woopies" (well-off-older people). Seniors constitute a distinct and increasingly important submarket worthwhile to study. One aspect still inadequately researched seems to be the rates of participation of elderly in tourism. An early indication of a significant evolution in this respect are the findings of C. Weiss that the US and Canada are the only countries where the participation rates of elderly exceeded the overall figures for the population as a whole. (Weiss, C., 1974, 1:2). The same development took place in other DCs in 1980s and the image of an average older person as non-participant surely belongs to the past. (Feige, M., 1988:17; Datzer, R., 1988).

The findings of the same researchers indicate that "older tourists travelled farther than younger ones." They also tend to stay longer at destination and are "attracted to relatively developed nations" (Weiss, C., 1974). This last observation may be partially explained by quest for comfort and superior health services in the DCs. Another research finding indicates that senior tourists tend to purchase more package tours (Sheldon, P., Mak, J., 1987,). This trend illustrates the desire for security, comfort and social contacts. Seniors prefer not only package tours but also choose more frequently air travel than other age groups.

In recreation the "golden age" people show their preference for less strenuous activities. Thus e.g. their demand for urban recreation, particularily for urban parks, is increasing. They also tend, caeteris paribus, to spend relatively more than younger people for their vacation, they enjoy luxury and pampering if they can afford it. However, as a whole, the elderly spend less per capita than the middle age group. Therefore, the tourism industry attempts to tap the seniors market by offering special fares and various discounts, many of them in off-season and shoulder periods. Indeed, the participation of senior citizens in tourism and recreation is not limited to seasonal peaks when the majority of seniors do travel. However, increasingly they travel in off-season and shoulder months (spring and fall). Especially favoured time for older people is spring, a kind of celebration of survival and victory over death of winter (Weiss C. 1974:2). The fact that the share of off-season travelers among seniors is higher than among other age groups is highly positive for tourism business because it decreases seasonality. Also, the cold winters of the northern latitudes motivate older people to seek the milder southern areas not only as a place of temporary but also of permanent residence (Florida, California, Spain etc.).

A good deal of travel by retired people constitutes the seasonal migration in the form of sizable southward movement in late fall and early winter with a return northward flow in the spring. This seasonal migration of seniors is not only characteristic for North America but also for Western Europe e.g. the number of West Germans spending winter or part of winter on Spanish Mallorca is growing by 8% annually (Spiegel 1988, 10). German retirees constitute the majority of winter tourists in Mallorca. Although the water temperature is only 13°C and the air not much more than 15°, the seniors enjoy the off-season rates and prices (often 50% lower than in season) and also intensive social contacts among themselves.

Another form of seniors' participation in tourism is education. There are a number of educational programs, especially tailored for this sub-market. Seniors travel both domestically and internationally to take part in mostly non-credit courses offered "off-season" by universities, colleges and other educational institutions worldwide. The seniors stay at campus dormitories and attend lectures and discussions on various subjects.

All these forms of tourism frequently mean a psychological boost for retirees, a sort of ego-enhancement. And old people need it often badly. The aging process is associated with loss of social status (identity crisis) culminating in retirement. Hence the need for compensation. The status of old people is enhanced by participating in tourism. Another motivation for old people to participate is an escape from inactivity and monotony and their wish to act "young" again.

Evaluating all the age-lifecycle submarket one may state that in relative terms the seniors' submarket is developing most dynamically. The baby boom submarket (people born between 1945-1965) is at present the most important, the youth submarket has lost its relative prominence so distinct in the 1960s and 1970s because of decreased share of young people in total population. In many developed countries this age group i.e. the "baby bust" generation born after 1965 has declined not only relatively but also absolutely.

The research into the age structure of tourists provides the tourism industry with important data invaluable for a more balanced approach to various age sub-markets and in practical terms for marketing. E.g. it is well known that US travellers are "consistently more elderly than the norm" (Pearce, D.G., 1978:6). Also certain destinations draw their clientele from specific age brackets e.g. Nepal appeals to the young. The elderly for reasons of safety and comfort prefer rather DCs as destinations. Marketing tourism for them should be focussed on newly retired and "younger" seniors because the participation rates drop among people older than 74 (Blazey, M., 1987).

Sex

In not too distant past tourism and outside-home recreation were almost exclusive domain of men. However, in the DCs this situation has changed enormously in the 20th century, especially after World War II when radical demographic shifts have occurred. One of the most important phenomena has been the emancipation of women and their massive entrance into the work force. In the US the number of working women was almost equal to the number of non-working housewives already in 1973 to exceed them subsequently. Today one could generalize that in the DCs more than 50% of women are working. Another demographic fact with consequences for tourism/recreation is the growth of unmarried women with no children. In the USA this group exceeded already in early 1970's the women living with husband and children.

The impact of these demographic changes on tourism/recreation was analyzed by researchers. They came to the conclusion that "working women have more money available to them than women who do not work" (Bartos, R., 1982:5). Therefore, their participation rates in tourism are higher. The large number of unmarried and married women with no children means also more customers for tourism.

All these developments indicate that tourism marketing should take the quickly expanding women market more and more into account. Indeed, in terms of total participation rates the differences between sexes are nowadays almost nonexistent with slightly higher male rates for all DCs. Thus women participate at near equal rates as men. Differences appear by analysing specific tourism/recreational activities. There are also variations in terms of specific tourist origins and destinations. Men prevail overwhelmingly in such traditionally masculine recreational activities like hunting and fishing, hockey. The rule of the thumb is that the more active form of recreation, the more physical effort, more risk-taking, more roughness the greater the male prevalence. There are, however, some exceptions to the rule like fishing or spectator sports which are dominated by males although in most cases they are rather passive forms of recreation.

Business travel still is dominated by men although the share of women in this field is growing (21% in the US according to Waters 1988:138). Females, at least in some DCs, tend to have a slightly greater share in pleasure tourism than males (55% for air pleasure and personal trips in the US - ibid.). One could only speculate about the reasons: perhaps males are more interested in sports, have more job obligations (although more than 50% of women are working) and live shorter. Women seem to be more interested in educational and cultural tourism, single women in such forms of tourism (e.g. group tours) which bring them in contact with opposite sex. In male dominated societies like Japan males constitute still an overwhelming share of tourists, not only in business but also pleasure tourism. Destinations also differ in terms of sex composition of tourists: female tourists choose mostly destinations offering stability and personal security. Therefore, they tend to travel more to established safe tourist regions located mostly in the DCs. Some destinations attract more males for various reasons: challenge of adventure, participation in vigorous and dangerous sports like alpinism, sex tourism (South Korea, Thailand) etc.

As a whole one could generalize that from all the other demographic variables sex is probably the least important in determining of participation rates because of equalizing tendencies characteristic for modern times. Nevertheless, sex is still an important variable in market segmentation.

5.7.2.2 Socioeconomic Segmentation

The group of socioeconomic variables used for market segmentation contains first of all income, education and occupation. There is a lot of overlap between them and the demographic variables (e.g. income and age) but it is possible with varying degree of difficulty to isolate their correlations with leisure patterns. The results of research in this field confirm the hypothesis that such a correlation is strong with respect to income and education and more elusive but still evident as far as occupations are concerned. Therefore people working in the field of tourism and recreation research follow closely the changes in socioeconomic structure of our society: long term trends to increase the discretionary incomes, to improve the standards of education and to change the occupational structure of society towards high participation occupations augur well for the future of leisure time pursuits.

Income

Income has a definite effect on participation rates. The availability of financial resources for the discretionary expenses connected with different forms of tourism and recreation constitutes an important prerequisite for participation. Income makes the transposition of needs into effective demand possible. Naturally, some of the differences between income groups are related to age, education and occupation. Nevertheless, the income level seems to be a pretty reliable yardstick for participation and expenditures, especially with respect to activities requiring substantial outlays, e.g. overseas travel, water skiing, boating. Generally, the correlation between income and tourism seems to be much stronger than between income and recreation, because the economic factors play a greater role in tourism than in recreation which is often provided free or almost free by the government. In other words, income is a primary determinant for tourism and less important for the choice of various forms of recreation. However, it would be incorrect to use income as the sole explanation for the choice of particular activities. E.g. a young couple may have the same income as a retired couple but spends it for different activities. To use solely income data would disguise this difference. After reaching a certain minimal level of disposable income when all the basic needs (food, shelter, clothing etc.) are satisfied the consumer starts to spend more of the discretionary income on leisure time activities. From this point on the positive income elasticity of demand for leisure time activities comes into play i.e. the expenditures in this field are growing more than proportionately than the disposable income.

After reaching a certain relatively high level of income the consumer's expenditures for leisure time start to grow less than proportionately to the income. Indeed, with respect to some recreation activities, especially those which are subsidized by the government, there is a drop in participation rates of rich people.

Education

The second socioeconomic variable is education. Its positive relationship with the volume and type of participation is strong. Education widens the horizons, brings an increasing awareness of the world, stimulates the interests in seeing new lands and experiencing other cultural environments. Its influence on participation in tourism is undeniable and increasing with improving educational levels of the population. Also the choice of recreational activities may be influenced by education. "Better educated groups ... spent less of their leisure time in passive pursuits, did more do-it-yourself work, watched television less ... were much more active in sailing, golf, walking and mountaineering" (Dower, 1970:254). Interesting ideas about the influence of education on leisure patterns have been developed by J.K. Galbraith (Galbraith, J.K., 1967:365-366). However, the relationship is only partially direct because of its overlap with income. Better education means more income and this is important for shaping the demand. Nevertheless, it is possible in many cases to establish direct relationships between education and leisure patterns e.g. participation in wilderness area activities is not expensive. Thus the income factor may be ignored. Therefore, the disproportionately large participation of well educated people in these pursuits proves the point. Another example: caeteris paribus less educated tourists spend money more freely than educated. They seem to equate fun with spending money.

Occupation

The third variable "occupation" is so strongly associated with income and education (Woodside, A., 1987) that it is difficult to ascertain its influence on leisure patterns. It comes to the point that some researchers deny the existence of any correlation between them. It seems, however, that some cautious generalizations are possible: civil servants, teachers, students, professionals, managerial and technical staff show high participation rates whereas farmers, farm managers, small shopkeepers, laborers and unskilled workers participate below the average level in out-of-home activities. Sedentary occupations of the former group increase their propensity to active recreation, whereas occupations requiring physical exertion are mainly associated with relatively passive behaviour in leisure time. Hunting - dominantly a sport of blue-collar workers, may be regarded as an exception.

Racial, ethic and religious variables

The racial, ethnic or religious minorities in many countries have had historically lower participation rates and different leisure time patterns as compared with the population as a whole. However, as these groups assimilate, emancipate and urbanize the differences tend to diminish as they approach the national socioeconomic standards. The underprivileged increase their participation, change traditional leisure styles and attitudes as economic and social barriers tend to disappear, as religious regulations and customs start to be applied more liberally. However, at least until this "catch up" process is completed the minorities constitute separate submarkets not to be ignored by professionals of the leisure research field. Thus e.g. a new "black market" appeared in the US in the 1960s and was reinforced in the 1970s in connection with the publication of Alex Hailey's book "Roots" and the television series of the same name. Since, special tours for black Americans have been organized to Gambia, Senegal and Ghana to give them opportunity to visit their ancestral places.

Ethnic and religious submarkets are widely used all over the world in organizing tours to the countries of ancestry or religious pilgrimages. The future for ethnic tourism seems to be bright as ethnicity is very much in fashion and religious fervor increases. To give a few examples, the North Americans of Irish or Italian ancestry participate in trips to their "old countries," the Catholics conduct pilgrimages to Lourdes, France or Santiago de Compostella, Spain, the Muslims to Mecca etc.

5.7.2.3 Geographic Segmentation

Various geographic regions or countries are often regarded as submarkets for a particular tourist destination. In such considerations special attention is paid not only to intrinsic characteristic features of various regions like e.g. emissiveness or travel intensity (the percentage of outbound tourist to the whole population in a given year) but also distance between tourist origin and destination. In this respect researchers often use gravity models to determine the degree of distance friction. These models take into account not only distance but also the size of population and the levels of discretionary incomes at origin. Thus, the US is focussing on researching Canadian, Mexican, West European and more recently on Japanese submarkets as the main sources of incoming international tourists. On these countries the US marketing effort is concentrated. The tourism marketers also know that certain cultural environments are favouring greater participation than others, e.g. Northern Europeans tend to travel more internationally than the Mediterranean people.

In addition to location the size of community or settlement constitutes a geographical variable which allows the discovery of distinct submarkets. The relationship between the size of settlement and effective demand for tourism and recreation is obvious: the higher the degree of urbanization (and consequently the size of locality) the greater the participation in recreation and emissiveness in tourism. People living in smaller communities and especially in rural areas reveal much lower degree of effective demand. There are a number of reasons for this situation. First of all a generally higher level of income with increased size of communities (overlap with the variable "income"). Also different life-styles and attitudes may play a role. The rural population, especially employed in agriculture has closer ties with their environment, enjoys an everyday contact with nature and has to cope with the peak labour period during the summer season. These reasons decrease mobility. The population of urban areas especially urban agglomerations working in rather unhealthy and congested natural environment (push factor) is eager to escape from the dull depersonalized daily routine during the weekend and summer leisure time. The urbanites seek compensation in extraurban recreational activities and tourism thus showing a much higher mobility than the rural population or even people living in small towns. The size of the family may also play some role; the families of inhabitants of metropolitan areas tend to be smaller than those of smaller cities and rural areas, where the large number of children may prove to be an additional constraint.

5.7.2.4 Life Style or Psychographic Segmentation

The use of demographic, socioeconomic and geographic variables has provided some useful tools for identification of submarkets. As a result of such research we have learned that the highest propensities to participate in tourism/recreation is characteristic for people with good incomes and education, living in large urban centers, young, members of small households, working in managerial and professional positions. The lowest participation rates are among the poor and uneducated, living in rural areas, very old and sick, single parents, parents with large families, especially with small children and working in unskilled, semiskilled occupations and agriculture. However, this information provided by researchers working with these "traditional" market segments, sometimes called inprecisely "demographics," proved to be insufficient as prediction of leisure time behaviour. Indeed, individuals identical according to traditional "demographics" frequently have different leisure time patterns. Therefore researchers had to reach "beyond demographics" in order to obtain a deeper understanding not only of general participation rates in leisure activities but also of demands, preferences, expressed by market segments for concrete leisure time activities, including tourism and recreation. This task is being pursued by life style research or psychographics which matches various life style submarkets with leisure time behavioural patterns. It does not mean that psychographics focusses entirely on psychological attitudes, perceptions, images. Sometimes, also traditional "demographics" is taken into account. Psychographic research uses combinations of a multitude of variables in order to provide detailed profiles of submarkets for the marketers who are trying to reach their clients in a most efficient way. Thus, the researchers identify combinations or clusters of measurable interacting and integrated variables (not just a sum of them) to arrive at psychographic submarkets. These life style groupings are being correlated with certain leisure time patterns which also constitute clusters of preferred activities. All this requires the application of advanced quantitative methodology. Various computer programs intended to digest and integrate available data have been used, numerous multivariate and clustering techniques have been developed. All these achievements in applied mathematics enables the researcher to come to more meaningful conclusions. Some of the practical applications of these methodologies lead to "computerized vacations": the prospective vacationers provide answers on many questions pertaining to their "demographics" and life styles and the computer matches these variables with various destinations.

Fortunately, there is already some empirical evidence that life-style variables are better predictors of tourist preferences and behaviours than "demographics" (Abbey, J. 1979). This explains the frequently differing leisure time behaviours of individuals of the same age, sex, profession, education, etc.

As far as concrete psychographic market segments are concerned one is faced not with "the theory" but rather with a number of theories. Therefore, in a textbook presentation it seems most appropriate to select only few segmentations as examples. Analysing these categories discovered by various researchers one may find out that they are somewhat similar but may be labelled differently.

The point of departure for some submarket classifications is the psychological division of people into introvert and extrovert categories. Thus behavioural groupings into psychocentric and allocentric have been used by some researchers (Plog, S.C., 1987)). Psychocentric (Greek psyche-breath, soul) means centered on ones thoughts, overconcerned, overprotecting, self-inhibited, passive not venturesome person with restricted lifestyle, preferring TV watching to more active out-of-home leisure patterns. This is a typical nonparticipant in tourism. Allocentric (Greek allos-other, different) outgoing, eager to mix with others, self-confident, adventurous, interested in various activities including extraurban recreation and tourism, at home selective TV watcher, print media oriented. There is a continuum between these two extremes which allows for classification of people into five categories: 1) psychocentric 2) near psychocentric 3) midcentric 4) near allocentric 5) allocentric. The distribution of these categories is unequal which is demonstrated by a bell-shaped curve (Fig. 5.2. after Plog 1987).

Figure 5.2

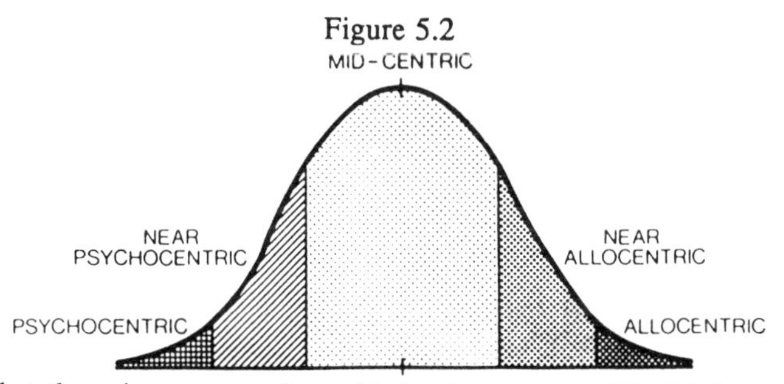

It is true that there is some overlap with the income variable (high percentage of low income earners among psychocentrics and high income among allocentrics) but for in-between people the correlations are only slightly positive. As recreationalists and tourists psychocentrics and near-psychocentrics are characterized by low relaxing activity level, prefer car to air travel and familiar destinations and atmosphere, choose rather group than independent travel. Allocentrics are looking for new experiences, prefer air travel and high activity level. S. Plog suggests that various tourist destinations occupy different positions on the psychocentric-allocentric continuum. For New Yorkers this position would move from Coney Island through Miami Beach, other US destinations, Oahu, Caribbean, Great Britain, other Western European destinations, outer islands of Hawaii, Orient, South Pacific, Africa.

As an example of Plog's research in correlating the psychological make-up of people with their leisure time behaviour serves the distinction between golfers and tennis players (Plog, S., 1987: 205). They "are demographically very much alike" (high incomes, many executives and professions etc.) but their personalities differ: golfers are very much social interaction-oriented, they need friendship, affiliation. They spent more on food and entertainment. As big spenders golfers are most desirable guests at resort location. Tennis players are competitive, achievement-oriented. "After match socializing and drinking are limited." The spending patterns of tennis players are more Spartan: they spend less for food and entertainment. The implications for tourism business of such behaviour differences are evident.

The psychographic market segmentation studies range from psychological submarkets like these presented above to more descriptive behavioural like the following one based on a study conducted under the auspices of Harper's and the Atlantic magazines (Bernay, E., 1971). A random sample of readers of these magazines and their neighbours non-readers has been investigated. Computerized statistical clustering procedures were used to classify the respondents of the questionnaires into homogenous groupings or types. For the programming the Q mode factor analysis technique was applied which allowed condensation of data derived from correlations of one hundred variables selected from the survey as relevant to marketers.

As a result of this study four life styles have been distinguished:

1) The movers and shakers. Active mobile types participating in many activities including recreation and tourism, involved both politically and socially. They are well educated, working in professional occupations and are well paid. The author of the study report regards this life style as indicating future developments in our society: their share is going to increase with the time.

2) Homebodies. They are less experience-oriented as the movers and shakers but more material-oriented homely people who invest their money and time in equipping their house with wall to wall carpeting, TV, home repairs etc. These people were less likely to recreate or travel outside their home area.

3) The older folk. All over 50, in lower income groups, rather inactive in all respects. (This evaluation is now certainly obsolete - Z.M.)
4) The established. Mainly affluent businessmen. In their marketing behaviour they were most similar to the "Movers and Shakers" and shared with them a heavy incidence of among others of travelling by air and car rentals. They differed, however, in that they were considerably less involved politically than "Movers and Shakers."

Market segmentation at the stage of antecedent conditions is instrumental in establishing generalized consumer groupings. Although it is still far from concrete consumer behaviour in tourism and recreation its utility for marketing strategies is beyond question. The three first "traditional" market segmentations (demographic, socio-economic and geographic) use single variables at a time. These variables are relatively easy to determine, therefore, they are called "hard variables." The psychographics operates with clusters of variables in order to arrive at certain consumer profiles. However, because these variables are difficult to determine and complicated they are called "soft variables."

5.7.3. The Motivational Market Segmentation

5.7.3.1. Motivations for tourist participation

The antecedent conditions, discussed above, influence the general behavioral patterns of the participants in tourism/recreation. These patterns are called motivations. Thus motivations are products of the interaction between demographic, socio-economic and life style variables which result in a particular propensity to participate, the behavioral patterns during leisure time. The market research brings interesting findings in this respect by establishing four basic motivations. The first two motivations are of mainly social and psychological character with prevalence of push factors, the two latter are of chiefly educational and cultural character with pull factors prevailing. The four basic motivations are as follows:
1) The desire to change the environment.
2) The desire to change the role.
3) The desire for new experience.
4) The desire for rest and relaxation.

To sum up the motivations: the purpose of vacation is contrast, change, break of pace, relaxation.

The desire to change the environment is mainly connected with the push factor of industrial-urban environment. The necessities of the industrial age with the overriding importance of economics of scale, technological and economic cooperation, transportation and communication facilities etc. lead to rapid growth of high density urban areas called urban implosion. This process of spectacular concentrations of people in urban centers, conurbations and megalopolis created a number of severe problems resulting from congestion (lack of space will be according to Fourastié the worst scarcity of the 21st century - Fourastié, 1965), air, water and noise pollution, decay of urban cores, etc. The flight to the suburbs, this "dispersed congestion" has not solved these problems but has contributed to further decrease of recreational resources close to the city which are disappearing, devoured by residential and transportation sprawl.

Not only the physical but also the social environment in the city exerts a number of stresses and pressures which contribute to the push factors causing the desire to escape the urban environment. Also the social links between the inhabitants of the city tend to diminish up to the point of "the lonely mass."

All these negative aspects of physical and social environment of the city constitute the push factor which motivates the people to escape, albeit temporarily, in order to enjoy more amenable physical conditions, in order to meet other tourists and/or local people at destinations.

However, the desire to escape from the everyday environment is not only limited to urban areas but to any environment. Obviously people become "fed up" with even pleasant everyday conditions and require change, variety, transitory escape. Of course, the urban environment, as the least agreeable one, causes the highest tourist departure rates (emissivness).

Obviously, the push factor is dominating this motivation. There is no pull to any specific environment. The only condition is that it should be physically and socially different.

The second and third motivation - the desire to change the role and to gain new experiences-results from another negative aspect of contemporary urbanized society - the mechanical, monotonous, routine, character of most jobs, especially industrial jobs. It is the consequence of the division of labor, specialization and mechanization of work. Gone are the days of creative endeavours of artisans and craftsmen. Although working hours are relatively short, most jobs do not satisfy, do not contribute to the development of personality. As a result people long for a connection to a larger world, outside their day-to-day lives. There is an urge for at least temporal evasion, escape from tedious daily routine, a desire to "get away from it all," to compensate by assuming another role, the desire to be an individual and not a depersonalized part of a smoothly running organization. Breaking out from a tedious daily routine prompts tourists to participate in adventure travel where they seek self-reliance, challenge, recognition of their own potential and individual confidence in themselves. Thus our time is characterized by increasing pluralism and individualism in life styles, retreat from traditional sex roles (sometimes even marriage) less conformity, more societal segmentation, more focus on oneself ("me-generation"). Whatever the value judgments over these developments - they certainly are good for tourism and recreation because increasingly self-centered people looking for "instant gratification" here-and-now expand their rates of participation. Psychologists and sociologists argue about the reasons of these trends. It seems that they result from cumulative impacts rather than from simple causation. For our purposes the impact of trends towards self-realization, self-fulfillment and self-improvement are essential. These trends impact on leisure time patterns in three respects:

a) In physical terms it is the tendency to healthful living which restores the validity of the Roman motto: "Healthy mind in a healthy body." People are nowadays preoccupied with their bodies by means of nutrition and exercise which requires a radical change in lifestyles. The fitness boom which started in 1970s established new standards of beauty: slim, muscular, healthy human beings. The leisure time is aimed at achieving these ideals by increased activity, i.e. participation in active sports like jogging, swimming, cross-country skiing, tennis, soccer. This trend is reinforced by ecological awareness, appreciation of nature and its aesthetic values which leads to forms of recreation in wilderness environment. These activities mostly require skills which have to be learned. Thus we identify a trend away from passivity, idleness towards active participation in contrast to work which nowadays requires hardly any physical endeavour from most workers. Because of increased individualism a slight trend may be observed away from group team games towards small groups and individual forms of activity like tennis, hiking, bicycling, swimming and other water recreational activities. The trend to physical fulfillment occurs sometimes in forms which have been criticized as aberrations: the tendency to nudity and increased sexual activity while on vacation.

b) Human desire of self-fulfillment is not limited to physical activities. It encompasses also growth of human intellectual potential. Its impact on leisure time activities appears in the form of cultural and educational tourism, in creative activities including various hobbies, in growing desire for social contact and adventure during groups travel to compensate for living among "lonely masses" in big cities. Many tourists travel to other regions of their own country and abroad in order to learn about the world, to acquire experiences, knowledge, information and skills. In practical terms leisure time professionals are responding to the demands by providing more interpretive services, more special interest tours. Alex Hailey's book "Roots" stimulated tremendous interest in heritage, visits to far-away archives, ethnic roots explorations in countries of the forefathers etc. Many people pay for participating in various research projects like archaeological digs, nature study expeditions, adventure trips often connected not only with high expenses, but also with physical hardships and dangers. The trend toward these forms of "challenging tourism" gains increasing popularity worldwide.

c) The leisure time activities and especially tourism contribute not only to physical and intellectual self-realization but also spiritual self-fulfillment and improvement. This aspect is characteristic not only for religious pilgrimages. Many tourists, especially young people, look for self-discovery in the new environment. In tourism, especially international tourism, one is given the opportunity to lose illusions, preconceptions one gains consciousness about existence of an individual in the global context, and better and deeper understanding of the world. Remember the slogan: "Explore a part of Canada and you will explore a part of yourself." Tourists gain more than ephemerical satisfactions provided by most products. Tourists gain insights into their personalities - a self-exploration, a self-evaluation and possibly new orientation. This potential for a permanent spiritual reorientation resulting from tourism, constitutes an effective antidote for the depersonalization of our "post industrial" society. Unfortunately, there are very many instances in modern mass tourism when these expectations remain in the sphere of theory. The practice is that tourists return home without having experienced any of these beneficial impacts.

The last motivation - the desire for rest and recreation constitutes the basic reason for taking vacation. Here not only the teleological reasons of restoring the working capacity and health (curative health tourism) play a role but also the simple desire to have fun.

One encounters sometimes pessimistic views that modern tourism, or at least some of its forms, frustrates the realization of the four basic motivations. According to these opinions tourists are unable to find their identities in the anonymous vacation crowds because they transfer from one anonymous big city to another anonymous tourist conglomerate. Instead of enjoying rest and relaxation they function during their vacation in another mobile and achievement-oriented mass. These pessimists should be reminded that contemporary tourists can fully adjust their vacation plans to their motivations by finding "their niche" in the vast array of supply: gregarious types are contact and experience-oriented, others avoid contact and seek solitude. All rest and recreate according to their own likes and there are no rigid behavioural patterns for everyone during vacation. The choice of vacation is individual.

5.7.3.2. Motivations and trends in societal values

The trend towards self-actualization, self-realization, self-fulfillment, has been strengthened by conventional materialistic values epitomized in the traditional work ethic. However, more and more people nowadays do not find self-fulfillment in work. Indeed, they turn away from work for material success, from work for work's sake. Their self-fulfillment lies not in careers or accumulation of wealth. They find that buying more and more goods brings less and less enjoyment and that goods are relatively short-lived. On the other hand collecting experiences brings great satisfaction. Experiences cannot be lost or stolen. They last for life-time. These trends in motivations prompt many people (but not the majority) into action: early retirement, resignation from high status and income executive positions in favor of increased leisure time are good examples. Indeed, nowadays thousands of people turn away from the compulsion to acquire vast amounts of money, from materialistic greed recognizing that the pursuit of economic self-interest is not a panacea for human fulfillment and happiness.

This trend away from materialism favors also aesthetic and environmental values, stresses the importance of healthful simple life often away from city rather than materialistic extravagances with its development for development's sake, waste of resources and environmental deterioration. These modern trends, however, do not mean anti-consumption but rather adjustment to scarcity. They are aimed more towards material sufficiency than material abundance. They promote only a kind of more austere consumption which differs from materialistic consumerism. They do not set long-range goals with attendant savings which are being quickly eroded by inflation. People subscribing to this philosophy save less, spend less on consumer goods. It is not investment saving model but a model of frugal modest consumption sparkled with joy of life here and now. In this sense this model of voluntary simplicity is neither aesthetic nor altruistic but rather hedonistic. Such a model of life tends to center on leisure time rather than work. Unable to get satisfaction on the job people turn to leisure time activities. At the same time it does not mean that people do not want to work. They only want a better balance between work and leisure shifting their values from the former to the latter.

5.7.3.3 Motivations: quest for variety

A glance at the four basic motivations for tourism, discussed above, brings one to the conclusion that change, variety is the pivot of tourism. Would the world be homogeneous tourism would not exist. Fortunately, this is not the case. The world may be becoming more homogeneous and standardized in technical matters which is a positive phenomenon for tourism. But what the tourists want conforms with recent social trend: the turn from familiar to variety, novelty, heterogeneity, individualism both in personality and lifestyle. This trend has three consequences 1) It is more difficult to arrive at generalizations pertaining to specific nature of social changes. 2) There is more tolerance towards diversity of behaviour, towards alternative lifestyles. This is our democratic "signum temporis." 3) There is much more interest to break out of parochial shell and make the life filled with a variety of pleasurable activities, especially during vacations. For a variety of demand tourism offers an incredible assortment of choices out of diversity of options. Diversity becomes the most important aspect of human life. Indeed, as J. Bronowski tells in his "Ascent of Man" - "diversity is the breath of life, and we must not abandon that for any single form that happens to catch our fancy ... of all animals, man is the most creative because he carries and expresses the largest store of variety. Every attempt to make us uniform, biologically, logically, emotionally, or intellectually is a betrayal of the evolutionary thrust that has made man its apex." (Bronowski J. 1973:400)

5.7.3.4. Motivational submarket research

The motivation research segments the market into various submarkets according to motivations. And again here, similar to the lifestyle segmentations, there is no widely accepted system but rather a number of systems. However, all of them have on common characteristics: one can always recognize that these submarkets contain at least some elements of the four basic motives discussed above, particularly the two latter ones. Secondly, one could generalize that all these motivation systems focus on benefits sought by tourists. Therefore, one could call them "benefit market segmentation."

We will open our examples of motivational market segmentation research for tourism and recreation with submarkets suggested by Cask: 1) Rest and relaxation vacationer 2) sightseer 3) cost conscious/attraction oriented vacationer 4) sport enthusiast 5) camper. These groups have been related to magazine readership (Cask, M., 1981). Another researcher distinguished 6 submarkets: 1) Young sports type 2) outdoorsman-hunter 3) winter/water type 4) resort type 5) sightseer type 6) nightlife activities (Bryant, B.; Morrison, A., 1980). And here are some of the submarkets distinguished by Barnett: 1) Museum market 2) Music lovers market 3) Sporting market 4) Back-to-Nature market 5) Market for the elite (luxury resorts closed to general public) (Barnett, E., 1971,3:107). Hahn distinguishes following motivational submarkets 1) Rest-relaxation 2) experience 3) health 4) family 5) entertainment (Hahn, H., 1971:24). Tourism Canada in its studies of Western European markets (Tourism Canada 1988, 1989 A,B) segments tourists by product-orientation by travel philosophy and expected benefits. The result is the identification of 5-7 market segments, like Cultural Touring, Sports and Entertainment, Developed Resorts, Rural Beach etc. Unfortunately, these motivational segmentations are not comparable from study to study and there are also too many internal overlaps between segments within one study.

These examples of motivational market segmentations represent closed systems. This author suggests a motivational ("benefit") segmentation of tourism market in the form of an open-ended list which seems to be more flexible.

1) Business. One of the earliest prehistoric motives for travel.
2) Meetings (including conventions and congresses) Preferably treated separately from business travel because of different nature and dynamics.

3) Visits of friends and relatives. Especially important for people with modest financial means, large families, senior citizens, inhabitants of smaller communities, people who feel insecure, need psychological and organizational props.
4) Pilgrimage. This religious motivational submarket is especially important in the Catholic and Islamic world.
5) Restoration of health. Today it lost is former importance, especially in North America. Europe has still a large number of health resorts.
6) Recreation, rest, relaxation (physical and mental).
7) Education, culture. Results from intellectual curiosity. Experience oriented. Characteristically for tourism an element of novelty is involved. Normally consumers display brand loyalty. Majority of tourists motivated by education - culture prefer a previously unvisited destination. Habitual destination tourists may be motivated by avoiding novelty because of fear of unknown. This fear may be removed by participating in group tours. Sometimes formal education takes place on trips: cruise ships have been used as floating universities.
8) "Ethnic tourism." Visits to country of ethnic ancestry.
9) Entertainment e.g. theater, concerts, "gourmet tourism," etc.
10) Spectator sports.
11) Shopping.
12) Ego-enhancement, snob appeal. Quest for prestige connected with visiting unusual places e.g. the South Pole.

In most cases an individual is motivated by not one but several motives, one of them dominant. Of course, this one plays a decisive role in tourist preferences. Most motives are containing both push and pull factors in differing degrees. However, they are not destination-specific. Nevertheless, they provide useful market segmentations for marketing purposes because the market segments identified by research reveal the benefits sought by potential tourists. Therefore, these segments may be targeted by marketing in the most efficient way.

5.7.4. Attitudinal Segmentation

Motivations constitute more or less generalized desires detached from specific products (in our case destinations or recreational activities). Therefore, a consumer in order to establish a direct relationship with the product has to reach the attitudinal stage. "An attitude can be defined as a mental state that causes a favorable or unfavorable inclination towards the product." In other words "an attitude about a product is an evaluation of the expected utility that it contains." (Boerjan, P., 1974, 3:87). An attitude is connected with an image of the product which epitomizes all the former stages of demand formation acting as sorts of filters. The attitude to the product ranges in a continuum from very favorable to very unfavorable. Along this attitudinal continuum the market may be segmented. The products in tourism and recreation are destinations and specific activities. Therefore, the researchers are trying to relate various market segments to these products and establish the degree of attitude stability. In other words they are exploring to what degree relatively stable attitudes to certain products can be changed by appropriate marketing. Hence the studies dealing e.g. with the images of USA as perceived in Western European markets. Such studies aim to determine the size of a potential tourist market for the USA. Some images are essentially correct e.g. the Western Europeans think that Canada is not well endowed in warm beaches (see Market Assessments). This image is close to reality. However, an image of Canada as containing almost exclusively natural and rather inadequate human resources has to be corrected by appropriate marketing. For the market segments with unfavorable perceptions (images) of the potential destination market strategies aimed at changing them into favorable are being developed. An example of such strategy of molding potential markets attitudes is the promotion of island vacations in the Bahamas as complete perfect relaxation as opposed to allegedly hectic active experience of non-island vacations. "Non-island vacation tends to be educational, the island vacation therapeutic." (Chib, S., 1968:116). The West German researchers found out that the Baltic Sea enjoys an image of relaxing tranquility for family beach vacation as opposed to adventuresome roughness of the North Sea. These findings were instrumental in devising an appropriate marketing strategy for this destination. Another example is the formerly negative attitudes of the public towards winters in general, and specifically to Canadian winters, in terms of tourism/recreation. This attitude is changing into positive and marketing is to a large extent responsible for it.

The marketing efforts aimed at establishing desired images of tourist products (destinations) lead, unfortunately, too often to distortions differing greatly from reality. The beauty of nature is represented mainly correctly. However, the host population is frequently represented as carefree pleasure-seeking people ready to serve the tourists and prepared to do absolutely everything to please them. This is an undesirable propaganda and the respective governments should exercise control over marketing to prevent its occurrence. It is also advisable that tourists establish their images of destination on the basis of less biased sources like opinions of friends and relatives who visited these destinations or serious travel literature. This is the best method to avoid misrepresentations of reality.

5.7.5. Intervening Variables

The attitudinal stage in market segmentation studies brings the researcher close to the product: with the images of a particular destination established the act of consumption, in our case - participation, is imminent. However, at this last stage in demand formation the potential demand formed at the attitudinal stage has to pass through constrains or barriers to participation, called by B.L. Driver "intervening variables" or "deterrent external factors." If the intervening variables prove powerful enough the potential demand will never be transformed into effective demand. It will be frustrated. The list of these barriers or deterrents to participation is open-ended: lack of money and time for travel, family obligations, poor health, government restrictions (passport, visa, currency, customs and other restrictions) fear of unknown environment, linguistic difficulties, safety risks (accidents, robberies, diseases, epidemics, strikes, riots, hostility of host population, terrorism, wars, etc.). In many cases the barriers are only apparent caused by desinformation like in the case of the West Germany newspaper Bildzeitung which carried an article about Yukon's murderous wolf population which eats people. The story was inaccurate since no one has been eaten by wolves in the Yukon. The Yukon tourist authorities were upset about this unfounded sensationalism because West Germany is the Territory's biggest overseas tourist market (Winnipeg Free Press 8-1-1983). Similar unfounded barriers for tourism are being erected by people who say: "I am not going to Africa. There is war."

However, most of the barriers are real, not perceived. Therefore, the researchers are trying to match various market segments with most likely intervening variables. Here are some examples:

Market segment	Intervening variable
young	money
old	money, health
jet setters	none
professionals	time
business people	time
low education	language, culture shock

poor people	money
women	family obligation
	safety

5.7.6. Market Segmentations in Perspective

To sum up the discussion on market segmentations and the stages of demand formation one may make several general observations and conclusions:

1) The transitions from one stage of demand formation to another brings the consumer closer to the product. It marks the corresponding gradual decrease of push factor inputs in relation to pull factor inputs. Push factors are instrumental in the wish of the consumer to get out of something less desirable, pull factors attract consumers to products. In the attitudinal stage almost entirely pull factors are at play.

2) Most market segmentations are interesting but they differ in the degree of their utility which increases with the transition from one stage to the next. In other words the closer the segmentation to the product the greater its utility for direct application of marketing strategies aimed at matching the demand segments with real or perceived attributes of tourism destinations and/or recreational activities. These matching endeavors are usually accompanied by attempts to mold the attitudes of consumers with respect to destinations and activities. Such image manipulations meet often with success e.g. under the influence of appropriate promotion the image of harsh Canadian winter may yield to the fun and gaiety of winter festivals and games.

3) Market segmentations constitute useful analytical tools, not ends but rather means. There are segmentations which are completely useless e.g. astrological segmentations or physical typologies (Fitzgibbon, J.. 1979). However in reality skillful marketers use them to sell their products to the naive.

4) Market segmentations constitute not only a useful tool for marketing strategies but has also important implications for the supply which may be adjusted in response to demand.

5) Research in demand (including market segmentation) belongs to behavioral studies as compared to economic or geographical ones. It may be conducted in various scopes and forms integrating their findings at a wide ranging continuum of conceptual frameworks or models. For purposes of convenience one could following Ward and Robertson (Ward, S.; Robertson, T., 1973) distinguish among

three levels of conceptual theories, models or frameworks. The first level - general or comprehensive theories relate all the variables concerning a given phenomenon such as behaviour of tourists. In other words models at this level examine the total pattern of tourism demand of the entire population. Such general approach is conceptually useful but difficult to apply on operational level because, among other reasons, these models overlook smaller submarkets and a host of other detail which brings the models closer to reality. At the other extreme of the continuum are highly detailed specific behavioral models (e.g. specific recreational activity, time and site such as Casino game preference market segmentation for Las Vegas in 1985) which are more useful operationally but "without some broader framework for relating such studies to other research, there is a risk of contributing to the chaos of unusable results." (Ritchie, B., 1975:179). In between these two extremes lie middle range theories isolating certain submarkets or other subsectors of tourist behavior and dealing with them on both conceptual and operational levels.

6) The behavioral research, to which tourism and recreation demand studies belong, has another operational drawback which should be taken into account by researchers in order to maintain the degree of utility required from any research. Frequently there is a significant gap between people's perceptions, motivations, attitudes and actual behavior. In other words people by far not always follow on operational level what they indicate they would like to do if they are asked by researchers about it. This lack of translation into action (effective demand) constitutes a caveat not to be ignored by research.

7) There are two characteristics of the tourism demand which merit a separate discussion: elasticity and seasonality.

5.8. ELASTICITY OF DEMAND

5.8.1. Introduction

There is an old clichee that expenditures for leisure time activities and particularly for tourism and recreation are very strongly tied to economic cycles. According to these views they constitute a pure luxury bound to be affected first of all in the time of economic stress. Allegedly, people do not regard these expenditures as essential and sacrifice them first if their incomes shrink. In other words, the demand for leisure time activities, including tourism and recreation, allegedly reveals a high degree of sensitivity (responsiveness) to the changes in incomes and prices: the demand for tourism and recreation is highly price and income elastic, higher than in other branches of economy. Are these statements true? Closer observation of this matter brings two confusing answers: both yes and no. In order to understand this question it is advisable to start with the explanation of few economic terms.

There are two categories of demand elasticity: income and price elasticity. The income elasticity of demand pertains to the impact of rising or falling incomes on demand. The price elasticity is the reaction of demand to rising or falling prices. According to this criterion of demand dynamics one may look at demand elasticity as either positive or negative. The positive demand elasticity occurs when incomes are rising and/or prices falling, the negative demand elasticity when incomes are falling and/or prices rising.

Elasticity of demand is measured as a ratio between changes in demand and corresponding changes in incomes or prices. In other words the ratio measures the degree of response or reaction of demand to changes of incomes and/or prices. The changes of all variables are expressed in percentages. Thus, if e.g. a 1% growth of income causes 2% growth of demand, then the income elasticity of demand (or more precisely coefficient of elasticity) is 2. This may be expressed in a formula:

$$E = \frac{Q_2/Q_1}{Y_2/Y_1} = \frac{\%\ \text{change in demand}}{\%\ \text{change in income}}$$

where E = income elasticity of demand

Q = quantity demanded (dependent variable)

Y = income (independent variable)

If $E > 1$, like in the preceding example, then the demand for a given product is income elastic. If the value of E is between 0 and 1 it is income inelastic.

Graphically the increase of income may be illustrated by changing the positions of the demand curve which moves farther off the point of intersection of the axes x and y. C_2 illustrates the new demand schedule after the increase of income. The price is assumed stable.

Figure 5.3

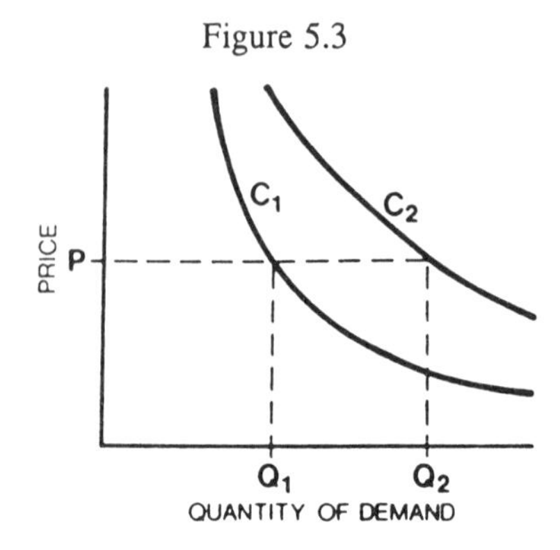

5.8.2. Income elasticity

Demand for tourism and payable recreation is regarded as income elastic. The negative income elasticity of demand was especially evident during the great depression of 1930s. After the end of World War II in times of economic upturn the positive income elasticity of demand came to the fore: the expenditures for tourism have grown more than proportionately in response to the growth of GNP. This is well documented by research for most DCs on the macro-economic level (primary elasticity). Because tourism expenditures as a rule have grown at an accelerated rate as compared with the growth of GNP, they have increased their share in total consumer expenditures. One could generalize basing on various sources that the positive income elasticity of demand have been about 1.5-2.0. On a more micro-economic level (secondary elasticity) there have been heterogeneous developments in terms of demand dynamics because in this respect many variables like e.g. changes in motivations and attitudes of consumers resulting from publicity, new trends in fashion, changes in transportation, in local supply patterns, etc. have worked in various directions in space and time. These spatial and chronological changes have affected the demand in certain tourist regions and centers in differing ways, either positively or negatively. Also demand for some particular forms of recreation and tourism services responds in various ways to economic situations (changes in incomes and prices), e.g. negative income elasticity of demand for hotel accommodation may be associated with tourists switching from hotels to cheaper forms of accommodation, a positive income elasticity of demand for air travel may correlate with the public turning to air travel from other means of transportation.

The phenomenon of highly positive income elasticity of demand for tourism would indicate that there is an equally strong negative reaction of the people in case of a decrease in their real incomes. The situation in this respect is, however, much more complex. The falling real disposable incomes result in 1) the decrease of discretionary expenditure, including leisure time expenditures if the basic spending for necessities of life (e.g. food, shelter) remains unchanged or 2) discretionary incomes remaining at the same level if there is a decrease of basic quality standards in expenditures for necessities.

It seems that the first scenario is prevalent. However, there are also indications that people lower their basic standards in order to maintain their treasured leisure lifestyles. In times of relatively mild economic downturns in 1950s and 1960s expenditures for tourism grew even absolutely in real terms although at a lower rate. Only very serious world recessions like those in 1973-75 and 1980-1983 when real disposable incomes in most DCs declined conspicuously did the demand for many forms of tourism and recreation decrease to rebound quickly when the recession was over. Tourism has shown a remarkable degree of resiliency. The negative income elasticity of demand proved to be weak sometimes even to a point of inelasticity (Bleile, G., 1976). This apparent invulnerability of tourism and recreation to economic crises has been explained as a result of changes in social values: leisure time pursuits lost their luxury character and became deeply ingrained part of life, a necessity rather than luxury, consequently relatively impregnable to business cycles. Some economists suggest the relative income hypothesis (Bleile 1976:2) stating that the households are inclined to hold to once achieved standard of living even at times of dropping incomes. This resistance to give up once achieved standard is called catch (Sperrhaken) effect by Meinke (Meinke, H.. 1968:31).

The concrete impact of the decline of real incomes on the tourism and recreation expenditures may be as follows:

1) No reaction at all. The negative income elasticity of demand is close to zero. Consumers are making cuts in spending for essentials lowering their standards with respect to accommodation, food, dress etc. and spending the same amounts of money as previously for leisure time activities. In certain instances they may cut savings or go into debt to achieve this.

2) Consumers may react by adjusting their leisure expenditures to the diminished level of discretionary incomes. The spending cuts tourists undertake rarely mean that they will cancel their vacation. What they do means simply that they will rather

substitute domestic for foreign vacations, travel shorter distances, take shorter vacations, use shorter slots of free time for travel (free days, weekends), take one-stop vacations rather than multistop (touring), turn to cheaper accommodation (e.g. camping or B & B), resort to self-catering instead of eating in restaurants, travel during shoulder months or off-season instead of high-season etc.

It is interesting to note that these spending cuts are normally less than proportionate to the decline of the real incomes. Therefore, one could generalize that such behaviour indicates relatively low negative income elasticity of demand for tourism in contrast to relatively high positive income elasticity. In these spending patterns lies the strength of tourism discussed above.

5.8.3. Price elasticity

For computations of price elasticity of demand two indexes are used for tourist destinations: Consumer Price Index and still better Travel Price Index which specifically relates to tourist expenditures. Travel Price Index is a composite measure consisting of the cost of transportation, accommodation, food, shopping and other services used by tourists. If the Travel Price Indexes are not available then one has to rely on Consumer Price Indexes for comparisons between various destinations. If the comparisons are international then foreign currency exchange rates have to be used additionally in order to compare the level of tourist spending between various countries and cities. One of the best sources used by tourists is the monthly publication "Maximum Travel Per Diem Allowances for Foreign Areas," published by US government. This publication is compiled in $US for daily expenses on lodging and meals for the use of travelling US officials. The information pertains to about 1000 cities worldwide and is extremely useful for tourists.

There is enough proof that both negative and positive price elasticity of demand for tourism is relatively high. Indeed, tourism business is highly competitive: potential tourists pay attention to statistics comparing relative costs of tourist services (accommodation, food, entertainment, shopping etc.) all over the world. Although there are changes in the positions of various places one can arrive at some generalization valid for a number of years. In the statistics of the most expensive countries (and cities) in the world Japan (Tokyo) occupies in the second half of 1980's consistently the first place. Also consistently expensive is Saudi Arabia. In Europe most expensive are Scandinavia and Switzerland and the cheapest are the southern countries (especially Portugal and Greece).

Potential tourists are monitoring the price levels in various places and taking them into account while preparing their travel plans. This is especially important with respect of travel in the LDCs: some of them may be expensive (e.g. Zair or Nigeria), partly because of the overvaluation of their currencies. The majority, however, is cheap (e.g. India, Thailand, Turkey) and this is a major incentive to visit them despite the generally longer distances to travel. Travel agents are frequently ill-informed with respect to these matters.

The Travel Price Index is subject to fluctuations and this frequently results in swings in demand. The reasons for those fluctuations vary. An example could be currency revaluations and more importantly devaluations in host countries which greatly enhances tourism (e.g. Mexico, Argentina). The post World War II history supplied many examples when tourists shifted from the Caribbean to Europe or from Italy to Spain as a result of price differentials. There are some exceptions in this respect though. The research results of Baretje and Defert indicate that destinations lacking attractions and/or facilities cannot count that lowering their prices will increase demand for them. Baretje and Defert found also that price reductions do not increase tourism to countries located off the main routes, far from tourism generating countries where travel expenses outbalance the price advantage. Instead of reducing prices such countries should concentrate on advertising. (Baretje, R.; Defert, P., 1972). Also well established destinations usually enjoy lower negative demand elasticities by rising prices. But any destination may price itself out of the market if the prices increase unreasonably.

Another area of price elasticity of demand is tourist transportation. In this respect the technical and organizational developments in the post World War II period has opened new vistas for tourism. Not only speedier and more efficient means of transportation but also special fares, charters, various forms of package tours have won widespread acceptance by the public because they contribute to fare decreases in real terms and consequently result in positive response on the demand side. The progress in airline deregulation in North America and Europe has also stimulated potential tourist demand. Extremely price sensitive segments of the market which would not participate at higher price levels have been tapped. In this way positive price elasticity of demand is instrumental in opening up entire new submarkets.

The majority of modern tourists is not only price-conscious and looks for better deals but also frugal while travelling, which is in concert with the trend to simple living away from extravagant luxuries. Thus the broad masses of tourists tend to spend relatively less per trip and per day. Especially young people travel cheaper. But the public at large is also following the price sensitive trends e.g. to budget accommodation. Another method to save on tourism expenses is to participate in trips organized by non-profit (tax exempt) organizations like universities, churches, museums for purposes of study, pilgrimage etc.

Elasticities of demand are not equal for all forms of tourism and recreation. Relatively high elasticities are connected with foreign pleasure tourism, especially if associated with certain forms of recreation like beach activities, or skiing. Slightly lower (but still high) elasticities characterize international sightseeing and cultural tourism because here demand is for concrete destinations looming high on the preference scale of tourists. Domestic tourism has mainly lower elasticity than international. Visitors to friends and relative show moderate elasticities. The lowest elasticities are associated with demand for business travel and local urban recreation as at least part of the expenses is paid by the company or is subsidized by the government.

As far as individuals are concerned the demand elasticity is certainly not equal at all income levels. Moreover, it may not exist at all at certain levels. Indeed, there is a certain minimal threshold level of disposable income which has to be attained before discretionary expenditures for leisure start to increase. Disposable incomes below this threshold are used totally to buy the necessities of life like food, clothing, accommodation. At this level there is no discretionary income. Once this threshold level of minimal disposable income has been attained the individual is able to take decisions as to the proportion of the discretionary income spent on leisure time expenditures, including tourism and recreation. As stated above, these expenditures tend to rise more than proportionately with rising real incomes or with decreasing prices by stable incomes. Such positive demand elasticity is characteristic for majority of people in the DCs. Increased income inequality would in such countries reduce expenditure for tourism. The situation in LDCs is different: here only relatively small minority has reached the threshold level of disposable income enabling them to participate in tourism. Unequal distribution of incomes in developing countries plays a stimulating role for tourism because equality in income distribution would not allow anybody to participate.

With rising incomes the positive demand elasticity does not stay high indefinitely. The high income elasticity is limited not only be a minimal threshold value but also by a certain upper level. By incomes higher than this level the expenditures of very rich people for leisure tend to grow at a slower rate than the increasing incomes. Thus the demand of the very wealthy becomes income (and price) inelastic. Many wealthy individuals have reached this upper income level but, of course, no country as a whole has done it. Thus the demand elasticity for tourism pertains to the vast majority of residents in DCs. Only two groups of consumers are excluded: the very poor and the extremely rich whose demand is inelastic.

To sum up the discussion on demand elasticity: The statement that the success of tourism and recreation business is a sensitive barometer of general economic situation, that it is susceptible to business cycles can be only accepted in a qualified way. The demand is positively income and price elastic in times of economic upswings. However, in times of mild economic downturns the demand proves to be, to the surprise of many, relatively income inelastic (showing low degree of income elasticity). Only a very severe recession with significant drop in incomes may seriously affect this demand. This proves that the old clichee of expenditures for leisure time as luxury is not true anymore for most people in the DCs. As for price elasticity it remains high at all times because of the extremely competitive character of tourism industry. However, the uniqueness of tourism product tends to reduce this clasticity in some cases e.g. a tourist paying a stiff price for visiting the South Pole will not decide on a cheaper vacation in Jamaica.

5.9. SEASONALITY OF DEMAND

5.9.1. Introduction

Similarly to some other industries the demand for tourism is characterized by uneven temporal distribution i.e. a peaking pattern in time. These seasonal and weekend peaks are, however, more pronounced in tourism and out-of-home recreation than by most other products e.g. food, clothing, housing. This temporal peaking pattern of demand is called seasonality. The normal reaction of any industry to seasonality is a response on the supply side i.e. the corresponding adjustment of output to the level of demand. Another response is that the production level remains relatively equal and seasonal surpluses go into inventory to be sold during peak periods e.g. at Christmas.

The things look differently with respect to tourism industry. The supply side of tourism has two features resulting from its service-orientation which set it apart from other industries with important socioeconomic consequences. The first feature is that supply is relatively inflexible. This means it is difficult to modify the output in response to demand fluctuations. For instance, if a hotel has 100 rooms, it is impossible, like in the production of most other goods, to change the output from day to day e.g. to 110 or 90 rooms. Similar is the situation with seats in a restaurant or in an aircraft. The second feature is that tourism industry is by and large unable to store its products. What has been produced must be consumed almost immediately. The tourist product, an empty hotel room, an unoccupied jet seat, perishes totally with zero economic value for the time it is unused. The relationship between supply and demand is direct and this results in frequent temporal disequilibriums between them and creates a number of severe problems: low productivity of capital, seasonal unemployment and other diseconomies. Therefore, the tourist business is naturally interested in keeping its occupancy rates and passenger load factors as close to 100 per cent as possible. Unfortunately, such optimal situation normally happens only during tourist peak or high season. Most of the annual demand is concentrated in this time, relatively short, not only for the temperate zone but also for many other parts of the world. However, even this optimal tourist period may be a mixed blessing for a destination area. First of all, there is a loss of potential business due to the full utilization of capacities. The surplus demand which exceeds the capacities cannot be satisfied and thus is lost. If the capacities are used to the utmost and, what happens frequently, overused, then the tourist industry and in fact the whole destination area suffers from a number of negative phenomena. The overused and overcrowded facilities cannot maintain high standards of service and this causes dissatisfaction on the part of visitors. Other problems include various impacts detrimental to the environment like air and water pollution. Frequently also there are problems with food and water supplies for the masses of tourists (good example: French Riviera). Traffic congestion in the peak season not only causes inconvenience for the tourists but also contributes to dramatic increase of accidents and jumbo traffic jams like in France during "Le grand depart" for vacation every first of August and "La rentreé" (return) in early September.

Some negative consequences of the peak tourist season are connected with its impact on other economic sectors. Many enterprises, and certainly the national economy as a whole, suffer from the disruption of the production process as a result of the seasonal outflow of employees taking vacations almost simultaneously. Especially drastic is this in France in August where some enterprises work on reduced shift and others close completely. France in August is economically paralyzed. Such an uneven use of capacities and labor force means substantial diseconomies for the whole country. Only in some rather exceptional cases such shutdown may be justified as the best way to do maintenance and repair work without unacceptable slowdown of assembly lines.

The underutilization of capacities in tourism industry may occur occasionally also during the peak season, caused by strikes in transportation or hotels or by prolonged bad weather period. Losses to the tourism business in such situations may be partially offset by insurance. However, paying insurance premiums increases operating costs.

The seasonal character of tourist demand is most damaging to the business during the off-season and intermittent shoulder months (in the temperate climate usually spring and fall). During this period the tourism industry is afflicted with capacity underutilization. The off-season is a critical period when the fixed costs (at least some staff wages, depreciation of buildings, heating or cooling, property taxes etc.) have to be incurred. And tourism industry has a high ratio of fixed to variable costs. Such a situation leads to wasteful use of resources, to low productivity of capital and labor and consequently to low profits.

A separate mention deserves the employment situation. The seasonality of demand results in seasonality of employment. The inability to hold a stable job exerts a negative impact on the labor force, its motivation, level of training, fluctuation. During the peak season lack of labor causes hiring of additional staff (if available at all) with lower qualifications. During the off-season even the highly qualified staff has to be laid off.

5.9.2. Measuring Seasonality

The seasonal disequilibrium between demand and supply in tourism business can be illustrated graphically by a hypothetical demand curve for 12 months: (mid-latitudes, northern hemisphere)

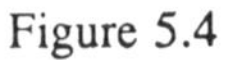

Figure 5.4

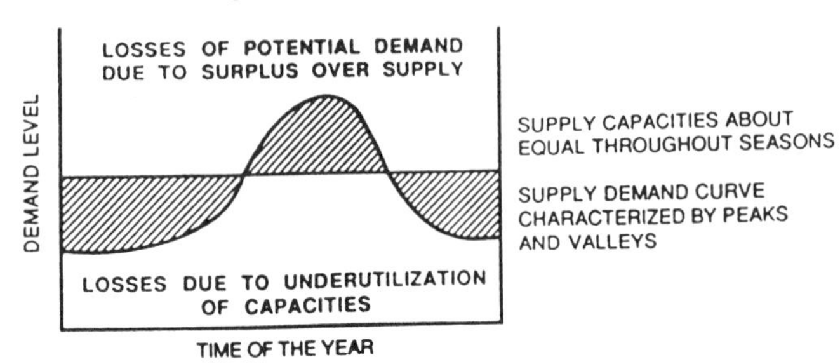

This curve shows that in off-season and shoulder periods (spring and fall in temperate climates) the demand is smaller than supply causing the utilization of capacities at under 100 per cent level. In the peak season the capacities are used 100 per cent but there is still surplus demand which cannot be satisfied. This potential demand cannot be translated into effective demand due to lack of capacities. Thus seasonal surplus and lack of capacities constitute the perennial problem of tourist industry.

Quantitatively the seasonal disequilibrium between supply and demand in tourism may be illustrated by three indicators expressing the relationship
1) between month of maximum and month of minimum demand i.e. the maximal demand in the peak month and minimal demand in the off-season month.
2) between month of maximum and annual average demand.
3) between the annual average demand and the month of minimum demand.

5.9.3 Natural and social seasonality

There are two categories of reasons causing seasonality: natural and social. The natural reasons pertain above all climate and especially the air and water temperature. In temperate climate the maximum in demand is directly correlated with the temperature maximum. Humans are unable at the present stage of economic and technological development to influence climate in a desired direction. Thus the natural conditions have to be accepted as an independent variable, a fact of life subject to no change of human origin. The influence of natural factors on seasonality of demand seems to be an almost universal phenomenon although there are certainly differences in intensity between various geographical regions.

There is no universal notion of seasonality. Various tourism/recreation activities require different natural conditions which are deemed as optimal for the satisfaction of participants. Such optimal seasons for groups of activities in mid-latitudes may be distinguished.

1) General recreation season which is identical with growing season (between start of blooming flowers and fall of the leaves).

2) Bathing season when water temperature is over 18^0 C or daily average air temperatures over 15^0 C.

3) Winter season when daily average temperatures are below 0^0 C.

4) Hunting or fishing season usually conforms with the reproductive cycles of the trophes. Frequently regulated by the government.

Optimal tourist/recreation seasons vary in their distribution over our planet. Thus e.g. the summer vacation season in the Mediterranean is long, the recreational activities are mainly water-oriented. Therefore, tourism concentrates mainly in coastal areas. The mountainous part of Europe enjoys two seasons - summer and winter whereas North Western Europe, having a short summer season, experience heightened day and weekend recreation periods. Summer is off-season in the Caribbean (high temperatures combined with high humidity, increased rainfall, less sun, hurricanes, insects, stinging jelly-fish in water) whereas winter is high season (dry, sunny, warm). The weather in Singapore is about equal all the year round.

The natural seasonality has been regarded as stable and unchangeable. However, recently some researchers are engaged in useful speculations on how the greenhouse effect will impact on tourism in the mid-latitudes: shortening of skiing season, extension of summer season etc.

The natural seasonality takes place at tourism destinations. The second category of demand seasonality is mainly associated with tourism origin zones and more exactly with social processes occurring in the areas of tourism origin. Social seasonality is more interesting for tourism planners because as human-related it is subject to influence and change. These are the people's attitudes to climate, to their traditional vacation and recreation patterns. Here belongs also the so called institutionalized seasonality like school vacations, religious holidays or any other institutionally regulated holiday seasons. Some authors emphasize the importance of human factors and make the point that school vacations constitute the determinant primary element in winter seasonality, natural factors playing only a secondary role (Escourrou, P., 1977). Indeed, there are many examples that social seasonality overrides the natural seasonality. Christmas occurs during natural off-season. People spend their money for travel and shopping and do not want to go anywhere in January. And January is the natural season in Florida and Caribbean. As a result this area suffers a sort of "off-season during peak season" and has to resort to discounting in order to get tourists.

The seasonality of demand varies in the degree of its impact on different types of tourism. Business travel is least affected and this is beneficial for tourism industry. Business travel (but not convention travel which is regarded by this author as a separate category of tourism) is virtually unseasonal and drops only during the institutionalized holiday periods and weekends. Tourism for personal reasons is intermediate in its seasonal variations. Sometimes it may be little affected by seasons (e.g. travel to funerals or weddings). Visits of friends and relatives display a higher degree of seasonality. Pleasure travel is highly seasonal and here lies mainly the "seasonality problem."

5.9.4. Response strategies to seasonality

Tourism industry has only limited possibilities of responding to demand fluctuations by modifying supply. As mentioned above, the rate of production can be altered with difficulty. Such adjustments, if possible at all, require longer periods of time in order to correspond more closely with fluctuating demand. Therefore, the most effective method seems to be modifying demand to approach as much as possible the ideal situation when supply equals demand. The measures undertaken in order to equalize the demand for tourism throughout the year are called "response strategies". There are two categories of these response strategies.

1) Primary response strategies influence the demand by government legislation which regulates e.g. the school year, government promotional programs or lower taxation of the tourist industry in off-season thus decreasing prices. The existing pattern of vacation (including school vacation) is heavily oriented to July and August in most DCs of the Northern hemisphere. The question is, what measures should be undertaken in order to change this situation, to extend the season. A group of measures which may be taken, albeit reluctantly, are administrative measures staggering the school and employees' vacations. Some people strongly oppose such primary strategies, especially staggering vacations. This author heard a French expert's opinion like this: "The vacations are for the people and not for the economic benefit of the tourism industry." However, the expert did not take into

account that seasonality is not only the scourge of the tourism industry: tourists are also suffering from congestion, high prices and environmental deterioration. Indeed, some governments have taken appropriate measures, e.g. staggering of school vacations according to a number of zones in which a country is divided. Thus France is divided into 3 zones, Poland into about 6.
2) Secondary response strategies are carried out by individual tourist enterprizes frequently under the guidance of national tourist offices. The best measure to boost off-season demand is the introduction of diversified pricing policy: discounted transportation fares and accommodation rates in order to woo the off-season market under the ad slogan "Off season - value season." There are many examples to prove the effectiveness of such policy. Winter ITCs (Inclusive Tour Charters) from New York City to various European capitals particularly to London, Paris, Vienna, and Leningrad for theater-opera and museum fans. Some of them are of very short duration ("Weekends in Europe"). Indeed, the great European cities offer a number of indoor cultural attractions which are patronized by thousands of winter visitors. As a result, winter is becoming the cultural, social and ...shopping season of Europe. Especially energetic with respect to marketing off-season tourism has been The British Tourism Authority (BTA) with its campaign called "Operation Off Peak." (Tourism Intelligence Bull. 1987, May) "Forty percent of French visitors to Britain now come in the six months from October to March, lured by Christmas shopping, January sales, discounted hotels..." (Tourism Intelligence Bull. 1987, Oct.). During the Caribbean off-season (15 April - 1 December) the hotel prices drop significantly to the level of 35% to 70% of winter (high season) prices (computed from New York Times 28-3-1988). Another example of price policy may be the establishment by IATA (International Air Transport Association) of three seasonal fare levels since December 1971: peak (summer), winter and shoulder. An interesting measure has been introduced in 1984/85 season in Greece: free off-season tourist transportation between small Aegean islands.

Besides price differential policy another strategy is frequently pursued in order to flatten the seasonal demand curve. It is multiple attraction strategy e.g. organization of event attractions during the off-season. These Hallmark Events, built around a major theme successfully woo tourists to fill the seasonally empty facilities: all kinds of festivals like Shakespeare, rock, folk, art and winter festivals proliferate to the benefit of tourism industry. Examples are Mardi Gras in New Orleans and Quebec Winter Carnival in Quebec City, Snow Festival in Sapporo,

Japan and Festival du Voyageur in Winnipeg, Canada. Also conventions, congresses, professional meetings take place normally in off-season or shoulder months between season and off-season. For this purpose costly multiple-use facilities are being built.

Other forms of multiple attraction strategy is the development and promotion of some formerly under-estimated and/or underutilized attractions like spring and fall fishing season, photographic foliage tours, "Indian Summer" tours etc. But most important of all is development of winter vacations.

After World War II the custom of taking winter vacations grew substantially and this development has undoubtedly alleviated the seasonality problem. There are two types of winter vacation: warm winter (winter-summer) and cold winter (winter-winter). The warm winter vacations turn winter into summer by utilization of zonal climatic differences using the possibilities for southern winter vacation in Southern United States, Caribbean and Mexico for North Americans and in the Subsaharan Africa for the Western Europeans not to mention the increasingly popular Southern hemisphere destinations. These policies certainly contribute to the newly established custom of taking vacations twice a year - once in summer, once in winter. The transportation and financial conditions for foreign winter vacations in the South have considerably improved since 1950s. In this respect fare policies, the volume of demand and distance play an important role.

The popularity of cold winter (winter-winter) vacations has also increased since the second World War especially in the Rockies and in the Alps. Here similarly like in former case not only the increase in discretionary incomes has been instrumental but also technical progress in transportation including the transportation of skiers up the slopes and the development of snowmobiles. These two winter sport activities, especially skiing (both down-hill and cross-country) have contributed to tremendous winter season boom in a number of resorts which have formerly known only the summer season. Because of skiing many mountain resorts became, according to the French expression, polyvalant, i.e. open year-round. During the summer peak tourism is more dispersed. The winter period is characterized by concentration in a few leading resorts. The development of winter recreational clothing and equipment (wet suits, floodlighting, snow making machinery, artificial ski slopes, helicopter skiing etc.) has helped not only to make a number of activities enjoyable but also extended the time of use. Some enthusiastic (and moneyed) skiers go during our summer to the Southern hemisphere to extend their time of enjoyment.

Many people still are not taking winter vacations during the Christmas-new Year festive period because they are reluctant to forgo the celebrations and their exciting atmosphere. However, it seems that a change in consumers attitude towards Christmas is in making. More and more people want to escape the commercialized festive season at home and celebrate it abroad on southern beaches or in ski resorts. Perhaps, to some extent, lower birth rates (less children) and the increased family disintegration have also contributed to spending Christmas outside home. In fact, this tendency has become so strong that another winter peak season during the Christmas-New Year period has been created. The trend has been reinforced by the fact that many enterprizes close for the festive period to reopen only after New Year. Thus the expansion of winter tourism has not only contributed to the alleviation of seasonality but ironically has created another "seasonality problem."

Some destinations have to devise place specific strategies in order to cope with the seasonal drop in demand. Bermuda is a good example. The off-season lasting from December 1 to March 15 is a time when this almost totally tourism dependent destination is suffering substantial economic losses. Although the air temperatures vary between 10-18^0 C the water is too cool for swimming. Therefore, Bermuda advertized its splendid golf courses in winter season. Also event attractions and lower hotel rates are being promoted. The winter off-season has been officially declared as "Rendezvous Time" since 1960 in order to woo tourists. If all these measures have brought rather disappointing results, lack of cooperation between the Bermuda Department of Tourism and the island's hotel industry is to blame (Lewis, R.; Beggs, T., 1982).

There are situations, however, that no measures to increase demand can be taken because natural seasonality excludes many destinations for part of the year, e.g. ski resorts located at high elevations (in the Alps even below 2000 m) have a too cool summer with little tourism. In such cases careful planning is required in order to cope with unemployment. Some countries have been able to acquire considerable experience in this respect. For example, in the Austrian Alps the labor force is employed during the off-season and the shoulder months not only in routine repairs but first of all in the production of sport articles (especially ski equipment) and souvenirs. Some men find employment in other tourist regions, in mining, and lumbering, women in handicrafts or textile industry. There are also some opportunities in agriculture although in most agricultural branches the peak seasons coincide with the tourist seasons. Also some employees may continue their education during off-season.

The extension of the weekend and the increasing popularity of long weekends also contributes to seasonal evening out of demand. People are participating in tourism and recreation more frequently but for shorter periods of time, including weekends, and this development is viewed positively by the tourism industry. However, the weekly fluctuations of demand bring their own problems not unsimilar to seasonal fluctuations. Especially severe are the traffic congestion problems connected with outbound and return traffic on weekends. It seems that these problems will be alleviated with the widespread introduction of compressed (4 day) working week. It appears almost sure that the four day working week for all in some DCs is on the horizon, hopefully even still before the year 2000 although this seems to be rather unrealistic at present.

A combination of weekend with seasonal demand fluctuation has proven to be a matter of concern. Excessive tourist flows cause congestion of transportation and facilities. The measures to cope with these problems, tested with full success, are as follows:

1) Staggering the begin of school vacation preferably starting in the middle of the week.

2) Staggering the vacations of employees especially avoiding their start on the first day of the month and on weekend. Vacations should be computed not in weeks or months but in working days. In this way the Friday afternoon rush may decrease and the vacationers will start their trip on an off-peak day of the week.

5.9.5. The results of response strategies

The positive results of the strategies discussed above are at hand: in almost all DCs there is a long term decrease in the proportion of vacations taken during July and August. This trend has been supported by research results, e.g. Drakatos (1987:584) found that within the Greek high season of May-September the share of May, June and September has increased at the expense of July and August. However, there is no possibility and perhaps also no need that the "seasonality problem" disappears completely. Some of the reasons for seasonality of tourism demand are subjective connected with biological and psychological needs of an individual and as such difficult to change by using merely economic incentives. From the biological point of view the human organisms and its functions are to a certain extent depended on the natural environment. This environment changes seasonally its impact on the vegetative nerve system and the hormonal functions. Thus these seasonal changes to a certain degree determine human needs for recreation - a process aimed at restoring human capacity to live and work. This need for seasonal change of work and rest for the organism is still not adequately researched by the medical science. Especially interesting would be to find out to what degree this biological mechanism could be subject to outside manipulation.

The other subjective reasons for seasonal demand fluctuations are psychological and are easier to change by news media, books, educational processes marketing etc. Human perceptions, images can be to a certain extent manipulated with respect of seasonality. The aim is to explain to the public the assets of off-season vacations and this is being done with a degree of success. Nevertheless, the concentration of tourism in time has proved to be more rigid than the concentration in space which is more successfully changed by diversification of destinations due mainly to the progress in the field of transportation.

5.10. SOME TRENDS IN DEMAND

In this section some of the trends in demand with respect to purpose of travel will be discussed. These trends are keenly observed by tourism planners involved in market segmentation.

The most important post World War II trend in demand is the increase of the share of pleasure and convention tourism at the expense of business travel and Visits to Friends and Relatives (VFR). Especially significant is the relative drop in business travel although the pace of change is only slight.

Business travel includes travel on company, organization and government business. Sometimes convention travel has been included. It seems, however, that it is more appropriate to treat it as a separate category mainly because of a large pleasure component and higher seasonality.

The importance of business travel for tourism may be summarized as follows:

1. Business travellers spend more per capita than pleasure tourists.
2. Business travel is income and price inelastic because the costs are paid from company expense accounts.
3. Business travel reacts much less to changes in economic situation than pleasure travel.

4. Business travel shows relatively little seasonal fluctuation with some drop in summer and short stop at Christmas-New Year time.
5. Business travel share is higher in domestic than in international tourism.
6. The participation of women in business travel is increasing and this has an impact on tourism e.g. on hotel room design (e.g. full length mirrors, hair dryers, electric curling irons).

Shortly after World War II the share of business travel amounted to over 50% of total travel in the DCs. However, since then there has been a slow gradual decline to 10-20%. Pleasure tourism dominates the contemporary travel scene. The reason for this development is the progress of computerization and telecommunication. Audio-visual equipment enables executives to hold even teleconferences (teleconferencing or video-conferencing) and engage in telemarketing. As a result many companies cut costs by saving on business travel. International business travel seems to be more affected by these developments than domestic. It has decreased to below 10% of US overseas travel. However, increasing world economic integration and globalization of trade certainly works in opposite direction and slows down this process.

Despite the relative drop in business travel it still grows in absolute terms and dominates some (mainly domestic) air routes and also the supersonic travel. In addition business is directly involved in the expansion of pleasure tourism by according fringe benefits to the employees in the form of so called motivational or incentive travel to mainly domestic destinations (typically: Las Vegas). These performance awards for employees constitute in some countries important tax-free fringe benefits. However, there is a tendency to tax them. Companies contribute to tourism also in other ways sponsoring picnics, excursions, tours, training courses in attractive places, etc.

Convention or congress tourism, which should be regarded as a separate category and not a mere segment of business travel, is booming and the projections for the future are excellent. The reason for this is not only the fact that the major costs are paid not by the participants but also the importance of personal contacts and pure pleasure aspects and the attractiveness of the venues.

There are less visits to friends and relatives. Reason: decreasing family cohesiveness. But not only loosening family bonds play a role - also the desire for personal freedom, and last but not least the fact that people in the long range have more discretionary money than in the past. This long-range trend to decrease the VFR travel pertains mainly to domestic tourism. In international tourism an opposite tendency may be observed: growing migrations from the third world to the first world countries stimulates not only ethnic tourism i.e. visits of former emigrants to their native countries but also VFR from the native country to the country of immigration.

As stated above pure pleasure tourism is the winner with respect to relative growth. But even here there have been important shifts: e.g. family travel has decreased causing the drop of average party size. There are several reasons for this development: smaller families, decline of family cohesiveness, teenagers going on separate vacations with their peers. The most important reasons appear to be more frequent separate vacations of husband and wife either as a result of difficulties to arrange simultaneous vacations for working couples or for psychological reasons: vacation from marriage connected with too much togetherness. Nevertheless, husband and wife vacations are still used as rehabilitation and renewal of marriage. Other trends in pleasure tourism demand include taking more than one vacation in a year, including "mini-vacations" during long weekends and shorter average stay at destination.

CHAPTER 6

SUPPLY

6.1 INTRODUCTION

The other side of the demand-supply equation - the tourist supply - is almost synonymously called the "tourist product" or "destination." In the history of tourism development the product/destination-oriented approach predates the more recent demand approach. Therefore, it is called "classical" or "historical" (World Travel, No. 129, March-April 1976:17). The product/destination oriented approach in tourism research, financing, development, management, advertising etc. focusses on the product as an independent variable shaping the demand patterns. Little attention is paid to the demand i.e. the motivations, attitudes, desires, etc. of prospective tourists. One cannot deny that frequently the supply constitutes the leading factor exerting its impact on demand by developing new resources and substituting some resources for others. As mentioned above, one of the proponents of the idea that in our modern times the stimuli mostly come from supply and not from demand is J.K. Galbraith who thinks that the market in classical sense, where demand (consumer) is leading, does not exist. According to him, supply is the decisive market element.

Many researchers, among them most geographers, have followed the traditional product-oriented approach to tourism development focussing on tourism resources, mainly natural resources which are more tangible, better measurable and more stable than the somewhat fluid and unpredictable demand. However, the supply orientation proved to be too one-sided. In reaction to it the demand-oriented approach has developed, especially after the World War II when the emphasis in research shifted to the "marketing approach". Psychologists, sociologists and some economists concentrate their research in this field. Thus tourism research and development as a whole is becoming more balanced in taking both sides of the demand-supply equation into account. The task is to harmonize the supply with the demand patterns both structurally and spatially. The development of supply has to be shaped in order to:

1) absorb the demand in a smooth and organized way to the full satisfaction of consumers

2) preserve the natural and social environment without deterioration or damage to the quality of consumption processes.

Thus, there is a need not only for systematic and regional inventories of supply factors but also for study of capacities to assure that their ceilings will not be surpassed. Also existing and potential user conflicts have to be researched in order to avoid them.

The terms tourism/recreation "supply" and "product" are used almost synonymously with "resources" and "destinations." The difference between them is rather contextual than conceptual. "Recreation resource ... used in its full dictionary sense to mean a source of supply of recreation opportunities, with no limitations on type or location. Books and television programs ... automobiles and aircraft ... golf courses and tennis courts; shopping centers ... homes and hotels ... museums ... restaurants ... streets and roads ... rivers ... park naturalists recreation program leaders ... are all included" (Chubb, M.; Chubb,H., 1981:287).

One could group the resources into following categories:

1) Attractions

2) Infrastructure or Facilities (including tourism facilities)

This general categorisation may be subject to still more detailed subdivision:

I. Natural resources (attractions) i.e. created by nature without human interference: land, topography, bodies of water (including hot and mineral springs), climate, flora, fauna.

II. Human (social, anthropogenic or man-made) resources developed by humans:

1) people and their culture, folklore, customs, religion, etc.

2) cultural, historical, architectural attractions (museums, cathedrals, theatres, monuments, festivals, art, crafts, etc.)

3) infrastructure (facilities)

The infrastructure means about the same as facilities, the other points pertain to the attractions.

This classification will constitute the basis for our further discussion. One has to note, though, that there is no clear-cut distinction between natural and human resources (take e.g. an urban park). The resources differ from each other in the degree of human modification on a continuum. It explains the conceptual validity of this classification although, not unlike with many other typologies, drawing a line at operational level may be impossible.

The terms "landscape" or "scenary" are designating combinations or integrations of various natural and human resources as elements of a whole.

Alongside with the above classification there are also other groupings of resources which use various criteria. Here are some of them:

Classification A - According to their relationship to tourism/recreation.

1) Primary or inherent.

These resources have in their essence no direct relationship to tourism and recreation. Tourism is unique in that it alone can turn to productive use many of these seemingly "useless" natural and man-made resources. Some of these resources may even constitute obstacles to other economic activities like rugged mountains. Some resources like lavish architecture of the past built at great costs without apparent economic return may only now - because of tourism - become sources of considerable income.

2) Secondary or derived.

In their origin and function connected with tourism and recreation. Artificially created by humans for this specific purpose e.g. an urban park or Disney World.

Classification B - According to the spatial extent of the market:

1) local 2) regional 3) national 4) international

Classification C - According to their evaluation

1) Absolute

2) Relative (comparison of resource endowment between various areas and destinations).

Classification D deals with the character of attractions.

1. Site attractions

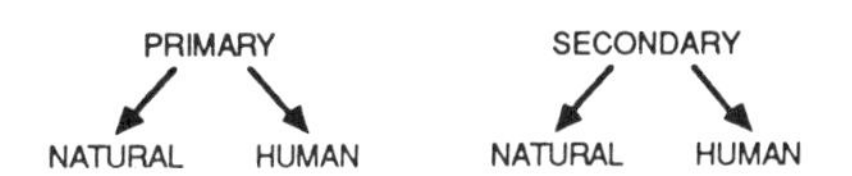

2. Event attractions

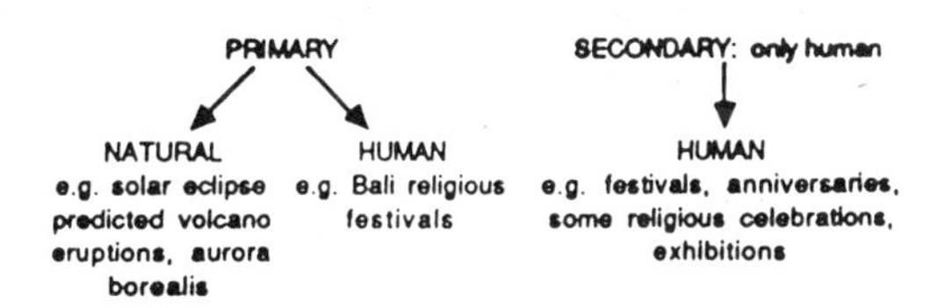

Classification E deals with typical and unique attributes of attractions:
1) Unique e.g. Taj Mahal, Grand Canyon of Colorado River, Ayers Rock, Niagara Falls.
2) Typical e.g. National Parks Act in Canada stipulates that "representative samples of Canada's natural landscapes, seascapes and ecosystems are acquired and maintained" (Environment Canada, 1981:3)

Geographical approach to supply

Geographers approach supply from their specific point of view encompasses two elements:
1) Space. Geographers investigate spatial distributions of resources and attempt to develop concepts of their regional differentiation and interaction, concepts of physical geographic, human geographic and integrated regionalization of resources and facilities.
2) Interrelation or better interaction between people (consumers) and their social and natural environment.

Some aspects of supply (e.g. supply for home-based recreation, hotel management issues) remain outside focus for geographers.

6.2. NATURAL RESOURCES

6.2.1. Introduction

Not all natural environments can be regarded as resources. E.S. Zimmerman expressed it succinctly: "Resources are not: they become". In this connection Zimmerman saw resources as "expanding and contracting in response to human effort and behaviour". (Zimmerman, E.S., 1951). Only such part of natural environment which is potentially useful becomes a resource. What cannot at the present time be used for whatever reason does not constitute a resource. The same pertains to tourism/recreation resources. Only what is subject to existing demand is a resource. "There is nothing to the physical landscape or features of any particular piece of land or body of water that makes it a recreation resource; it is the combination of the natural qualities and the ability and desire of man to use it, that makes a resource out of what otherwise may be a more or less meaningless combination of rocks, soil, and trees." (Clawson, M.; Knetsch, J., 1966:7).

Following this line of thought one could divide resources into existing or available and potential.

Resource is nowadays an important word in economic-geographical books and articles. There is always an abundance of discussion on natural resources for industry and agriculture but seldom for recreation and tourism. Also applied physical geography's contribution to evaluation of natural resources for tourism and recreation is very modest. These oversights may be attributed to certain traditional attitudes on the one hand and the newness of the question at the other. Recognition of this situation explains why relatively less research has been conducted into natural resources for tourism/recreation as compared with resources for other branches of economy. The matter is complicated by the fact that the latter are products of some particular part of the natural environment e.g. forests supply timber, rocks-minerals, the former in contrast constitute integrations, combinations of many environmental elements. These resource systems (integrations) are called natural landscapes or sceneries with single components less distinguishable and more difficult for analysis than the resources for traditional industries.

Another reason of rather scant attention paid to natural resources for tourism/recreation is the fact that they are partially of marginal importance for other industries and the competition for resources between tourism and other industries is to a certain extent limited although in recent years increasing, mainly because of the competition for land between tourism and some primary industries like mining, lumbering and in some LDCs agriculture. Tourism uses mainly renewable resources which are in a sense "free" or almost free (e.g. beautiful landscapes, good climate, fresh air) because the tourist industry has not paid for them and they are not used or little used economically. Frequently their economic value without tourism would be small. This is certainly an economic asset. The fact that the tourism resources are largely renewable and of permanent economic, social, cultural and environmental value, puts tourism in favorable position - it can flourish in areas of low economic productivity e.g. mountains where traditional economies have collapsed in competition with new technologies. A good example is tourism development in the Alps where traditional agriculture cannot compete with modern agriculture. A number of "dying" fishing villages or mill towns in northern New England and in Canadian Maritime provinces have been given new economic opportunities as a result of tourism development.

There are instances, especially small overpopulated islands, when practically any conventional resource base (energy resources, minerals, raw materials) is absent but there are such "free" tourist resources as mild climate, beautiful beaches, clean warm ocean waters, coral reefs, impressive volcanoes, etc. which are not coveted by other industries. Thus tourism seems to be the only alternative in addition to subsistence or plantation agriculture in most of the Caribbean islands.

Of course, tourism uses not only renewable resources, but also some non-renewable like oil, coal, gas. However, tourism is relatively energy-efficient and does not require too many raw materials as compared with other economic sectors.

However, the way of thinking that resources for other industries, particularly non-renewable resources, are limited whereas tourism/recreation resources are mostly renewable and thus inexhaustable natural gifts, proved to be incorrect. Only a relatively short time ago, we learned the bitter lesson of decreasing quality of our existence. We learned that the resource abundance (including even renewable resources) is illusory that all resources are limited and require most careful planning, management and protection. We overlooked the fact that human action may have not only positive but also more often a detrimental impact on resources decreasing their qualities for healthy and pleasant living.

By taking the tourism/recreation resources for granted and concentrating on the development of agricultural, industrial and other resources our society neglected also another important fact: the physical environment cannot be regarded exclusively as a source of increased output of raw materials as an arena for economic expansion. It is the challenge of our time to appreciate this environmeent also as a stage on which we spend our leisure time. Thus natural or physical environment should be regarded not only as a source of raw materials but also a source of enjoyment.

6.2.2. Classification of natural recreation resources

In 1950, Marion Claswon proposed a classification of natural resources for recreation which was later developed in a book coauthored by him and Jack Knetsch (Clawson, M.; Knetsch, J., 1966:37). This simple division , today regarded as classical, may be applied to resources or areas. The three categories are user-oriented (intensive or concentrated use of resources), resource-based (extensive or dispersed use of resources) and intermediate. The total spectrum of resource use should be regarded as a continuum in real world conditions. The characteristics are contained in the following table.

TABLE 6.1

GENERAL CLASSIFICATION OF OUTDOOR RECREATIONAL USES AND RESOURCES

	Type of Recreation Area		
Item	User Oriented	Resource Based	Intermediate
1. General location (accessibility)	Close to users; on what-ever resources are available	Where outstanding resources can be found; may be distant from most users	Must not be too remote from users; on best resources available within distance limitation
2. Major types of activity	Games, such as golf and tennis; swimming; picknicking; walks and horse riding; zoo, etc. playing by children	Major sightseeing; scientific and historical interest; hiking and mountain climbing; camping, fishing and hunting	Camping, picknicking, hiking; swimming, hunting, fishing
3. When major use occurs	After hours (school or work)	Vacations	Day outings and weekends
4. Typical sizes of areas	One to a hundred, or at most to a few	Usually some thousands of acres perhaps many thousands	A hundred to several thousand acres
5. Common types of agency responsibility	City, county, or other local government; private agencies	National parks and national forests primarily; state parks in some cases; private, especially for seashore and major lakes	Federal reservoirs, state parks; private

(Source: Clawson, M. Knetch, J. "Economics of outdoor recreation" 1966:37).

6.3. NATURAL RESOURCES: SYSTEMATIC APPROACH

In real world natural resources for tourism and recreation occur always in integrations (systems). However, for analytical purposes they may be discussed topically in the following order:

1. Space (land)
2. Topography, geology, soils
3. Water
4. Climate
5. Bioresources (vegetation and wildlife)

These resources should be evaluated both as factors attracting and stimulating recreation and tourism as well as impeding them.

6.3.1. Space

6.3.1.1. Space as a resource

The term "land" "territory" "area" used almost synonymously imply certain geographical concreteness as compared to abstract geometrical "space". However, land besides being a carrier of a variety of natural resources constitutes also a resource by itself as finite geographical space. Although some authors do not regard space as a recreation resource (M. Clawson 1972: 442) the empirical fact of space shortages and the necessity of spatial planning and management dictate such a recognition. In other words: the land is a finite limited resource, the demand for it is not limited. The result is shortage of land for humanity in general and for tourism/recreation in particular. Indeed, Fourastié was probably right predicting that lack of space will be the worst scarcity of the 21st century (Fourastié,J.,1965:204). Another futurologist, Herman Kahn, expresses similar opinion predicting that tourism by the year 2000 will have a growth rate that may be "self limiting" because of "too many tourists and two few sites." That the problem of space is not entirely new one could facetiously draw on a story from Antiquity: Diogenes lived in a tub and when Alexander the Great, in gratitude for an interesting talk with him, told him to name his wish, he simply asked the emperor to move out of the way, not to block the sunlight. In a more serious vein, the modern problems of space are epitomized by the pressure of infinite demand on finite space resources which are in increasingly short supply. The acuteness of this phenomenon differs from place to place, from region to region but it is widely recognized as a world problem. In North America space is especially scarce in the East which is most urbanized. The densely populated urban areas, suffer most from this problem as a result of conversion of rural land to urban use in the urban fringe areas. Western Europe and especially West Germany and the Benelux countries are, of course, still more afflicted than North America. In West Germany over 10% of land is covered by settlements. But even in less urbanized areas land resources are subject to intense competitions and conflicts of various uses, for instance mining and lumbering encroach on wilderness areas.

The amount of space devoured annually by humanity is staggering e.g. in the US already for may years more than 150,000 sq. kilometers of cultivated land are lost annually to settlements, industrial and transportation use. The "space devouring" function of human settlements has been reinforced by the development of modern transportation systems, especially highways. The cities produced by the automobile are less dense than those produced by railroad and street car. The increased mobility associated with the automobile requires more space for highways, sprawling suburbs and last but not least - parking space: Disneyland in California occupies less than 34 hectares (83 acres) but has a car parking of 44.5 hectares (110 acres).

Thus the supply of land decreases, the demand increases and the result is overcrowding, deteriorating the quality of land whose carrying capacity has been overtaxed by intensive use. The area of pristine wilderness shrinks rapidly. Even in less populated LDCs the land use patterns change under the pressure of rapidly increasing populations. Especially characteristic is the expansion of crop cultivation at the expense of forest and of grazing areas, at the expense of savannahs with all the associated negative environmental impacts. In the DCs the forested areas sometimes even increase as a result of the retreat of agricultural margins following the conversion of existing farm operations to other economically more suitable uses such as forestry and recreation. At the same time, however, frequent failures to regenerate cut-over land or to take care of wooded areas that were burned or ravaged by insects, decreases the quality of forested land for both recreational and non-recreational uses.

Not only space in general but also tourism/recreation space is increasingly in short supply. Extraurban sprawl of second homes, retirement and other communities is associated with land price inflation even in areas relatively distant from metropolitan centers as the developers often leapfrog the more expensive suburban lands and invade the usually more environmentally attractive exurbia.

This "land rush" in turn increases the use of energy (automobiles) and sometimes contributes to despoiliation of valuable recreational resources. As a result the availability of recreational land for urban population is seriously affected.

The availability of recreational land differs spatially: the scarcity is mostly acute in urban and near-urban areas whereas the longer the distance from densely populated metropolitan areas the more abundant is recreational land and available at reasonable prices. However, the increased distance-decay function and the associated incremental costs of travel from the city to the destination constitutes an important negative factor. This spatial dichotomy between center and periphery is especially strongly pronounced in North America. In Western Europe land is in short supply and expensive at local, regional and even national levels, but the distance-decay function plays a lesser role than in North America because of small areas of these countries.

6.3.1.2. Space requirements

The human response to the problems associated with land as recreation resource has to recognize the fact that specific activities within the recreation spectrum require different qualities and quantities of land. Some activities can occur almost on any land (e.g. hiking), some are location-specific (e.g. skiing, beach activities). Availability of mountains with reliable snow cover in winter, clean water bodies, sandy beaches, game and fish etc. relate to various recreation activities. Thus the resource endowment of a particular area - the qualitative dimension of its space - make it more or less suitable for various types of recreation activity.

In addition to these qualities of land as resource carrier there are differences in space requirements. Indeed, the amount of space required for various types of recreation varies greatly. Generally, the user-oriented activities located at relatively close distance from city require less space than the resource-based activities (regional or extraurban destinations). Some of the user-oriented activities have strictly specified space requirements like tennis, soccer or football, some less precise like golf. These activities not only require space monopoly associated with exclusive use but also exclusive property rights. A really good golf course will not tolerate goats or wildlife grazing on it or people walking through it even when snow-covered because it ruins the insulating properties of snow and damages the grass. On the other hand, such activities like swimming or boating require the temporary exclusion of other uses but they mainly do not infringe on property rights needing only the privilege of access. Most resource-based activities do not require monopoly of space. Certain kinds of tourism/recreation, especially less formal and organized activities do not require any special provision of space and it is impossible to determine any specific space standards for them, e.g. sightseeing, walking, driving for pleasure etc.

Another group of activities have actually very small space requirements but they need a surrounding "accessory space" which could be called "social space" which varies dependent from the activity and the preferences of the user. "At one end of the scale the amount of space may be so restricted that the conduct of the activity and the value of the recreation experience are seriously impaired; at the other the marginal value of additional space to the user for the activity is negligible. Family picnicking may be a frustrating experience if strangers are similarly picnicking 10 feet away, yet a family exclusively occupying an acre of ground for its picnic would probably not value an additional acre very highly." (Perloff, H.S.; Wingo, L., 1962:90).

Clawson and Knetsch illustrate a hypothetical example of user satisfaction in relation to intensity of recreation area use by the following graph (Clawson, M.; Knetsch, J.L., 1966:168).

Figure 6.1

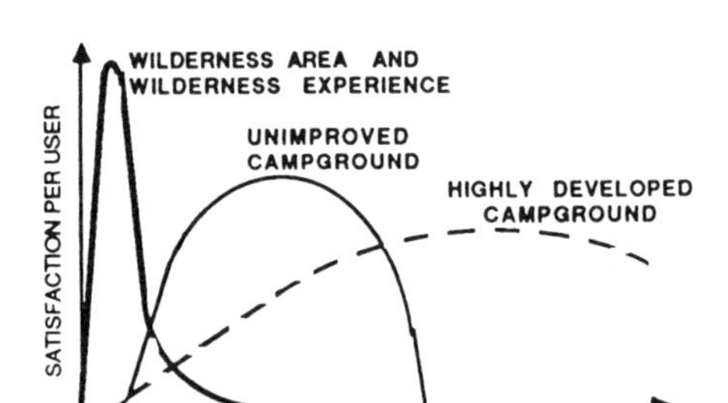

The point here is that few people want to be entirely alone. However, on the other hand there also exists a certain threshold or point of saturation within the "social space". Exceeding this point by reducing the available space causes a decline in quality of recreational experience and thus diminishing satisfaction of the users (social carrying capacity). This threshold value of space use differs (fig. 6.1) for various types of recreation. It may also vary among individuals. Clawson and Knetsch comment on the quality of wilderness experience: "Forest Service regards the encountering of two other parties per day as an informal standard for a tolerable level of intensity of use within wilderness areas. Whatever may be the actual limits, there would be general agreement that the tolerable intensity of use must be relatively very low if the wilderness experience is still to be maintained" (op. cit:168).

The urgency of the provision of local recreation space increases with the size of the community. It is also a well known fact that travel intensity (emissiveness) grows more than proportionately with growing population of cities. Lundgren observes that this "runs contrary to what one would expect, namely that the larger the urban economy, the greater would be its ability to provide its own recreational requirements. This may be true for typical "urban" recreation such as theater, concerts, museums, dining, spectator sports; however, for any activity requiring outdoor open space the situation is quite different. Usually one finds that the urban economics have priced themselves out of open space allocation for recreation, which makes a high outbound travel intensity inevitable for the urban population" (Lundgren, J.,1974:171). Thus the high emissiveness of large metropolitan centers is at least partially caused by lack of open space for recreation. The problem should be viewed in light of the fact that vast majority of out-of-home recreational activities takes place locally. Therefore, urban and near-urban recreation space is subject to greatest demand pressure and land use conflict. Hence provision of this space for easily accessible user-oriented activities should loom high in the priorities of planners and politicians.

6.3.1.1. Factors moderating the space requirements

The discussion of space requirements for recreation may suggest that there are a number of factors which moderate these requirements. First of all, as stated above, there is a group of activities which practically do not require any space and frequently even no access. Such activities like sightseeing, walking, driving for pleasure are to a large extent associated with enjoyment of landscapes as a visual and aesthetic resource. In such cases the main requirement is the scenery visible from the road.

Secondly, even in cases when tourism and recreation requires space this land is to a large extent less desirable for other branches of economy, particularly agriculture. Agriculture covets flat, well drained fertile land, recreation and especially resource-based recreation on the other hand uses agriculturally marginal lands, particularly areas with some undesirable features for agriculture like steep, rough, rocky, sandy, wet or highly erodable. Indeed, in the long range, despite some economically or politically motivated fluctuations in this respect, agricultural production in the DCs concentrates on better lands leaving economically marginal areas to other uses, mainly forestry and recreation. In the American geographical literature the recreational potential of these "empty areas" was discussed for the first time by Lester Klimm (1954), later by Dennis Durden (1966). They developed the typology of "empty areas" and indicated that their extent is subject to a dynamic process both expanding and contracting their acreages at various times and in various places. The general long range trend is that the recreational lands are spatially expanding. Examples abound not only from North America but also from Western Europe where West Germany is a good case where abandoned fallow land called "Sozialbrache" constitutes about 3 per cent of agricultural area, especially in environmentally most attractive parts of the country thus extending the available recreation land. The farmers have abandoned these marginal lands to seek employment in the city and all this in a country whose leaders short time ago complained about lack of "living space."

The third factor moderating the space requirement for recreation is the principle of multiple use. As indicated above, most of resource-based activities do not require complete monopoly of space i.e. exclusion of other uses. They function according to the principle of multiple use i.e. complex, integrated land use with other branches of economy. Examples of multiple use management are many forest areas which provide timber, forage, water, wildlife and recreation. Moreover, well planned and properly managed multiple land use is not only compatible but may also enhance the recreation experience and decrease the pressure on land. For the benefit of wildlife watchers and hunters properly managed logging operations may increase browse for deer, provide space for camping and hiking. Grazing, farming and logging is compatible with skiing, stock raising with dude ranching etc. However, some of the recreational activities require at least temporary single use pattern or monopoly of space e.g. hunting, skiing. In this connection improper management may cause land-use conflicts between various incompatible forms of recreation. The conflicts are frequently asymmetrical i.e. the negative impact is only in one direction whereas the culprit may be little affected by competitors. Here are some examples of such conflicts: hikers versus horseback riders, or off-road vehicles, anglers and swimmers versus motor boat users and water skiers, cross country skiers versus snowmobilers etc. Especially water oriented recreational activities like swimming, water-skiing, sunbathing, boating, sailing, canoeing are susceptible to use conflicts. Here not only spatial incompatibility plays a role but also quality of water. Proper management techniques (zoning) prevent such conflicts.

Fourth factor pertains also to resource-based activities. These activities are undertaken largely in peripheral zones distant from the urban centers. Their participation rates, due to mainly distance friction (called also distance decay function) are lower than those of resource based urban or near urban activities. The relatively low participation rates for such activities like wilderness hiking, canoeing, mountaineering etc. indicate lower demand for space in these remote areas although also here conflicts occur with increasing demand pressure.

The fifth factor is diverted demand. The "land rush" for second homes diverts the demand of the owners from heavily overused public recreation areas but at the same time it creates its own environmental problems of congestion and drives the land prices up to the level as to make it unaffordable for locals if they would wish to return to agriculture. Large agriculturally marginal areas in the Alps where many farmers have totally or partially given up their farms, have been taken over by recreation creating a number of severe environmental problems (Karp, G.. 1981:57). Thus this diversion of demand serves at best as a palliative and, similarly to the other factors, does not solve the problems of demand-supply disequilibrium for land. Indeed, despite the modifying factors land for recreation still becomes less available and less accessible necessitating appropriate human response to this problem.

6.3.1.4. Response strategies to land use problems

Human response strategies to the problem of lack of space as a resource present an infinite array of measures at the government level and at the private sector level.

The government intervention dates back at least to the 19th century when city governments started to provide scenic beauty and relaxation for the inhabitants in the form of urban parks. In the 20th century with the growth of land use problems, the intervention of governments on all levels has played a very significant role. One could generalize that this role has been increasing in the long range despite some short range policy fluctuations. This trend is evident especially since 1945.

The intervention of the government in land use matters, including recreational land use has expanded substantially.

The forms of government intervention are varied. In urban areas the policy aimed at provision of open space for the population continues. As there is less space available than in the past, the government agencies sometimes in cooperation with private ones introduce such innovative and cost-effective methods as conversion of vacant land into park, conversion of unused industrial facilities into recreation areas, use of abandoned railroad corridors and utility rights-of-ways as gardens, playgrounds, and bike/hike trails, use of covered dumps and sanitary fills for parks, use of abandoned buildings and old schools, etc.

Other forms of response to land use problems are zoning regulations trying to separate various incompatible land uses including land use for recreation. Zoning regulations assign their own space to incompatible uses in order to avoid conflicts. As stated above, such conflicts may even arise within recreation compelling the government to intervene.

However, the highest degree of direct government intervention is connected with land ownership. In this respect one has to indicate the difference between North America and Western Europe as far as direct government land ownership is concerned. In the US and Canada the government, especially the federal government, owns a substantial share of land and thus is able to influence the land use patterns directly. The government land ownership is essential for certain types of lands where the preservation and recreation function have to be coordinated in a rather "monopolistic" fashion on relatively large areas. Wilderness preservation requires such a "monopolistic" approach to land. Here, of course, so much space and such long periods of time are involved that public land ownership seems imperative. In this respect U.S. and Canadian National Parks records are the most outstanding. Examples of multiple use of land for recreation and other branches of economy like forestry are the US National and Canadian Provincial Forests, US State Parks and Canadian Provincial Parks. The governments of the US and Canada at all levels have also a long history of public land acquisition from private ownership to reserve them for present and future enjoyment of the population. The US Department of the Interior through Land and Water Conservation Fund has bought thousands of acres of private land for preservation and recreation e.g. 18,000 additional acres were acquired for Florida's Everglades National Park in 1972.

The development of public lands for purposes of recreation in US and Canada has a long history reaching colonial times. However, more recently the governments in realization that they cannot alone supply for all out-of-home recreation opportunities, started to intervene in the private sector to extend these opportunities. Another reason for government intervention has been the situation of private lands as a finite spatial resource: land use conflicts between various branches of economy and recreational use and also within the recreational spectrum. These conflicts are sometimes difficult to resolve on the basis of cost-benefit calculations. Rush to buy recreational land contributes to pressures associated not only with environmental deterioration but also with speculation, misrepresentation and outright fraud.

All these reasons make government intervention indispensable. Especially valuable as tourism/recreation resources are undeveloped private lands. The government intervention in private land use occurs not only in the form of fiscal policy (taxation) or in legislation (e.g. zoning). Sometimes mere exhortation indicating both private and public interests will suffice. However, there are many other methods like the purchase by the government of scenic easements (development rights) to protect the view from scenic roads, rivers and urban landscapes, subsidisation of recreation on private lands in the form of cost sharing privileges to private landowners for practices beneficial to wildlife or in the form of long-term loans for recreation development (e.g. the US Food and Agriculture Act 1962). Another form of public sponsorship of recreation are tax credits for protection of scenic resources. However, there are various obstacles which act as deterrents to provision of recreation opportunities on private lands e.g. liability, vandalism and compliance with health standards which have to be overcome by proper planning and management. The form of government intervention most resented by owners and supported by the public, is legislation pertaining to maintenance of public access to recreation land or outright expropriation.

The issues connected with the government intervention into private ownership is a controversial one and depends on political philosophy. The right of the center political philosophy opposes government intervention on constitutional grounds. The US constitution states that "Nor shall private property be taken for public use without just compensation." Therefore even land and water surfaces zoning regulations interfering with the value of private property constitute sensu stricto a violation of the constitution. More centrist political philosophies, which have gained gradually the upper hand, advocate a rather substantial degree of government interference. Indeed, the concept of land as a purely private commodity to be regulated only by economic forces has been giving way to the concept of land as a national resource in which all citizens have interests. This new "land ethic" philosophy does not propose nationalization of private lands but advocates the shift of power from the individual owner to the government as far as land use planning and control is concerned. The present situation in North America is charcterized by rather large amount of regulations, restrictions and other forms of government interference which significantly affect land use and land values. On top of this, local governments in the framework of comprehensive land use plans increasingly establish population ceilings in a number of communities, especially recreation and retirement communities although sensu stricto this is another violation of the US constitution guaranteeing to all citizens the right to travel and inferentially the right to settle where they wish.

On the state and provincial government level a number of regulating steps have been taken in the framework of "land ethic" policy which have a significant beneficial impact on environment and recreation. It seems that the trend is in the direction of taking some land use authority from local public agencies by the mid-level governments facing a number of serious problems transcending purely local issues. Examples of such problems include e.g. disappearance of shoreline and other recreation areas available to the public, destruction of ecologically important wetlands, dilemmas about locating airports and other infrastructure facilities, to name only a few issues. Dealing with them requires a machinery with authority at all levels of government which would reconcile the conflicting interests of different levels. And such a machinery does not exist in North America.

The steps aimed at land use regulation fall into two categories: one is measures dealing with geographical entities like local zoning, shoreline, wetland regulation or scenic river-preservation. The other category is functional controls like erosion control, regulation of real estate (including recreation), siting of the transportation network etc. All this, increasing government influence on land use, constitutes a controversial political issue between preservation of environment and development. The interests of tourism/recreation industry lie mostly, but far from always, on the side of preservation, on the side of integrated land use planning which, however, at the present moment is a goal yet to be achieved both in US and Canada.

The land use policy in the EC countries is rather difficult to generalize but two characteristic features ought to be mentioned without going into details which differ from country to country: first is the relatively small amount of land in public ownership; second, is a rather stronger degree of government interference because space in Europe is a more scarce commodity than in North America.

On the global level, the UN resolutions, which are not binding for the members, go far to the left advocating a land use policy stopping short of private land nationalization.

At present, there is virtually no economically developed country where the government would not interfere at least in some degree into land use. The response strategies to the limitations of space and environmental problems necessitates the development of government research, planning and management apparatus in order to implement and put into action policies which would benefit the public. Such an apparatus deals with a wide spectrum of factors like physical, economic, social and even psychological.

Response strategies at the private sector level (private land owners) are very modest as compared to government level and rather limited to urban open space. Here are some examples:

1) A new branch of geography called necrogeography ("nekros" means "dead" in Greek) investigates the role of cemeteries in our landscapes. Necrogeography among others studies the space requirements of cemeteries which constitute such a conspicuous feature in some landscapes. In this respect one has to note following trends:

a) Increasing tendency for cemeteries to take less space by various methods including body cremation or vertical rather than horizontal storage of graves (high rise mausoleums)

b) The modern cemeteries, called "memorial parks" or "memorial gardens" are frequently beautifully landscaped. Some are simple, easy to maintain flat lawn-type monuments flush on the ground. In areas in critical need for urban open multiple use of cemeteries is spreading i.e. use for walking, picnicking and other forms of recreation. The proponents of this idea argue that it is ecologically sound, enjoyable for the living and decreases vandalism. The opponents insist that it is insensitive and desecrating. The decision in adopting a specific policy lies in the hands of local population, and such a policy has been approved in many cities, especially in some US cities, where the provision of open urban space has been neglected. In the U.K. still more radical steps to save space have been undertaken: complete removal of some cemeteries or their total conversion.

2) The modern hotel design makes better use of space: the rooms are smaller, ceilings lower, corridors narrower, lobbies better designed than in old hotels. Japan, a country where the problem of space is very acute, introduced in 1980 a new type of hotel with no rooms but only "life capsules" of about two square meters arranged vertically.

6.3.2. Topography

The topography on our planet ranges from Mt. Everest, 8,848 meters over sea level, to the Dead Sea, 394 meters below sea level. In terms of topography tourism and recreation certainly competes little with other branches of the economy. Agriculture prefers flatlands because they usually have better soils and allow the operation of modern machinery on large fields. Manufacturing also locates on flat terrain but location on the mountain valley bottom frequently aggravates the problems of pollution and access. It also limits the potential for territorial expansion. Tourism and recreation mostly prefers location in areas with at least some topographic variation. One could generalize that the greater the relief energy the better the location. This trend has been especially pronounced since the beginning of modern tourism in the 19th century. It follows the changing tastes of the people who even in this century switch their preference from smaller topographic forms to the grandeur of Alpine landscape. This tendency has been confirmed e.g. by German research which found out that German domestic vacationers in the GFR are increasingly turning to Bavaria from Mittelgebirge ("Urlaubsreisen 1968":31-32). Indeed, rugged mountain landscapes are regarded as beautiful and preferred as a background even for recreation activities requiring small pieces of flatland. For some rapidly expanding recreation activities like down-hill skiing, mountain hiking and climbing, land with high relief energy is, or course, essential. Spectacular mountain peaks like Mt. Everest or Matterhorn focus the attention of all, including those who admire them from the distance only. Also the health aspects of clean fresh mountain air have gained wide recognition. Sightseeing in conjunction with rigorous recreation activities focusses on volcanoes (Etna in Italy, Mt. Saint Helene's in the US), remarkable rock formations, caves, waterfalls, scenic outlook points, etc. As a result of these trends, tourism and recreation is gradually gaining space in the mountains at the expense of agricultural and pastoral land use. Of course, these developments are not only caused by purely techno-economic factors like inability to use machinery or steep slopes, but mainly by socioeconomic push and pull forces which have caused the migration of rural population to cities. Recreational land use takes over vacated land and increases so much that it sometimes surpasses the saturation point like in the case of the Alps or even in some areas of the Rockies.

Whereas in terms of macrolocation for tourism and recreation mountains are practically always regarded as favorable, microlocation or siting on steep slopes may involve some hazards or at least inconveniences. In this respect dangers of erosion, landslides, mudflows, avalanches should be always taken into account while planning any siting of facilities. Sometimes high relief energy may cause additional costs in transportation (rail, road and even trail construction). One has to build serpentines or tunnels. Tunnels are extremely expensive and also involve the sacrifice of views. Box canyons with only one exit should be avoided for siting facilities, especially in Western and South Western parts of US because of forest and bush fires and also flashfloods. Siting of campgrounds and other facilities among mountain rivers should consider the possibility of flooding and washouts, siting on flatland must take into account the proper drainage. Campgrounds require a good drainage on one hand but should avoid slopes more than 10% on the other. If there is a lack of slopes flat enough for construction of facilities then additional costs must be incurred for building level sites.

Downhill skiing must be planned first of all from the point of view steepness of the slopes which was formerly measured in degrees and at present almost exclusively in per cent of incline which indicates the relationship between the rise of the terrain and the horizontal plane. Thus e.g. 5 meters of rise for 100 meters of horizontal distance means 5% slope. Consequently, 5° translates into 9% slope, 10° into 18%, 15° into 27%, 25° into 58%, 40° into 84% and 45° into 100%. Beginners normally use slopes up to 25%, intermediate skiers 25-35% and experts 50-70%. A successful ski resort should offer a variety of slopes adjusted to the skill levels of its market. This circumstance in addition to vastly increasing capacities of uphill transportation of skiers created the necessity of slope designing to increase the capacity and variety of slopes and also to smooth out some "rough spots" on the ski runs. In this connection a number of specialized firms like e.g. Ecodesign Mountain Recreation Planners Ltd. which designed the Mount Allen for 1988 Calgary Winter Olympics, are working at "sculpturing" the mountains for creating ski slopes better suited to modern requirements. At the same time the firms stress their commitment to environmental conservation (New York Times 20-10-1986).

The downhill ski planners take into account not only the steepness of slopes but also a number of other parameters like vertical drop: in North American West 800 meters vertical drop is regarded as a minimum for a ski run, in North East however only 400 meters is acceptable. Roughness of the area must be taken into account by recreation planners in connection with vertical drop. It indicates whether the slopes are long and even or short and irregular. Rock outcrops and erosion gullies while sometimes spectacular may inhibit the construction of facilities. Other important factors in downhill ski run planning are aspect, exposure and prevailing winds. Aspect means orientation of the slope towards the sun. Generally speaking northern aspects are preferred because snow keeps best on such slopes. Southern aspects shorten the season and create some hazards, especially for late winter downhill skiing: the snow melts during the day to freeze up at night time thus creating ice conditions for morning skiers. Exposure is the ability of the sunlight to fall on an area and must be taken into account while planning downhill runs. There may be cases when another mountain blocks the sun from the ski run. Ski runs facing prevailing winds should be avoided. This creates some problems in areas with prevailing NW winds. In such cases NE aspects of ski slopes should be given preference if possible.

6.3.3. Geology and soils

Geology has to be taken into account when selecting sites for tourism facilities and recreation areas, e.g. in earthquake areas is better to build a hotel on hard rock like granite than on sand or sandy soils. To reduce the danger of earthquakes hotels should be built on solid steel pipes anchored in bedrock. This, of course, increases the costs. The best beaches have white fine textured sand rather than black one of volcanic provenience because the latter absorbs more sunrays and becomes hotter. Natural hazards like avalanches, seismic areas etc. are to be avoided for tourist location but they themselves become tourist attractions sometimes long after the disasters have struck.

In site selection one has to take into account not only slope and drainage but also soil characteristics like susceptibility to erosion and compaction. Thus the best siting for tourism/recreation facilities is a gently sloping area with deep well drained soils of medium texture like e.g. loams. Very coarse textured soils like sands or gravels have certain advantages like lack of dust or mud but they are prone to erosion (especially sands) and do not give enough support for vegetation because of droughtiness. Heavy soils like silt or clay may allow too much surface runoff and are susceptible to pudling (loss of structure resulting from use when too wet or too dry). Soils prone to mudflows are very frequent in the mountains the best example is the destruction of the Medeo outdoor speed skating rink near Alma-Ata in the USSR. The rebuilt rink is now sheltered by a special dam.

6.3.4. Water

Water always has held a special attraction for people as an essential element of visual landscape. For tourism and recreation it constitutes an extremely, some say, the most important resource. In summer it is water on which recreation concentrates. Winter recreation activities like skiing, skating, snowmobiling, ice fishing focus also on water, although in a different form. The by now classical US Outdoor Recreation Resources Review Commission report from 1962 states that "probably the major portion of outdoor recreation is in association with water areas" (ORRRC study report no. 4, 1962:3).

The same report suggests that "there are two types of uses of water for recreation:

1. Water used in recreation as a medium on which or in which the recreation activities are being carried out e.g., swimming, boating, hunting, fishing.

2. The second general type of use, if it may be called a use, is that in which water provides principally the background or setting enhancing the intrinsic satisfaction to be derived from any recreational activity. This general type includes all recreational activity for which water as a scenic asset is involved, or where it adds to the emotional quality of any personal experience. For instance, camping primarily for nature study, photography, or just plain contemplation, requires a natural, unspoiled setting" (idem:3). The outstanding water resources in this sense are waterfalls, geisers etc. used for sightseeing.

3. An additional type of water use could be distinguished that as a health resource: mineral and thermal water.

As a result of these functions, site locations for facilities and activities tend to concentrate on water or at least with the view on water. Site location directly on water's edge may spoil the attraction.

The resources for water oriented recreation may be divided into two groups:

1) littoral resources of the sea
2) inland water resources.

The quality requirements for both groups are similar. One of the early discussions of water quality for recreation is contained in the ORRRC reports (ORRRC report no. 10, 1962: 14-20). The water should be free of sediments, waste, chemical and thermal pollution, algal growth etc. The report emphasises the importance of water quality research and deplores the non-existence of recreation hydrology. During the years since the publication of the ORRRC report there has been some progress in this field.

The water quality standards for recreation do not need to equal the standards for drinking water, however, some standards need to be established in order to protect the public. In terms of water cleanness the highest standards should be required from water used for swimming and bathing. It seems that 1000 coliform organisms per 100 ml of water should be regarded as maximum for swimming (Talaeva, U., 1979;209). Generally speaking the ocean waters are not as hospitable to pathogens as freshwater but unfortunately, many sea beaches around the world do not live up to the standards of cleanness, especially at closed seas like the Mediterranean e.g. Alexandria, Egypt (Cabelli, V., 1979:144). Also coastal waters in open ocean suffer from contamination e.g. the pollution on the US Eastern Seaboard is a well known fact (New York Times 24-1-1988). Besides purity, the water for swimming requires transparency (clarity), and also some minimum temperature, generally not below 18^0C. Of course, the water requirements differ for various recreation activities e.g. cleanness standards are lower for activities not involving direct contact with water (e.g. boating, sailing). There are also other qualities of water which should be taken into account like minimum depth, spread (distribution in space) - paramount prerequisite for sailing, absence of undertows endangering swimmers (e.g. in parts of the Gulf of Guinea coast in West Africa or East Australia) danger of shark attacks (Australia, Caribbean, US West Coast, South Africa). Examples of water related problems in inland areas are swift currents, rocky bottoms and too low water temperatures in the mountains, dangerous fauna (e.g. piranas in South America), disease contamination (e.g. bilharzia in Africa).

Planning for water related recreation activities must also take into account the character of the shoreline. Here the beach-type shoreline is better for siting than bluff or marsh for a number of reasons, one of them being access. Bluff location, however, offers views, sometimes spectacular.

Water related recreation activities

There is a wide spectrum of water related activities. One could divide them according to various criteria. One such a criterion could be historical.

1) Long established (traditional) activities: fishing, canoeing, boating, bathing, swimming etc.

2) Modern activities: water-skiing, scuba diving, hang-gliding, surfing, wind surfing.

These activities reflect the widening range, expanding diversity by introduction of new water related sports.

Of course, recreationists enjoy also other land activities near water of more universal character like hiking, bird watching, golfing, horseback riding, picnicking, camping, photographing, painting, sightseeing (scenic, scientific, historical), nature study (biological, geological, botanical).

Inland waters

The inland waters supply the recreationists with a wide variety of recreation opportunities ranging from opportunities identical to that provided by the sea to spas based on mineral water. This considerable diversity of water resources necessitates classification. Morphologically inland water resources may be divided into linear and enclosed. Linear forms comprise rivers, streams, canals and other waterways. Enclosed waters comprise lakes, ponds, reservoirs. There are several major concerns connected with the use of inland waters for recreation. First is the access problem which looks differently depending on activity. It is most severe for boating and sailing because in addition to access or property rights it normally requires construction of special launching facilities. Another problem is user conflicts, not only between recreation and other uses but also between various water related recreation activities. This necessitates planning and management of resource allocation to various users. Still another problem presents the preservation as opposed to use. Generally, recreation does not endanger the inland water environment but in cases of inadequate planning and management such dangers may arise. Therefore, a number of regulations limit some water related activities in the interests of conservation, e.g. water skiing has been practically eliminated from many water bodies as a result of the restrictions imposed on use of motor boats or the imposition of speed limits designed to prevent bank erosion and user conflicts.

The location of recreational activities and related facilities on inland waters follows certain guidelines which may be difficult to implement in certain specific situations. On rivers and streams in temperate climate areas, the best locations are mostly in the middle course because of availability of shallow sandy access to the water (beach type of shoreline) and slower speed of the water. Upstream sections frequently have mountain character with access problems, swift current, rocky shores and bottoms. The water is often shallow, cold and repeatedly changes its level. Such an environment eliminates many recreation activities like swimming and leaves some like white water canoeing which is thriving in such conditions. Location in lower parts of river courses may be associated with such drawbacks like mud, marshes, insects, increased pollution, too wide course.

A very special use of water is connected with its alleged healing capacities. The history of mineral spas reaches antiquity but only modern tourism contributed to significant development of spas. The medicinal capacities of mineral waters are generally less accepted in North America than in Europe where special branches of medicine have been developed: balneology - study of chemical composition of waters (and muds) and its capacity to heal certain ailments, and its application - balneotherapy. Hydrotherapy is the external therapeutic application of water in baths, whirlpools etc. In most spas people also use mineral water internally for drinking therapy. Medical uses of sea water have been first recognized in 19th century but there is relatively little practical application. The Institute de Thalassotherapie in Ouiberon (Brittany, France) uses water and maritime climate for cure. In terms of absolute volume of tourism the spas keep their place but their relative share in tourism has dropped to almost insignificance even in Europe where, like elsewhere in the world, the emphasis is on water-related recreational activities which dominate the scene of resorts. In North America the most popular inland water resort area is the Lake Sidney Lanier in Georgia, on an artificial lake.

Water constitutes not only direct resource for tourism/recreation but also indirect as "conditio sine qua non" for the existence of tourists at destinations. Acute shortages of fresh water in certain areas, notably at some subtropical islands seriously limits tourist development. Good examples are small Caribbean islands.

Therefore, in some cases measures are being taken like construction of expensive desalinisation plants or transportation of water in tankers, practiced between Spain and Canary islands. Another method, used more frequently, is simply to limit further hotel development, especially in areas where local population suffers as a result of lavish supply of water to 5-star hotels. An example of such situation is the southern part of Bombay, India which faces a water shortage problem of significant proportion. The ban on further hotel construction seems imperatiave.

6.3.5. Climate

Climate has been an independent variable since the dawn of history. Human society is able to exert its impact on climate only since the start of industrial revolution and especially during the present century. However, this impact is largely negative, also for tourism. It seems that our society will be unable to modify climate in a positive way for its benefit, at least in foreseeable future. Only indoor temperature and humidity conditions for recreation can be modified by heating, air conditioning, and more recently by freezing (artificial ski slopes under a dome near Tokyo).

Climate as a resource for tourism and recreation may be approached from three points of view according to dimensions of area under consideration.

1) Macroclimate pertains to an area with linear dimension of above 100 km, involves areas of thousands of square kilometers.

2) Mesoclimate: from 1 to 100 km, involves areas from few to several thousands square kilometers.

3) Microclimate: less than 1 km, involves small usually topographically well delineated areas. (Griffiths, J., 1976:55).

For the purposes of tourism and recreation macro- and microclimate are considered.

6.3.5.1. Macroclimate

People watching in the middle of a winter blizzard the sporting events on TV in warm and sunny Florida dream to be there. If they have time and money such a dream may become reality quite easily. Indeed, in the modern era of rapid transportation the opportunities to enjoy "good climate" at any time of the year are increasing quickly. Therefore, more and more tourists became interested in climatic conditions of their potential destinations. However, disappointment frequently awaits them if they try to get the climatic information from the travel agents because it is non-existent, or inadequate, and sometimes even misleading. (Mieczkowski, Z., 1983).

The tourism business and planning are concerned with macroclimate as far as the duration of tourism season is concerned because it determines the degree of investment utilisation. Also climatic extremes (especially in temperature) must be taken into account because of discomfort for long-stay visitors. In some cases the developers have to plan for indoor recreation facilities to create alternatives for periods with inclement weather. Examples: indoor tennis courts for rainy or excessively hot and humid weather, or curling rinks in Canada when outdoor conditions become impossible for recreation.

Climate is divided into component parts called elements: temperature, humidity, precipitation, sunshine, wind etc. Temperature in conjunction with humidity and precipitation seem to be the most important elements for tourism. To determine the most likely climatic conditions for potential tourists in various parts of the world data for climatic elements must cover long periods of time, preferably 30 years, in order to be reliable. Such multiannual averages for climatic elements are called normals. They are available for most DCs. However, the data for many LDCs are inadequate. Occasionally some climatic data may be difficult to obtain even for DCs like e.g. number of days with persistent fog, thunderstorms, snow cover, frost, etc.

Temperature combined with humidity is the basic component of human climatic comfort. For tourism the averages for daily maxima are more important in most cases than daily averages because during the night most tourists stay indoors. During the daytime when tourists spend time outdoors they would like to enjoy optimum thermal conditions suitable for their particular type of activity. Not favorable are too low temperatures as well as too high, especially when they are combined with high relative humidity which makes the conditions particularly oppressive. Annual average temperatures are not useful for tourism, the best would be the normals for each of the 52 weeks of the year but because of availability of only monthly data we have to use the normals for the maximum daily averages for each of the 12 months. With respect to other climatic elements we shall limit ourselves to a statement that generally tourists like sunshine and try to avoid precipitation and strong winds.

In the post-World War II the biomedical researchers have established certain standards of human comfort in various temperature-humidity conditions. On the basis of research conducted on people in a climatic chamber (climatron) one could establish that the optimum comfortable range of temperature for humans using indoor clothing is 20-27^0 C by 30-70% humidity. The degree of comfort decreases on both sides of this most comfortable conditions. These results in conjunction with the statistics for monthly normals of temperature and humidity have created the potential for evaluation of world climates for tourism. Thus monthly integrated indices of climatic comfort have been computed for 453 localities in the world. In addition to the evaluation of temperature and humidity conditions also weighted values for sunshine, precipitation and wind have been taken into account (Mieczkowski, Z., 1985).

6.3.5.2. Microclimate

The microclimate must be taken into account in planning for second homes, resorts and other recreation facilities. Indeed, recreation siting has to follow a number of rules and caveats in order to avoid costly mistakes. Some of them will be discussed here. Daily temperature ranges should be kept to a minimum by recreation sitings. In the mountains middle slopes have most uniform temperature and best ventilation (day updrafts, evening downdrafts) but difficulties of access and lack of flat lots for construction may constitute problems. Locations at the bottom of mountain valley give better access and flatland for construction of tourist facilities but are climatically unfavorable because of air drainage problems: cool air sinks to lower levels, collects at the valley bottoms frequently causing radiation fog which may be persistent in the absence of wind. High sites in the mountains offer spectacular vistas but allow exposure to cooler temperatures and winds. Best aspects or orientations for down-hill skiing are North and North East (especially by prevailing North Western winds) because of snow protection against melting. Campgrounds are best located with eastern aspects to utilize the warming sun in the morning and provide shade at noon and in the afternoon. Northern aspects for scenic overlooks prevent the tourists from blinding direct sun and provide for better photographing. The best aspects for beaches are South or West in moderate climates to provide maximum sun and warmth. The most frequent mistakes committed by architects in this respect have been high-rise hotels shading beaches (e.g. some hotels at Aruba, Dutch West Indies).

The consideration of exposure to sunlight has to take into account the specifics of the area and the type of recreation. High degree of exposure may be welcomed for beach activities and in the fall, winter and spring but too hot for summer rest and relaxation. Diminished exposure of a ski slope with southern aspect caused by a shadow of another mountain may contribute to better protection of the snow cover. Wind is welcomed for certain recreation activities like sailing and detrimental for swimming or skating. Outdoor speed skating rinks are known to have natural or artificial windbreaks. Lighter accommodations for tourists like tents, cabins, have to avoid placing the entrance against prevailing winds. Fireplaces also should not be located windward of accommodation. Marinas should not be placed at the end of long axis of a lake in order to avoid large waves. Resort microlocation should be in the lee of prevailing winds.

6.3.5.3. Climate and human health

There is a large literature dealing with the health aspects of climate. Almost every handbook on meteorology or climatology contains a chapter devoted to human health. Moreover, interdisciplinary fields of biometeorology and bioclimatology have been developed. Biometeorology researches the impact of various weather patterns on morbidity, mortality, crime, accidents, etc.). Bioclimatology focusses on such topics as acclimatization, human climatic comfort, the impact of climate on health, curative properties of climate (climatotherapy or climotherapy). Bioclimatology is of more interest for tourism than biometeorology. However, these new applied fields of meteorology and climatology do not enjoy equal development and recognition throughout the world. Generally they are more recognized in Europe than in North America. In Germany doctors often recommend their patients vacation in a stimulating (Reizklima) or sparing (mild) climate (Schonklima).

6.3.6. Bioresources

Bioresources for tourism/recreation will be discussed in two parts: vegetation (flora) and wildlife (fauna).

6.3.6.1. Vegetation

Most of the land is covered by some kind of vegetation either natural or cultivated although the line of division between them becomes increasingly blurred in modern times. From all the vegetation covers, forest is generally most preferred by tourists and recreationists. Meadows and pastures take second place, cultivated land the third. Densely populated human settlements take the last place in this evaluation (Kiemstedt, H., 1967:43-46). However, such a general statement requires clarification. Uninterrupted blanket of dense forests loses much of its attraction even if of high quality. Tourists and recreationists love variety of land use. The more variety in vegetation cover the more valuable the landscape. In fact, many researchers found out that there is a definite tendency for recreationists to congregate at the boundaries between various land uses. This tendency is called "edge-effect" and the length of the "edges" is measured quantitatively. The longer the border line of different land uses in an area, the higher the evaluation of such landscape for tourism and recreation (idem.:37-38). Certainly, such a landscape would not get high marks for agricultural use, especially for modern agriculture with large regularly shaped fields most convenient for machinery.

The importance of forests for the survival of humanity is widely recognized. Its influence on micro- and macroclimate is especially important now when enormous quantities of carbon dioxide (CO_2) pour into the atmosphere Forests use them for photosynthesis and production of oxygen thus diminishing the threatening greenhouse effects and other environmental disturbances. Other not directly measurable environmental benefits of forests include protection against flooding, soil erosion (especially watershed protection), landslides, avalanches, provision of habitat for wild animals and areas for recreation. The areas covered by forests in DCs have not changed much recently and even slightly increased in some regions (e.g. Alps) as a result of abandonment of agriculturally marginal lands but their quality has deteriorated due to various factors like mismanagement, acid rain and other pollutants. The Waldsterben (death of forest) has reached alarming levels in most valuable recreationally mountain areas of West Germany, Switzerland and Austria. At the end of 1987 about 60% of forests in West Germany were "sick" (Spiegel 1987, 37). The culprit is a mixture of various air pollutants with the dominant impact of acid rain. In North America the most affected are the forests and lakes of the eastern part of the continent.

The tropical forests of Africa, Asia, Central and South America are even more important environmentally than the forest of the temperate zone, mainly because of their genetic variety. They are also threatened not so much by air pollution but rather by quickly progressing deforestation. They are disappearing at a frightening pace of about 10-12 million hectares annually due mainly to exploding population pressure and resulting expansion of cultivated areas. Exports of cheap beef from cattle raised on cleared forest land constitutes another factor. Tourism and recreation plays a neglibible role in the deforestation process and on balance plays a positive role in the preservation of flora. The culprit is development of other economic sectors: agriculture, ranching, hydrostations etc. Saving of tropical forests is in the interests of entire humanity. It seems, therefore, appropriate that the DCs should pay at least part of the opportunity costs of preserving the tropical forests. A good example of this new approach is the new trend called debt-for-nature swaps: the LDCs are forgiven some of their debts to the DCs in exchange for preserving specific parts of their tropical rain forests (Costa Rica, Bolivia, Brazil).

Forests, and particularly tropical forests, play an important role as gene pools, as regulators of global climate (e.g. moderating the global temperatures), as "factories" of oxygen and reservoirs of moisture. They are returning up to three quarters of precipitation they receive to the atmosphere in the form of evapotranspiration and thus increase the rainfall as source of fresh water invaluable also for tourism development. Deforested areas lose the rain water to runoff which moves relatively far from the site, frequently causing disastrous floods, and is not easily returned to the atmosphere. This role of forest as fresh water supplier is important for tourism because, as indicated above, lack of fresh water constitutes a constraint for tourism development in some areas, particularly in some subtropical island locations.

The forest constitutes a valuable resource for recreation. Perhaps its value is highest in lower mountain areas and lowlands especially if connected with water surfaces. High mountains have large areas without forest cover and also dense monotonous stands of conifers. Similar is the situation on lowlands in higher latitudes.

The therapeutic influence of forests has been discussed by some researchers (Kostrowicki, A. 1970:641). Forests provide shade, fresh air and lower the wind speeds. The best for rest seekers are pine forests with their beneficial influence on breathing system, nerve soothing impact lowering the blood pressure etc. As far as microlocation is concerned a recreation planner should not only take the species composition into account (Mazur, D., 1975) but also the density of stand. e.g. picnic areas have to provide playgrounds. Therefore, they have usually less dense stands than camping areas where high crown closure with eastern aspect opening are preferred. Such areas should be also cleared of noxious weeds such as poison ivy or poison oak. The recreation planner should also make use of trees as protection against solar radiation, as windbreaks, as noise barriers (Heisler, G. 1977).

Another threatened bioresource for tourism and recreation are wetlands - the interface between land and water ecosystems. Wetlands are increasingly appreciated as important habitats for wildlife and ecological stabilizers retaining excess water and trapping silt and organic matter. They modify water chemistry by storing of environmental contaminants. Wetlands also reduce flooding and control erosion. Unfortunately many people think of wetlands as unproductive swamps valuable only if drained and "developed." The result is the widespread destruction and/or degradation of wetlands. The government agencies in North America have undertaken a number of measures to save the wetlands. Some of them involve close cooperation between the U.S. and Canada. The North American Waterfowl Management Program is a good example of such cooperation aimed at preservation of wetland habitats which constitute nesting grounds for waterfowl. This 15 year program started in 1985. At a cost of $ US 1.5 billion it assures US government funding for conservation of wetland nesting grounds in North America, 70% of which are located in Canada. The 1989 bill provides for the inclusion of Mexico into the plan. Tourism will be one of the beneficiaries of these measures because it is vitally interested in retention of wetlands for the consumptive and non-consumptive use of recreationists not to speak about other economic and environmental benefits of their integrity.

6.3.6.2. Wildlife

The human attitude to wildlife underwent an evolution in the recent two hundreds of years. The feudal royalty and nobility of Europe enjoyed recreational hunting in specially created preserves. However, the rest of population saw savagery and danger in the wilderness. Consequently, wildlife in the remaining areas was subject to indiscriminate destruction which continued in the settled or colonized areas overseas not only by way of hunting many wild animals to extinction but also by reducing the wildlife habitat, especially by spread of agriculture. Thus humanity formerly surrounded by wildlife radically reversed the situation and is threatening today the very survival of many species. The "civilizing" of wilderness by changes of land use patterns and overhunting has significantly contributed to a virtual disappearance of wildlife in many regions. For example, such species like zebra, rhinoceros and giraffe are absent in West Africa. The modern ideas of nature conservation have their roots in 18th century philosophy but had not developed until the 19th century and have been put into widespread practice only in the present century.

Although the habitat destruction still goes on there is enough evidence that the new preservation policy and management of wildlife and habitat is bearing fruit in a number of DCs. Many endangered species are making a comeback as a result of protective measures undertaken. However, the situation in the LDCs is, with some exceptions, disastrous both in terms of physical destruction of wildlife and what may prove more ominous on long range, habitat destruction. With respect to physical destruction of animals one has to express dissatisfaction that the tourist industry by its demand for wildlife souvenirs creates part of the market for poachers who are often not only tolerated but also aided and abetted by corrupt local officials. However, it must be emphasized that tourist demand for wildlife souvenirs constitutes only a small part of the total demand for parts of wild animal bodies (e.g. skins, ivory, rhinoceros horns) and for wild animals as pets. This demand is mainly non-tourist. There is a wide-spread international trade in these items. One of multiple sources of such demand is East Asia`s traditional medicine which uses parts of wild animals' bodies as aphrodisiacs or drugs allegedly curing various diseases. Rhino horns are used as dagger handles in Yemen. In order to match the supply with rapidly growing demand there is one only "solution": poaching. Poaching to cater to the needs of East Asia's medicine market is not only rampant in the LDCs but also exerts an increasing toll in North America.

Besides the international trade in wildlife, there is also a domestic demand for wild animals in the LDCs. Indeed, the physical destruction of wildlife is certainly enhanced by traditional attitudes of people in the LDCs, especially in Africa, that any bioresources without direct practical use for humans are useless. The Swaheli word for meat (nyamu) means also animal. Therefore, appropriate educational programs among the local population seem highly recommended.

In terms of habitat protection the acquisition of land to establish various types of nature preserves is the best method. There have been many examples of the costs of acquisition and management being covered by subsequent visitation fees collected from tourists and the creation of jobs in addition to obvious environmental benefits. Nature conservation requires close cooperation of government bodies at all levels including international forum where such organizations like International Union for the Conservation of Nature and Natural Resources, the United Nations Environment Program and the World Wildlife Fund are doing their best, as evidenced by their study published in 1980 entitled "The World Conservation Strategy." Threatened wild animals are also protected by the Convention on International Trade in Endangered Species of Wild Fauna and Flora, known as CITES treaty (joined by almost 100 countries) and a number of national legislations like US Endangered Species Act of 1973.

There are two uses of wildlife for recreation: consumptive (mainly hunting and fishing) and non-consumptive (bird watching, wildlife photography, game viewing, etc.). The consumptive use is older historically and easier to compute in economic terms. The licence fees and taxes on hunting and fishing equipment provide the bulk of money for wildlife and habitat management. Renting out shooting to tourists is a sizable business in Europe (e.g. pheasants in Great Britain). North American natives are selling their hunting quotas for polar bears to rich foreigners. More difficult is to compute the economic impact of non-consumptive use although it is nowadays generally higher than the consumptive use. The consumptive use is regulated by governments and has reached stable or even levels in North America and Europe in recent years. It has a "bad image" in many circles, an image which is based on a misguided emotional assumption that every species of wildlife is endangered and needs complete protection. The reality is different: scientific management based on sustained yield principle allows controlled sport consumptive use for the benefit of whole ecosystems. Wildlife populations must stay balanced with their food supply in order to maintain the animals and their habitat in good condition. Starvation and disease result of overpopulation and are cruel and wasteful alternatives to hunting. Also the balance between prey and predator should be maintained. The controversy between anti-hunters and hunters, vast majority of whom are recreational hunters, will never end. In recent years the anti-hunting camp has been strengthened by various sorts of militant animals rights movements e.g. Hunt Saboteurs Association in Great Britain which also fights against all fishing. Many of these organizations are against killing and consumption of any animals by people and also use of their furs, skins, etc. Some are using violence to promote their cause. Most of them are well funded which sounds as an irony in view of the fact that the "Digit Fund" for the preservation of Rwanda's mountain gorillas is broke.

Another type of conflict arises between consumptive use of wildlife for recreation and use for food by North American natives. Locally it leads to bitter disputes over native hunting and fishing rights established by 19th century treaties especially in connection with native land claims which are perceived by the tourist industry as a threat to its livelihood. Another example of conflict is the competition between commercial and sports fishing. In all conflict cases the representatives of the tourist industry argue that recreational consumptive use of wildlife is economically more advantageous than any other consumptive use. Indeed, recreational fishing and hunting is a multimillion industry. Its importance differs spatially e.g. it is very significant in Western Canada, Northern Ontario and Minnesota. It is a source of foreign currency for Canada. It creates employment for guides, outfitters, accommodation facilities, food outlets etc. Benson's 1961 study "Fishing and hunting in Canada" for the Canadian Wildlife Service established that 12.6% of the Canadian population over the age of fourteen were either hunters or anglers or both. Part of them travel within Canada as tourists. There is also a strong foreign tourist component in Canada's consumptive use of wildlife (US, Western Europe).

The economic and ecological benefits of consumptive uses of wildlife does not mean that it should go on uncontrolled. Proof for this is the relentless extinction of many species. Clearly, all consumptive use of wild animals placed on the endangered list should be forbidden. Such species like panda bears in China or mountain gorillas in central Africa are already for some time under protection but again in these as in majority of other cases habitat destruction seems to be the main obstacle in protecting these animals, still worse than poaching which is, of course, very harmful and illegal. In some countries poachers are shot on sight (Zimbabwe, Kenya). The enforcement of laws protecting a number of species will be impeded until there is a market for ivory, rhino horns, furs and skins of rare animals, rare animals as pets etc. In North America some endangered species (e.g. Florida panthers, mountain lions) are threatened by native hunters which indicates the necessity of the final regulation of the economic and legal issues with respect to aboriginal Americans.

In majority of countries the consumtive use of wildlife is regulated seasonally or forbidden with respect to certain endangered species which are on the endangered list. Rare are cases of total outright ban on all hunting like in Kenya where it was outlawed in 1977. This meant the end of tourist hunting called safari hunting. Such radical solutions mostly do not lead to desired effects. Tourist hunting in Africa is strictly regulated and economically extremely profitable. It is an established fact that the presence of hunters scares poachers off. Tanzania, Zambia, Central African Republic and some other countries allow still a very limited big game hunting for rich tourists. There are many other examples of ill-conceived conservation measures leading to undesirable results. What needs to be protected, e.g. rhinos, mountain gorillas, whales, condors and whooping cranes should be protected. However, protection of some seals, which are by far not endangered, leads to their proliferation harmful to the fishery industry. Grey seals are killed in Canada not only for their skins and meat but because they deplete fish stocks and spread parasites (New York Times 1-1-1988). Even some formerly endangered species rebounded so much that some "harvesting" may be locally considered: the proliferating alligators constitute a nuisance in Florida. There are also too many common leopards in southern and eastern Africa. Protection of elephants in Tsavo National Park in Kenya lead in the past to their periodic mass starvation and destruction of vegetation. In 1980s the situtation has changed: poaching for ivory became so bad that total protection is essential. The sea otter destroys the crab industry in Alaska Bay, moose in Sweden block airports, train tracks and destroy young trees. There are too many kangaroos in Australia destroying crops in the fields. The lack of natural predators is causing some of the protected species to proliferate to such an extent that they are not threatened by extinction anymore. Moreover, they expand often beyond their food basis necessitating human control measures to restore the ecological balance. In such cases regulated hunting, mainly tourist (sport) hunting, is the appropriate wildlife management technique, applied for a limited time.

It is interesting to note that in historical perspective, in the DCs, the share of consumptive use of wildlife resources is decreasing to the benefit of non-consumptive use. Initially this was only a relative drop accompanied by absolute growth of consumptive use. However, in recent years many sources report an absolute decrease of hunting (Winnipeg Free Press 31-10-1987). The fast growing non-consumptive use of wildlife surpasses that of consumptive use not only in economic terms and also in terms of user satisfaction (Vaske, J., 1982). It occurs almost anywhere (e.g. bird watching), in zoos, commercial game parks but especially in government owned and managed nature parks of all types. Game farms are places for both consumptive and non-consumptive use of various wild animals (crocodiles, ostriches, elks, elands, alpacas etc.). Also exotic tamed animals like camels in the Middle East and elephants in South and Southeast Asia constitute a tourist attraction for viewing, photographing and riding. Besides its economic value the nonconsumptive use of wildlife has also a significant social value which is more elusive though. One of these social values is the appreciation and understanding of wildlife by broad masses of population. The public learns to regard even such "ferocious" animals like wolfe, tiger or sharks as simply parts of ecosystems indifferent and mostly not dangerous to humans. Similarly like with vegetation cover, the tourists appreciate more variety of animals in the ecosystems than monotony although this does not exclude special interest tours destined to view only one species like e.g. whale watching on Pacific coast of North America or in Churchill, Manitoba, Canada. Typically, however, whole ecosystems focus the interest of tourists, the study of wildlife in Galapagos Islands, Ecuador being an outstanding example.

Non-consumptive use of bioresources by tourists has a powerful economic clout in the preservation of species and their habitats. In fact, one could generalize that it is due mainly to tourism rather than to the noble efforts of various conservation agencies that many species are saved from extinction. The reason is simple: tourism supplies financial means for conservation, and the international agencies are chronically short of money. Thus e.g. the delicate balance achieved in Central Africa's Vigunga Mountains, straddling Rwanda, Uganda and Zaire in the preservation of nearly extinct mountain gorillas is the result of money brought by 20,000 tourists a year. These tourists view and photograph a few habituated gorilla families in the Volcanoes National Park, Rwanda. The locals say: "tourism equals gorillas." Otherwise there would be no gorillas in this densely populated country (over 250 people per square kilometer) with exploding population (natural increase 3.6% annually with average 8.6 chidren per woman in her life time) where every hectar of arable land is an invaluable asset for starving people. The tourists indirectly pay for preservation of the gorilla habitat. The locals are proud of "their" gorillas which may become Rwanda's national symbol. Another example of non-consumptive use of wildlife as the main source of foreign currency for a whole country is Kenya. This is the reason why the Kenyan government embarked on such a radical policy against poaching, particularily against elephant poaching. The extinction of elephants would mean a threat to the whole ecosystem because the elephants transform the habitats by foraging. They create a mixture of savanna woodlands with grassland. These diverse habitats are used by a diverse array of other plant-eating species and consequently their predators. Thus practically most of Kenyan wildlife habitats depends on elephants creating the desired balance between wildlife and habitat.

The use of bioresources (both consumptive and non-consumptive) for tourism creates a multitude of management problems which have to be addressed by appropriate authorities and organizations. Without going into details one could indicate a couple of areas requiring constant vigilance in order that scientific management techniques may be applied if necessary: relationship between prey and predator and relationships between wildlife and its food base. Also important are the issues of resource use conflicts which are partially related to tourism:

1) When the recreational land use overlaps with wildlife habitats undesirable contacts between tourists and animals like bears, bisons, elephants, lions, buffaloes etc. may occur.

2) The life of motorized tourists and other travellers may be endangered within wilderness areas when wildlife crosses the highways. Fences or warning systems like Deer Alert, installed in cars, or simple advice to be more alert and drive slower may be considered.

3) Incompatibility of wildlife habitat with other land uses e.g. crop depredation by birds (solution - lure crops) bird hazards to aviation, damages caused by beavers, ungulates and predators (e.g. pumas, canines) to agriculture and livestock. Animal depradation constitutes a severe problem at the fringe of many National Parks.

The wildlife as recreation resource is appreciated also beyond the wilderness areas and national parks - in zoos and, more recently, in super-zoos. The management of wildlife in captivity gains in sophistication. There are a number of outstanding zoos in the world like San Diego, California, San Antonio, Texas, Toronto, Ontario, Bronx, N.Y., Sydney, Australia, Amsterdam, Netherlands, Hamburg, West Germany, West Berlin, Dresden, East Germany. The newest developments are the super-zoos under various names like "wilderness park" e.g. Apple Valley near Minneapolis, Minnesota where wild animals roam in semi-natural environment.

A separate category of wildlife resources which have earned millions of dollars for the tourist industry is a resource which - there is strong suspicion - does not exist at all. The Loch Ness Monster in Scotland is a classic example. The monster has already acquired a number of "brothers" all over the world. Although much less successful financially these monsters range from Ogopogo in the Okanagan Lake, British Columbia, Champa in the Lake Champlain, Vermont to a lake monster in Yakutia in the USSR. In the mountains and other wilderness areas of the world many costly expeditions have been chasing various varieties of "abominable snowmen" an over 2 meters tall ape-like creature with various local names like "Yeti" in Asia or Sasquatch (Big Foot) in North America. The results so far: numerous "sightings," absolutely no concrete scientific proofs, but what seems to be the name of the game is income to the tourist industry which uses this "wildlife resource" quite effectively. The existence of International Association of Criptozoology organizing expeditions to Central Africa in search of live dinosaurs gives a boost to these endeavours to find more about these "hidden" bioresources - mythical monsters, "prehistoric" serpents, mysterious primates, etc.

6.4. NATURAL RESOURCES: ENVIRONMENTAL SYSTEMS

The systematic (topical) discussion of natural resources for tourism and recreation is certainly useful for analytical purposes. However, one has to stress that in the real world natural resources do not occur separately. They occur in landscapes - units of environmental systems, in combinations, complexes, or integrations of elements discussed systematically above. Only the regional interactions of these resource elements, arranged spatially in a specific way, constitute a proper basis for research of recreation potential, and also for concrete planning, development and management of natural resources of a given area.

The following discussion of environmental systems of natural resources is based on the classification proposed by Pierre Defert in 1966. (Defert, P., 1966). Defert distinguishes three basic types of "areal integrations of natural resources" (which will be called "systems" here) for tourism which, adjusted for our purposes, will be discussed under following designations:

1) Seashore system
2) Mountain system
3) Inland rural system on flatland

The three systems represent a transition from the seashore system which is characterized by spatial and temporal concentration of tourism in the narrow coastal zone to the inland system on flatland where tourism is much more dispersed.

6.4.1. The seashore system

The seashore system represents an interface between land and sea, a junction of landscape and seascape. Therefore, it constitutes a special attraction in conformity with the "edge effect" idea. The natural resources for recreation are arranged in the littoral zone in a linear fashion, linearly following the shoreline with varying width (depth). This coastal area could be subdivided into four zones or sectors (Gunn, C., 1971: 6-7).
1. Marine zone 2. Beach 3. Shoreland 4. Hinterland

Marine zone

The marine zone covers roughly the area of continental shelf and does not reach deeper than 200 m. (continental shelf). It starts at the continental slope and reaches up to the limits of the beach. Recreational activities taking place in this zone encompass deep sea sport fishing, sailing, boating. The requirement for successful recreation is abundance of marine bioresources, especially fish (but not sharks) and no pollution.

6.4.1.1. Beach

Beach is the most important of the four zones. It is part water and part land. The seaward limits of the beach cannot be precisely determined because they depend on the extent of beach activities, mainly swimming. They normally reach beyond the low tide line. Thus the beach extends from its ill defined water limits with the marine zone up to a much more conspicuous line - the natural dike built from materials accumulated by the waves. In seaside resorts this natural dike is artificially reinforced and frequently accommodates the dike road or promenade. Beach is composed from products of rock decomposition of various texture. The best for recreation are fine textured pure sands in light colors enjoyed for play and building wind breaks. Black sand beaches (some of them of basaltic provenience) which may be encountered e.g. on southern parts of the Dalmatian coast (Petrovac) or at the Big Island of Hawaii absorb too much of sun heat and burn the skins of recreationists. Beaches covered with rocks, pebbles, gravel and other coarse textured materials are inferior for recreation. Examples: parts of French Riveria,

Crimea, Eastern Black Sea. They often occur in conjunction with a bluff type of shoreline which may offer some advantages, mainly vistas, excursion areas, scenic highways ("en corniche"). Nevertheless rocky shores do not induce large tourist developments. Marshy shorelines are to be avoided completely for tourist locations. However, their ecological value becomes now increasingly appreciated. Tourism should not invade these areas but may use them as bioresource following strict environmental guidelines. However, marshes do not constitute suitable tourist locations mainly because of wetness and accumulated materials like mud and clay, present also in river mouth sediments. Also various insects constitute a nuisance.

The beach is mainly composed of two parts: the foreshore (strand) and the backshore (berm). Strand is the intertidal zone between low and high tide. There are various types of strands depending on their slope and tidal range. The tidal range is the vertical range between low and high tide. It amounts on the ocean shores to between 4 and 14 meters and in mediterranean seas like Baltic or Mediterranean it reaches from few centimeters to a maximum of 40 centimeters which renders the maximum horizontal width (intertidal zone) of about 1 meter. The steeper the slope of the strand the narrower it becomes. Too steep strand constitutes a danger for nonswimmers and children, especially in combination with high tidal range. Steep strand also increases the danger of coastal erosion which may contribute to the destruction of vacation homes and other recreational facilities because it does not provide enough braking power against waves and ocean currents. Shallow bottoms provide such a power.

The ocean tides, generally speaking, are beneficial for recreation. They help the self-purifying process of water and make the beaches healthier. The recreationists enjoy the rolling wave crests of high tide, if not excessively large After the low tide bares the strand leaving, fish, shell fish, shells and other materials, they take pleasure in collecting the "fruits of the sea" (frutti di mare). The strand constitutes also an excellent place for play. If the sand is hard enough car racing takes place on it. Of course, this form of recreation at the beach is rightly criticized by environmentalists. The optimal width of strand is 200-500 meters. Wider strands are inconvenient because one has to walk too far to take a swim during low tide.

The backshore of the beach, called berm, is the most important part of the beach because here do the recreationists spend most of their time. It consists of materials deposited by the waves above the line of high tide and becomes inundated only during storms. Berm is commonly called beach. Although this is inexact, this meaning will be used in the following discussion.

Not only the width but also the length of the beach is important for recreation because it is a capacity indicator. A large beach resort should have at least 1 km length of beach. An example of long gorgeous beaches of 25 km lengths are resorts on Italian Adriatic coast. The Colva beach in Goa, India is almost 40 kilometers long and over 50 meters wide. However, some resorts may do fine with several smaller bay beaches (e.g. Costa Brava in Spain). Such beaches have lower capacity but offer more intime environment and are less exposed to wind.

The recreationists spent most of the day at the beach engaging in a wide variety of activities in this land-sea environment: sunning, building castles of sand, walking, beachcombing, swimming, wading, fishing, spearfishing, snorkeling, scuba diving, surfing, wind surfing, boating, etc. Important is also aesthetic appreciation of seascape, watching long vistas, observing fish and sea mammals.

6.4.1.2. Shoreland

The shoreland, starting seawards with the natural or artificial dike, is the place where tourists spend evenings and nights. This is the area where tourist facilities are located: accommodation, food and shopping services. It is visually connected with the sea, most of the time (except during heavy storms) but it does not have any physical contact with the sea. Winds bringing sand from the beach sometimes form sand dunes which constitute an excellent recreation environment requiring, however, proper management. Dunes protect the shoreland from erosion when storm waves encroach on the shoreline. They also provide wind protection. The dunes should be stabilized by vegetation, some of it like pine trees has especially positive impact on health. Sometimes planners complain about dunes blocking internal waters or isolating brackish lagoons and thus constraining tourist development but such situations may actually enhance the environment.

There are cases that inadequacies of the beach have to be "made up" in shoreland zone by placing beachlike facilities there. As an example may serve swimming pools constructed in some resorts as substitutes or supplements for beach activities. The reasons may be too cold sea water, sea pollution, sharks, lack of access to sea (distance, bluff type of seashore) etc.

6.4.1.3. Hinterland

Hinterland is the last zone located most landwards. Here lives the local population and supporting services are located. It does constitute a separate planning zone. In some cases hinterlands are followed landwards by mountain backdrops which usually enhance the beauty of landscape but may impede the development of hinterland. Such mountain backgrounds constitute scenic assets for coasts all over the world. Good examples are the Dinaric Alps at the Dalmatian coast in Yugoslavia and the Crimean Mountains at the Crimean coast in the USSR and sections of the California coast.

6.4.1.4. Morphological classifications of the seashore system

In addition to the characteristic features of the four component parts of the seashore system of natural resources also various types of physical (morphological) classifications of the seashore system should be taken into account for recreation/tourism planning, development and management.

The 1962 historical Outdoor Recreation Resources Review Commission report brought following morphological classification of the shoreline for the purposes of recreation. Three types of shoreline are distinguished (ORRRC,1962 No. 4:12-31).

1) BEACH - a wide expanse of sand or other beach material lying at the waterline and of sufficient extent to permit its development as a facility without important encroachment on the upland.

2) BLUFF Shoreline (falaise) - a bank, bluff, or cliff, immediately landward of a relatively narrow beach (if any) and varying in height from several meters up to mountainous elevations. Bluffs may be composed of either loose or solid material - from sand to granite. Cliff is rocky and of considerable height. The bluff shore provides a marine environment, scenic values of a high order, and frequently the isolation, many outdoor recreation seekers prize so highly. The drawback is frequent lack of suitable beaches or problems with access to the beaches, often necessitating construction of stairs, lifts etc.

3) MARSH - a marsh shore indicates the existence of tidal or nontidal marsh as the principal shore feature. The marsh shore may be both the most ignored and the most promising type of shoreline for future recreation use, as explained above.

Another morphological classification has been developed by Pierre Defert who distinguishes between four types of seashores for the purposes of seaside resort development (Defert, P., 1966:7-77).

1. Oceanic: continuous and linear.
2. Oceanic: discontinuous and concentrated
3. Mediterranean: continuous and linear
4. Mediterranean: discontinuous and concentrated

1. Oceanic: continuous and linear

Here the sandy beaches extend at considerable distances sometimes for many kilometers. Tourism facilities follow the shoreline.

2. Oceanic: discontinuous and concentrated

This type develops when as a result of various resistance of coastal materials to the sea erosion, the coastline is not linear but alternates in bays, promontories, peninsulas etc. Usually the best beaches are located in the bays, the promontories and peninsulas playing the role of wind breaks.

3. Mediterranean: continuous and linear

This type is especially suitable for tourist development because of the absence or very small tides, (outside the Mediterranean area represented e.g. in Florida). The width of the beaches is stable, the length sometimes very considerable (e.g. about 25 km at Rimini, Italy). There are excellent sites for seaside resort development often with the beach several hundred meters wide and in a gentle slope. Lidos belong also to this type.

4. Mediterranean: discontinuous and concentrated

With respect to the fourth type there may be two possibilities. First is a Mediterranean "garland" littoral with wide open bays bordered by promontories or capes. The beaches in the bay are generally good. In the background there are hills or mountains with distinct elevated terraces conducive for resort development (e.g. Taormina, Sicily). Sometimes this may pose some transportation problems (cable cars, funiculars, stairs, trails etc.). These elevations constitute also excellent shelter against wind.

The second variety is the mediterranean littoral divided into small coves with tiny beaches. This situation favors a dispersal of tourist facilities (e.g. Costa Brava, Spain).

6.4.1.5. The climate of the seashore

Microclimatic considerations are an indispensable part of discussion of seashore ecosystem. There is no question that the Mediterranean summer is most conducive for seashore recreation. It guarantees reliable sunshine, warm temperatures, calm winds and absence of storms. It is also relatively cool - the excessive daytime temperatures are moderated by sea breezes resulting from convection of heated air over land and the movement of cool air flowing in from the sea to replace the uplifted air. THe daytime sea breeze pleasantly ventilates the beach by its cool moist air. The influence of the sea breeze normally does not exceed few kilometers inland, however, examples of deeper penetration abound. The greater the temperature contrast between sea and land the farther the range of sea breeze penetration which reaches 60-80 kilometers in Egypt. Understandably such deep penetration is connected with strong winds which e.g. reach gale force at Valparaiso, Chile. Such situations are also encountered in Western Australia and do not contribute to enjoyable recreation at the seashore.

The sea breeze attains its maximum at about 3 p.m. and then settles down gradually to die out completely about the sunset to make it a more serene and relaxing view. The nights are clear and calm especially in their beginnings with breezes flowing from the land to the water as the air over the land cools quicker.

Thus the breezes from land (night) and sea (day) tend to decrease the effects of diurnal temperature and humidity variations. Without the breeze evenings and nights would be too humid and afternoons too hot and dry. These pleasant climatic characteristics contribute to the comfort of tourists who spend day at the beach and evenings at the seashore promenades.

6.4.1.6. Planning and management problems of the coastal ecosystems

There are a number of problems at the seashore which require attention of tourism and recreation planners as well as politicians. These problems are connected with physical changes of the shoreline not only in the beach zone but frequently reaching into the shoreland. There are cases in the Mediterranean area that resorts have to relocate as a result of sea currents expanding the lidos (sand bars) and closing the access from the resort beach to the open sea. As resort location on a closed lagoon is not suitable, a new resort is created at the lido and the old one has to change to another function.

The most serious problems, however, are caused by wave erosion. Some shorelines, like the US Atlantic seaboard, suffer very much from this phenomenon. The short steep winter waves erode the beach and wash it away and make the berm lower, steeper and harder than in summer. Severe winter storms destroy not only the berm and devoid it of protective sand but wash away scores of shoreland vacation homes built too close to the beach. Frequently also the protective sand dunes suffer from wave erosion.

The problem of wave erosion in coastal zones seems to be worldwide and relentlessly increasing in recent time. Alarming reports are coming not only from vacation areas like US Atlantic and Pacific seaboards or North Sea (e.g. the island Sylt, West Germany) but also from areas where is hardly any sea-related recreation (Bangladesh). The long range climatic change, suspected to be caused by humans, ("greenhouse effect"), results in gradual melting of mid-latitude mountain glaciers and perhaps even polar glaciers. According to various sources, the average increase of about one millimeter annually in the level of the world ocean during the last century is bound to increase to several millimeters annually making the problem worse in the future.

In some cases the reasons for coastal erosion are rather local than global. Take the example of the Black Sea. The Bulgarian and Romanian coasts located in the lee of prevailing Western and Southwestern winds are not being eroded. However, the Soviet Coast, particularly that of Caucasus suffer every fall under the onslaught of storm waves driven by winds up to 8 or even 9 force on the Baufort scale. The destruction of beaches, coastal forests and roads, damage to hotel buildings has been especially severe recently as a result of diminished quantity of sedimentary material brought to the seashore by mountain rivers and creeks. The reasons for this were the engineering works, mainly hydrodams constructed on these water courses. Especially threatened by the sea is the Georgian resort Pitsunda. Among other expensive measures undertaken is trucking of at least 0.5 million tons of gravel and sand to the beaches. Another example of tourist beaches and facilities endangered by shore erosion are the coastal regions of Ghana, Togo and Benin. The construction of the Akosombo dam on the river Volta in Ghana in 1960s deprived the coast of material (mainly sand) formerly deposited at the shore of the Gulf of Guinea. As a result the sea has destroyed sections of coastal road and several villages in these countries. Hotel "Tropicana" situated 10 km east of Lome, Togo is one of the facilities threatened by the waves. The measures undertaken consist mainly of costly engineering works aimed at accumulation of sand in the strand zone (Jeune Afrique 12-8-1987).

Examples of local measures only will be discussed here because global measures aimed at the reduction of "greenhouse effect" are beyond the direct interest of tourism. The best protection against wave erosion is plenty of sand in the marine and especially in the strand zone. In such cases when the strand and the immediately adjacent part of the marine zone are sandy and shallow, the waves are being slowed down by the friction on the bottom and are less likely to cause damage. Unfortunately, however, at the US Atlantic coast there are strong ocean currents parallel to the shoreline which carry the sand away (normally northward) thus exposing large sections of the shore to wave erosion.

The measures undertaken against shore erosion are costly and by far not always successful. They include various concrete walls, slabs of various materials (concrete, rocks). Also applied is bringing sand to the affected area by various means, mainly trucks and pipelines. Most successful are measures undertaken to keep the sand in place and to accumulate additional sand by construction of jetties from concrete or rock, using dams or curtains of various materials including sandbags, plastics, car tires etc. The accumulated material decreases the impact of the waves. Also important is stabilization of dunes by means of fences and vegetation.

The 1972 US Coastal Zone Management Act (CZMA) provides the organizational framework to save the coastline by regulating the land use in fragile coastal areas, providing for funds to acquire environmentally endangered lands including acquisition, development and operation of estuarine sanctuaries for scientific, educational and other uses. Also sections of the Nationwide Outdoor Recreation Plan 1973 provide for environmental protection of shorelands. Special protection measures are being undertaken within the boundaries of National Seashores and National Recreation Areas encompassing coastlands. Also marine parks spreading all over the world since 1960s serve the same purpose.

Unfortunately the measures undertaken by US federal government proved to be only a partial success: not all the coastal states took up the initiative in implementation of the guidelines and utilization of funds provided. Still many of environmentally detrimental phenomena are taking place on US coasts: overdevelopment, folly of building on protective dunes and barrier islands, destruction of dunes by four-wheel-drive vehicles and dune buggies, all forms of pollution including pollution caused by off-shore oil drilling, destruction of coastal wetlands by development. Part of the problems in implementation of anti-erosion measures is that about 90% of the US coast (excluding Alaska) is in private ownership and only 5% is accessible for public recreation.

Many issues are still controversial. There is a wide-spread criticism of government disaster relief and low interest loans to rebuild vacation homes damaged by the sea. The argument is that vacation homes should not be permitted in danger zones. Another issue is public access to the beaches. At this time, one could generalize that the courts tend to favor public access at the expense of private owners.

The problems of recreational use of shorelines are by far not limited to the US Atlantic and Gulf of Mexico seaboard. The oceanic waters are on the offensive also in California, and at the North Sea Coast. The shore erosion is active even at the Great Lakes and other inland lakes like Lake Winnipeg. Factors contributing to this worldwide devastation have been overdevelopment and pollution. Bad examples are the Mediterranean and Baltic coastlands. Only relatively recently steps have been taken to stop this environmental outrage. But even the ocean shorelines are deteriorating. An example of this is the degradation of coral reefs (Caribbean, Australian Great Barrier Reef) to a large extent as a result of human (including recreational) impact. Reefs not only provide shelter and food for fish, lobsters and other marine life but also protect the shorelines from waves.

An example of efforts to preserve the best of British coastline provide the endeavours of Countryside Commission aimed at designation of "heritage coasts" in England and Wales. As a result of these actions the National Trust takes custody of especially environmentally valuable sections of British coasts to save them from ruin observed at many Mediterranean coasts. Indeed, "the campaign to save 900 miles of Britain's best coastline from developers has been successful. Between 1965 and 1989 the National Trust has acquired 500 miles of the designated coastline, including the famous Dover white cliffs. In spring 1989 the Trust started the campaign to acquire the remaining 400 miles" (Economist 15-4-1989).

6.4.2. The mountain system

The most important physical factor limiting tourist development within this system is accessibility. This means that the permeability of mountains by tourists is limited mainly as a result of gravity i.e. there are problems with vertical movements. The linear or horizontal accessibility within the mountain valley is generally good for modern means of surface transportation. Because of the accessibility and the availability of land for construction of tourist facilities most of the mountain resorts are located in valleys. The problems of accessibility start when the traffic has to move from valley to valley through mountain passes or when tourists want to reach the mountains from the valley bottom. Here the forces of gravity prevent vertical movements. Only alpinists have almost total access in the mountains, sick or old persons if left on their own, are very much limited not only by horizontal but especially by vertical friction of distance. Therefore, for the convenience of tourists new means of mountain transportation have been developed since the 19th century. First were various forms of surface transportation like special mountain rail roads capable of climbing steep slopes e.g. the Rigibahn in Switzerland. Later came the funicular. The 20th century brought the cable car, chairlift, T-bar, Poma-lift etc. Finally, in recent years various forms of VTOL aircraft (Vertical Take-off and Landing), the helicopter most important of them, have been used to improve the accessibility in the mountain environment.

The physical impact of altitude is not limited to problems of access and earth and snow movements. It also necessitates the adaptation of human body which requires some time, and may cause mountain sickness. One also should keep in mind that with increasing altitude the temperature drops about 1^0 C for about every 200 m (variations depend from air humidity). This is a blessing for tourists visiting tropical highlands but an obstacle for alpinists. Moreover, the daily temperature extremes increase with elevation because heat is quickly absorbed and dissipated. Also the weather conditions in the mountains are less stable than in the lowlands, it is generally more cloudy and the precipitation is higher. Therefore, such locations for tourist facilities should be selected which minimize precipitation. In this connection preferred are locations in the lee of a mountain range to windward locations where the precipitation is brought by prevailing winds.

Certain microclimatic features of the mountains should be also taken into account while planning for recreation e.g. the system of local winds should be understood and utilized for location of tourist facilities, summer homes etc. On warm days the winds tend to blow up the slopes during the day time as a result of warm air convection. During the night the cool catabatic winds are flowing downwards carrying cleaner air towards the valley floors where the cool air accumulates. This pattern of winds indicates where are the favorable locations. The area of mountain peaks or plateaus at the one extreme and the valley floor at the other tend to be the coldest places. The warmest and best ventilated places are to be found at intermediate locations i.e. at the higher parts of the valley slopes where good insolation may prove as an additional warming factor. This warm slope (thermal belt) indicates the best tourist location with longest frost-free season. The availability of flat terrain for construction usually may constitute a problem but in many mountain environments glacial terraces (e.g. Mürren, Switzerland) provide for adequate supply of land for construction of tourist facilities.

The intermediate location is preferred also for other reasons. There is less likelihood of adiabatic fog (ascending air cooled by extension) developing on the slopes. This type of fog develops mostly on higher elevations. The air on the slopes tends to be clearer and less polluted then on the valley bottom. The polluted air (dust, smoke) tends to collect at the valley floor as a result of temperature inversion occurring with nocturnal radiation. The valleys also tend to develop more frequently radiation fog, normally up to 50 m thick, during the night and early day hours. The reason for the adiabatic fog is the accumulation of cold air which contributes to lowering of the ground temperature of the valley. Fog forms especially after the sun starts to warm up the air following the sunrise and the ground surface still remains cool. This process of fog formation is reinforced by the condensation nuclei (e.g. particles of dust).

The vegetation plays an important role to reduce the speed of the wind (the rougher the surface the thicker the film of air which clings to the ground) and thus reduce the wind-chill effect on the horizontal plane.

Mountain areas are being used for recreation both in summer and in winter. The use in summer is more varied in terms of activities and as a consequence the physical requirements differ: e.g. the requirements for mountain climbing, hiking, lake fishing, golfing certainly are not the same. The winter use of mountain areas for recreation is less structured and this makes the analysis simpler: the mountains are mostly used for skiing. One could formulate the physical requirements for a vacation oriented ski resort as follows:

1) Proper aspect (orientation) to retain snow and protect from wind. In North America and Europe the ideal aspect of ski slopes is North, and in case of prevailing Northwestern winds the best aspect for skiing is between Northeast and almost East. Ideally the ski resort should be located roughly to the North of a Northeast facing ski mountain. Southern aspects of ski slopes should be avoided because of melting snow during the day which refreezes during the night creating icy conditions for morning skiers. To complete the ideal requirements there should be a mountain on the luv (windward side) of prevailing winds for protection. To maximize the sun exposure there should be no large mountain casting shade at the resort. A lake would be desirable for skating and ice sailing and for water-oriented summer recreation. Also a golf course should contribute to the extention of the season. Of course, it is difficult to expect such ideal physical conditions in the real world, nevertheless the knowledge of advantages and disadvantages of various locations may constitute a useful planning tool.
2) Vertical drop (denivelation) of at least 400 m in US Northeast and 800-1000 m in the West. Size of skiing area at least 10-20 km^2 accommodating 3000-5000 skiers. Minimal size of single ski slope 1 km^2 (100 ha). Advisable to combine neighboring areas to form a ski circuit (e.g. "Ski Zircus" of Lech and Zürs area in Austria).
3) The best elevations in the Alps and in the North American West are from 1000 meters on slopes with North and Northeast aspects and 1500 meters for West and South aspects. The highest recommended elevations for skiing are 1800-2000 meters. Higher may be too cold and also smaller amounts of oxygen may cause some adjustment problems for the bodies of skiers who tire quickly in such conditions. However, in order to enjoy skiing at lower latitudes, including the subtropical zone, such high altitudes may be the necessary prerequisites: there are resorts in the Andes of almost 5000 meters elevation. Well known ski resorts at high elevations are e.g. Portillo, Chile - 2,800 meters and Oukaimeden (Morocco).
4) Snowfall for skiing should be abundant (about 5 meters annually) and coming at rather frequent intervals in order to support commercial skiing. Generally speaking eastern locations in North America should have an at least 100-day season to be commercially viable, in the West the minimum is about 120 days. Some resorts here are open for 130-150 days annually. With the wider application of snow-making equipment many ski resorts have extended their seasons substantially although snow-making requires temperatures below freezing and plenty of water.

5) A ski resort should have a variety of avalanche-safe slopes to accommodate skiers at any level of proficiency. The rules of thumb are for beginners up to 25% slope, for intermediate skiers 25-50%, for experts 50-70%. Some slopes require grooming with special machinery, including snowmaking, shaping, mulching, seeding etc.
6) Sunshine should not be excluded from the slopes entirely because it makes skiing more enjoyable although it also shortens the season. Ideally, the days should be clear with snow falling at night.
7) An adequate source of water for the tourist facilitates.
8) Adequate base terrain to accommodate the lift terminals, base buildings, parking, lodges and other amenities supporting the mountain.

All these physical requirements have to be taken into account by ski resort planners in order to ensure their commercial viability from which depends the rentability of capital investment in the ski resort.

Like the seashore resource systems the mountain systems have also their environmental problems. Although of different nature they are also partially associated with tourism overdevelopment. Tourism which has saved large mountain areas from socio-economic decay, when overdeveloped, has inflicted substantial environmental damage to the mountain resource systems. The damage is caused by buildings, roads, soil destruction and compaction and deforestation which increase the runoff and thus the danger of flooding. Wet soils deprived of binding vegetation tend to slide. Catastrophic landslides and flooding in Italian Alps and Swiss Cantons Tessino and Graubunden in summer 1987 are good examples of such environmental neglect leading to loss of life and millions of dollars in lost property. Among the most destructive land use changes one has to single out the deforestation. Additionally, the trees suffer from acid rain which is, of course, not tourism-related. Especially damaging is the loss of evergreens which are valuable because they break the impact of rain with their needles and store the moisture. Deforestation is to a large extent responsible for the landslides in summer and avalanches in winter. The dangers increase as more tourist facilities move into previously inaccessible high mountains closer to the recreation areas in order to minimize the costs and time for commuting. In the post-World War II much research has been done on landslides and avalanches and interested readers may find ample supply of literature (Brunsden, D., 1971).

The most severe environmental impact of tourism occurs in the narrow-valleys which serve as transportation corridors. The physical, environmental and socio-economic carrying capacities are under pressure all year almost without off-season respite as it is the case with the seashore system. The pollution of mountain valleys is especially severe in winter due to the combining effects of temperature inversion and heavy motorized traffic. In more recent time, with the advancement of vertical transportation technology also the higher altitudes are coming under environmental pressure of tourism development. New mountain resorts are being developed in altitudes which were not accessible before World War II. Characteristic in this respect are French Alpine resort complexes with totally integrated facilities (so called "total destinations." A good example is LaPlagne with 33000 beds, 100 lifts, 107 runs, processing 33000 skiers an hour (Winnipeg Free Press 10-9-1988). Their environmental impact is more significant than in the Austrian and Swiss Alps where small hotels prevail and tourism development is grafted into the existing settlements. One has to stress, however, that much of environmental deterioration in the mountain areas of the world is the result of global pollution (Sustainable Development 1988, August; Council of Europe 1988).

6.4.3. The inland rural system on flatland

This system of natural resources for recreation includes lands located outside the seashore and mountain environment. The tourism/recreation locations in this system focus on water-oriented activities associated with rivers, lakes and other inland water bodies. Physical requirements are few: the area should be well drained (not swampy or periodically inundated), geologically stable, some relief desirable, abundance of vegetation (mainly forests) is a must.

This third system has a number of characteristic features as compared with the former two:

1) Historically these are the oldest locations where people enjoyed their vacations in extra-urban setting. They developed in times when people feared the mountains and did not swim in the sea.

2) Good locations for tourism/recreation are less concentrated and more dispersed because the space in this system is more available and more homogenous and there are relatively few physical obstacles.

3) As a result of more abundant availability of land for recreation the land prices are generally more moderate especially on lands of lower agricultural quality (marshes should be never used for development of recreation facilities). Also, the

constraints, and regulations pertaining the land use for recreation are here less rigorous.

4) Cheaper land prices attract the recreation use by people with modest financial resources.

5) The environmental impact of tourism/recreation is less pronounced than on the seashore and in the mountains and does not cause, as a rule, deep structural changes like in the other systems mainly because of higher environmental carrying capacity. There is less congestion due to more dispersed character of tourism.

6) This system does not present problems in accessibility. Its physical setting allows good permeability of space by transport in all directions because there are few natural obstacles.

7) The transportation network is already in place, created for other reasons, before tourism/recreation comes to the area.

8) The relative homogeneity of land facilitates the application of abstract quantitative methods while researching the tourist locations.

9) There are practically no physical limitations with respect to land use. The limitations are mainly legal: private property, nature resources, military areas.

10) This system is in relatively better position climatically than seashore and mountains: precipitation, storms and other aspects of inclement weather are of less concern.

11) The inland rural location offers an interesting environmental potential for tourism/recreation, for example marshy areas although not recommended for direct facility siting, nevertheless contain many opportunities for both consumptive and nonconsumptive use of bioresources for tourism/recreation which may be associated with better economic yields than drainage and agricultural use.

The location factors for resorts and dispersed tourist facilities in rural flatland areas are few and not too restrictive. The existence of a major transportation route in the vicinity is helpful but not indispensable. Freeways speed up the connections with urban areas but their existence has very limited impact on local economy because the services provided on freeways for motorists are isolated from regional context and mainly derived from the metropolitan area and not from land they are traversing ("sterilizing effect" of freeways). Another location factor could be the existence of a natural or man-made attraction. Because of the prevalence of water-oriented recreational activities in this system the preferred location is on a lake or reservoir, close to attractive forest complexes. Often a cave, a castle, church, folkloric tradition may attract tourism development. Sometimes even this is not indispensable: just the existence of accommodation facilities in a village or town may suffice.

6.5. WILDERNESS AND PARKS

6.5.1. Wilderness

The systems (complexes, integrations) of natural resources could be approached also from another point of view - according to their location on the continuum of environmental modification. While discussing the three types of natural resource areas (resource-based, intermediate and user-oriented) we mentioned that in the real world these areas should be put on a continuum as there are no clear-cut division lines between them. Indeed, such a continuum of environmental modification, ranging from the primeval to paved, from wilderness to urban area, does exist. Some of the resource areas on this continuum are particularly interesting for tourism/recreation: wilderness, resorts, urban areas attract tourism, because of specific clusters of resources or a single attraction they offer. The resource-based and to a lesser extent the intermediate areas will be discussed in this section (6.5) on account of their natural resource orientation. Special attention will be given to the resource-based extreme: wilderness and near-wilderness. The remaining part of the continuum, particularly the user-oriented areas, because of their mainly human type of resources, will focus our attention in section 6.7 Human Resources.

The appreciation of wilderness is a relatively recent phenomenon in the history of mankind. Throughout most of the ages wilderness was associated with evil. The "howling wilderness" was feared and avoided. The relationship between man and nature was epitomized in the Christian beliefs that the earth should be peopled, cultivated and "conquered." This transformation of wilderness into civilization was regarded as the noble mission of mankind. Wilderness was in the imagination of people the habitat of demons, devils, robbers and vagabonds. Only the Renaissance with its Humanism initiated a slow gradual change of the people's attitude to wilderness. The 18th century deism as religion of nature reversed the long time associations of wilderness by linking it with God. However, especially important in creating positive attitudes towards wilderness was the 19th century Romanticism. Indeed, gradually people started to regard wilderness not as something to be conquered and dominated but as a source of inspiration and renewal.

The North American pioneers still belonged to the category of wilderness conquerors. However, in the second half of the 19th century the attitudes changed and conservationists started to advocate wilderness preservation. Nevertheless, it took time for these ideas to take root. Thus the beginnings of National Parks in North America resulted from practical needs such as promotion of private exploitation of some unique nature resources, tourist development or securing of water supply for urban centers. Wilderness was thus preserved unintentionally. Only later, in the 1890s did the wilderness preservation function of national parks become important when a number of US conservationists realized that the American wilderness might soon be on the brink of extinction. A milestone in this respect was the foundation of the Sierra Club in 1892 as a result of John Muir's crusade to preserve western wilderness areas on public lands. Thus the USA can be credited with being the forerunner in the development of wilderness conservation in the world. The 1916 National Parks Act officially recognized the preservation function of the parks as of equal importance to enjoyment of recreational uses.

In Canada although the preservation function has been enshrined in the 1930 National Parks Act, it was not until the 1960s and early 1970s that primary use of parks for wilderness protection for the benefit of future generations came to the fore. Prior to that development, including tourism development, had been the main focuss. It can be argued that this resulted from a perceived overabundance of wild land and the economic needs of a young rapidly developing country. Therefore, the awareness of the value of wilderness appeared much later than in the US. Increased awareness of the need to protect wilderness was demonstrated in the strong public support for the establishment of new wilderness-oriented parks like Kluani and Nahanni, the protests against the Winter Olympic Games in Banff National Park and the defeat of Lake Louise development project proposed in the early 1970s. However, the legacy of National Parks as tourism development attractions are still visible in Canadian National Parks as exemplified by townsites of Banff, Jasper and Wasagaming (Riding Mountain National Park).

The appreciation of the value of wilderness is growing as the remaining natural areas continue to shrink around the world. The degree to which this value is recognized by governments and the general public reflects on the scope of practical steps undertaken to preserve wilderness which survives unchanged only in a few areas of the world. The long-range socio-economic, psychological and biological benefits for humanity connected with wilderness preservation have been proven by many authors. The problem is one of convincing the owners of land that protection of wilderness is a legitimate land use.

One of the most important areas of action in wilderness preservation is the expansion of National Park Systems all over the world. However, not all wilderness is or will be part of National Parks and not all the territory of National Parks is wilderness. Therefore, some countries, notably the United States, embarked on a policy geared to wilderness preservation at large, both within and outside the national Park System. An extremely important initiative in this respect was the 1964 Wilderness Act. The Act defined wilderness as follows:

> "A wilderness, in contrast with those areas where man and his own works dominate the landscape, is hereby recognized as an area where the earth and its community of life are untrammeled by man, where man himself is a visitor who does not remain ... an area of undeveloped Federal land retaining its primeval character and influence, without permanent improvements of human habitation, which is protected and managed so as to preserve its natural conditions and which (1) generally appears to have been affected primarily by the forces of nature, with the imprint of man's work substantially unnoticeable; (2) has outstanding opportunities for solitude or a primitive and unconfined type of recreation; (3) has at least five thousand acres of land or is of sufficient size as to make practicable its preservation and use in an unimpaired condition; and (4) may also contain ecological, geological, or other features of scientific, educational, scenic, or historical value."
>
> (Quoted after: Outdoor Recreation Action 1977, 44:4)

On the basis of this Act and subsequent legislations, millions of acres in National Forests (part of "the roadless areas"), large areas within the National Park System and National Wildlife Refugees have been designated as components of the National Wilderness Preservation System. These designations have been frequently debated. It is important to note that nobody has questioned whether any wilderness should be preserved. The issue has been rather how much land should acquire wilderness status (at present only less than 2% of the land in the lower 48 states), where these lands should be located, what categories of natural environment should be preserved, how large a specific area should be etc. These wilderness-allocation problems constitute the most controversial and hotly debated issues of natural resource utilization because the areas designated as wilderness must not be economically exploited. Especially controversial is the exclusion of lumbering and energy (gas and oil) exploitation. Proponents interested in development of mining and lumbering resist the wilderness designation pointing to the economic losses resulting from locking up "useful land" and locking out thousands of potential jobs "for the benefit of the few." Environmentalists argue that only a very small percentage of land already designated or being considered for the National Wilderness Preservation System has any real mining potential. As far as lumbering is concerned the economic troubles of the industry can be traced to reasons other than inadequacy of the resource base.

Another controversial issue is that of wilderness management. Simply designating an area as wilderness does not guarantee its preservation. What is needed is an enlightened wilderness management regime which recognizes wilderness as a land use which must be integrated with all other land uses. Wilderness is only one kind of land use which occupies the extreme position on the environmental-modification continuum ranging from primeval to paved. Wilderness areas are generally regarded as those which have not been modified by humans. Apart from the fact that this may not be entirely correct from a scientific point of view, the issue of resource management in the wilderness has been closely linked with management of recreation opportunities. The limited wilderness areas, we have set aside, cannot provide for all of the demands for primitive forms of dispersed extra-urban recreation. Some of these have to be provided by semi-wilderness or intermediate back-country areas located farther form the wilderness extreme on the environmental-modification continuum. In the real world these areas are usually located between the road's end and the wilderness. Such semi-wilderness areas with primitive dispersed recreation (mostly hiking, canoeing and camping) constitute excellent buffer zones insulating the wilderness from other land uses. Within the "pure wilderness", conservation is supreme and recreation may be limited or even excluded because of its vulnerability not only to what is happening within its confines but also what's happening nearby.

But here starts the controversy as to wilderness management. Two opposing philosophies have developed with respect to recreational use of wilderness (Hendee,J.C.,1974: 29). The anthropocentric or man-oriented approach advocates enhancement of recreational use of wilderness with appropriate management techniques which are geared to a high degree of recreation opportunity development with some "wilderness enclaves." The opposing philosophy may be called biocentric or "pure" approach. The primary management objective is not recreational facility development but rather preservation of natural integrity of wilderness ecosystems at the expense of recreational and other human use. This approach while limiting the recreation opportunities in wilderness areas advocates the provision of a full spectrum of these opportunities in other non-wilderness areas. This is a counter-argument against those who criticize the biocentric approach of wilderness management as "elitist" providing access to a privileged few at the expense of the majority, including the poor, the very young, the very old and the disabled. Indeed, the protection of wilderness against the excessive physical and social impact of tourism requires the introduction of management techniques which not only regulate and control the recreational use of wilderness, but also, if necessary, limit access. One has to realize that not the number of users but their physical and social (social carrying capacity) impact is decisive. Thus first of all, uses causing the greatest impacts should be restricted or redistributed in time and space. Thus the greatest impact on wilderness is caused by large parties on horses camping overnight and building campfires as compared to small groups of day hikers. The 1964 Wilderness Act recognizes the need for visitation of wilderness areas for recreational, scientific and educational purposes. Banning visitation is ruled out.

There are many management techniques aimed at limiting the visitation impact on wilderness. Thus visitor management may include introduction of entrance fees, permits regulating the time of visit, length of stay, area visited, party size, method of travel and behavior, various types of reservation systems including lottery. However, rationing of wilderness use is only the last resort. First comes information and education of visitors.

Today management techniques with respect to wilderness ecosystems are trying to avoid extremes. Thus e.g. complete fire control is generally rejected. The same may be said about a hands-off approach to fire control. Complete fire protection changes significantly the natural ecosystems decreasing some habitats (e.g. for ungulates) and increasing the chances for disastrous wide-spread fires as the forest ecosystem in the climax stage starts to age and the amount of dead, easily inflammable material increases. Preserving the ecological status quo is impossible. Therefore, fires, sometimes controlled (prescribed) fires should be allowed. At the other extreme, non-interference policy may cause severe damage to our limited wilderness resources not only in the form of fire losses but also other impacts caused by visitors: e.g. sometimes relocation of trails is necessary to minimize erosion damage or adverse impact on wildlife. Thus the environmental management in the wilderness has to allow some interference to offset human impact and recognize the place of the wilderness area in a broader region.

6.5.2. Park Systems in North America

The most important mechanizm form devoted to conservation of wilderness resources are national parks. National parks contain resource-based areas of outstanding value. But they do not stand alone. National Parks represent only one type of park in a broader family of parks which is called park systems. Therefore, before focussing on national parks it seems appropriate to outline briefly the totality of park systems which spans the whole continuum of environmental modification. Moving along this continuum from user (facility) oriented through intermediate to resource-based parks, helps us to establish a hierarchy of park systems. A park in the most generalized sense is any open, mostly public, space devoted to two functions, recreation and preservation. More specifically, the relationship between these functions changes as one moves along the continuum from intensive to extensive resource utilization.

1. Playlots and miniparks are usually small units aimed at providing short time recreation near home mainly in urban setting. In most cases they are planned in conjunction with a housing development like apartment blocks, condominiums etc. Miniparks in downtown areas serve shoppers and employees.

2. Neighborhood parks serve several blocks and are easily reached on foot. Their size is normally over 2 hectares. They are "footloose" in the sense that almost any piece of land, even those with poor natural resource qualities may be used and developed.

3. City parks (community, district parks, central parks) serve larger parts of cities containing a number of neighborhoods. Their size is at least 20 hectares. In site selection for these parks, physical features (lake, river banks, interesting vegetation, topography) play a role although not an overwhelming one. The first half of 19th century was the period of vigorous park development in Europe. In 1635, the first public park in America, Boston Commons, was set aside but the Central Park in NYC was the first predesigned landscape park in America (1858).

4. Theme parks are the space-age descendants of the old amusement parks. This category is difficult to place on the continuum of environmental modification because of the greatly varying nature of these parks. Their common feature is that

the attractions are tied together as a coordinated package developed as a single theme. Otherwise they vary. They may be located in urban or extraurban areas but always with easy access from large metropolitan centers. They may be publicly or privately owned. They are designed around an unending number of themes. Their importance may be local, regional, national and international. Even zoos of various types and Skansens (e.g. Colonial Williamsburg, Virginia) could be counted within this category of parks. However, the most representative theme parks are the dream parks, technological wonders of the space age. The oldest theme park is Disneyland, California opened in 1955 which in August 1985 recorded its 250 millionth visitor. Other outstanding examples are Disneyworld and Epcot Center in Florida. The religious theme park Heritage USA in South Carolina received some publicity in 1987.

5. Regional parks are large parks located near large population centers (within or outside city limits). Minimum size 200 hectares. In regional parks natural resources play an important role.

6. National Forests and State Parks in the U.S. and Provincial Forests and Provincial Parks in Canada contain the intermediate type of natural resources for tourism/recreation in North America. Some of them vary considerably in their location on the environmental modification continuum and cannot be counted as intermediate category. Most of parks in this group are large (thousands of hectares). Their location is extraurban. A zoning system is used to manage these parks. The zones range from wilderness conservation to intensive concentrated use. Planning and management is based on multiple use principle. This means mining, lumbering, hunting, fishing, grazing, watershed protection and recreation are allowed. The resources should be used on sustained yield basis which has been legislated for them (e.g. the US Multiple Use-Sustained Yield Act of 1960). As in the case of national parks, the development and conservation advocates clash. Developers argue for resource extraction, conservationists for resource conservation, ecosystem protection and non-consumptive forms of recreation.

In the case of US State Parks the ecosystems are frequently manipulated (e.g. silviculture) to enhance the scenic beauty of the park for the enjoyment of visitors. Canadian Provincial Parks vary considerably among and within the provinces. The noun or adjective added to the name of the park often reveals its designation (classification): recreation, wilderness, heritage, historical, archaeological, enthnological, natural, marine etc. Some of them are multi-purpose, in some the emphasis is on recreational use, others focus on wilderness preservation like

Atikaki Provincial Wilderness Park in Manitoba established in 1985. Such a park certainly should be placed very close to the resource-based extreme on the environmental modification continuum.

7. National Parks and equivalent reserves. In most cases these constitute larger units and are located farther from urban centers than the former category. They are owned and managed by the federal government. Single use principle excludes all other economic utilization but preservation and recreation, giving preservation priority, especially recently. Strict zoning systems guard against recreational overuse. Wilderness zones may be of limited public access or no surface access (roadless areas). Intensive recreation and habitation zones are limited in area and in the scope of activities. Wherever possible, tourist facilities and intensive uses are encouraged to locate outside the park boundaries.

The history of modern park systems started in the first half of the 19th century Europe with the development of the first predesigned urban landscape parks for public use. The second half of the 19th century marked the establishment of the first national parks in North America. Since that time all categories of parks, especially resource based parks, expanded all over the world. The most modern addition are the theme parks which enjoy an almost exponential growth. The closer the position of park to the user-oriented extreme on the continuum of environmental modification the greater is the intensity of use per square kilometer. Parks positioned at this end of the hierarchy are user-oriented. Here belong all the urban and exceptionally extraurban parks (e.g. extraurban theme parks). The importance of natural resources increases as one moves to higher categories: one could place the state and provincial parks in the intermediate group although some of them, at least partially, belong to the resource-based category if they contain wilderness areas. The national parks must be inequivocally regarded as resource-based parks where the use is extensive rather than intensive. The identification of some specific parks with user-oriented, intermediate and resource-based groups may pose some problems. Thus today we have a system of parks composed of a number of subsystems like urban, state or provincial and national parks. All of them interact with each other. If a given area does not have a certain category of parks on the continuum, the user pressure turns to the next category in the direction from user-oriented towards resource-based. The intervening opportunity of user-oriented and intermediate parks decreases the pressure on wilderness areas in national parks thus contributing to its preservation. This is one of the reasons of the need for their development. There is an urgent need for coordination of planning, communication and cooperation between different agencies responsible for various parks and park systems and also participation of the public in planning and management decisions. All these endeavors are helping to establish a dynamic balance between conservation of natural resources and the recreation needs of the public.

6.5.3. National Parks

6.5.3.1. Terrestrial parks

The national parks constitute the largest (territorially) part of the park systems. They had their forerunners in Medieval Europe where royalty and nobility established game preserves for their hunting pleasure. Also ancient rulers in Asia had their hunting preserves. By protecting the game, introducing "closed seasons", eliminating the encroachment of outsiders, mostly poachers, whole ecosystems have been preserved, although unintentionally. Later these areas became gradually open for public recreational use and in 19th century the first conflicts developed between the two functions of these protected lands: that of preservation and of use.

The beginnings of national parks in 19th century North America were not necessarily associated with pure and noble ideas of preservation. The early beginnings in North America were associated with commercial interests. These interest, to a large extent connected with railroads spanning the continent at that time were to a large extent instrumental in the establishment of parks. The railroad companies were interested in commercial development of "empty areas" in which the new railroad construction was taking place. Exploitation of the beauty of the landscape for tourism and establishment of resorts, including hot spring resorts, was very much in their minds. This intention was obvious with respect to the 1885-1887 creation of Banff National Park in the Canadian Rockies in partnership between the federal government and the Canadian Pacific Railway Company. The Banff townsite was developed and all sorts of commercial use allowed well into the 20th century when the trend to phase out multiple uses became evident. Nevertheless, even today many national parks, especially in Canada, have to cope with facilities which are totally incompatible with natural ecosystems. The modern concept of national parks has been developed since the second half of the 19th century in the United States and quickly followed by Canada. This concept underwent a transformation during the more than 100 years of its history from initial stress of recreation gradually to the emphasis of the preservation function. One could generalize that conservation prevailed only since about early 1960s. It is important to mention that North America has retained a leadership in the development of the national park idea which has been quickly spreading all over the world. Today national parks and equivalent reserves are regarded as the most important form of nature preservation on earth. This attitude is reflected in the 1969 definition of national parks recommended by the International Union for the Conservation of Nature and Natural Resources (IUCN). The national parks have to fulfil following conditions:

1. large areas where one or several ecosystems are not materially altered by human activity;
2. contents of outstanding plant, animal or geomorphological features of value for special scientific, educative and recreative interest;
3. managed by highest competent authority of the country;
4. no exploitation or occupation;
5. visitors allowed to enter for educative, cultural and recreation purposes.

(McBoyle, G., Sommerville, E., ed. 1976).

The UN agency, International Commission on National Parks (ICNP) publishes periodically updated lists of national parks in the world called "the United Nations List of National Parks and Equivalent Reserves." It is a list of honor. To be placed on the list requires not only compliance with the above definition but also with the excluding criteria. Thus excluded from the lists are:

1. Scientific reserves which can be entered only by special permission (strict nature reserves).
2. A nature reserve managed by private institution or lower authority without some type of recognition and control by the highest competent authority of the country (central government).
3. A special reserve devoted to preservation of one feature and not the whole ecosystem.
4. An inhabited and exploited area where public outdoor recreation takes priority over conservation of ecosystems.

(World Centennial National Parks 1872-1972, 1974:7).

Today almost every country in the world, excluding ministates, contains areas under nature protection organized as national parks or their equivalents. These designations spread throughout the 20th century and especially after World War II. The United Nations List of National Parks and Equivalent Reserves contains over 1200 entries from more than 140 countries. The exact designations vary ("national park," "game reserve," "refuge," "sanctuary," "reserve" etc.) but the main function is preservation. Unfortunately, in some cases, this may be mainly on paper. If the preservation function is an exclusive one and tourist access is limited, then the term National Park seems less appropriate and should be substituted by "preserve" or"reserve." However, in Canada the term "reserve" has a specific meaning associated with native land claims as stated later in this section.

One of the main principles of conservation is the unity of ecosystems which is frequently disrupted by international boundaries running frequently on rivers or on ridge tops. In such situations international parks may be careated like Waterton Lakes - Glacier International Peace Park established in 1932 at the Canada-US boundary or the Pieniny and Tatra National Parks at the boundary between Poland and Czechoslovakia. In fact, the Pieniny National Park is the first international park in the world (established 1924). Another concept of international cooperation is national park "twinning" (e.g. Canada's Ellesmere Island N.P. and Northern Greenland N.P.) Similarily like twin cities, national parks of different countries cooperate exhanging experiences in problem solving.

Assuming proper planning and management there is no conflict between the two functions of national parks: conservation and tourist use. In fact, tourist use provides the necessary economic basis for conservation and pays opportunity costs of lumbering, mining etc. (This issue is discussed in detail in Bella, 1987). The national parks of the world constitute powerful magnets for national and international tourism with all the ensuing benefits and problems. North American national parks systems attract millions of people annually (in 1987 well over 300 million in US and over 25 million in Canada) and various revenues from these visitors constitute an important benefit for the national and local economies. National parks in developing countries constitute an important source of foreign exchange. Expenditures by international tourists in parks usually more than outbalance the costs of park establishment and management and also opportunity costs associated with their existence. Especially significant in this respect are the parks of Kenya, Tanzania, Zambia and Galapagos Islands, which prove that wild animals are more useful alive than dead. Kenya's tourism brought over $ US 300 million into the country in 1987, more than coffee, traditionally the number one export (Winnipeg Free Press 27-2-1988).

Besides the tangible quantifiable economic values of national, parks there are also other values which do not lend themselves to quantification. Indeed, social, educational and scientific values are often forgotten because they cannot be readily expressed in dollars.

It is difficult to generalize about national parks in Europe because of the diversity in their organizational and management makeup (e.g. the land in British, Italian and many other National Parks is privately owned). However, one generalization seems appropriate. Most of these parks are organized on the basis similar to early stages of North American parks development, i.e., multiple use is allowed and/or there is an emphasis on tourism recreation development and not conservation of nature. These charactertistic features are the result of the fact that unaltered landscapes in Europe were hardly existing when the concept of National Parks came along. Therefore, compromises were unavoidable. Nevertheless, some European parks are devoted to strict preservation. They are only partially accessible like Bialoweza, Poland, or accessible for very few, like the only National Park of Switzerland, Engadine where conservation and research are the main functions.

Some national parks in the DCs allow various forms of economic exploitation beyond tourism. These forms have no detrimental influence on the ecology of the parks in the opinion of authorities, responsible for them, and also yield a badly needed economic gain for the local population. Thus e.g. the Royal Chitwan National Park in Nepal provided about $1 million annually of thatch grass for local construction needs, while the Dumonga-Bone National Park in Sulawesi, Indonesia, is providing water for a nearby irrigation scheme as well as protecting a rain forest area (Land 1983, 1:11). Many parks are administered by sub-national governments or by private boards and not by the national governments ("highest competent authority of the country" according to ICNP). Some national parks do not allow public access. All these features make the inclusion of some parks into the United Nations List of National Parks and Equivalent Reserves questionable.

National parks constitute as a rule a very small percentage of the total area of LDCs (e.g. only half of 1% in Africa). Nevertheless even these are under increasing pressure of escalating visitations which necessitates the adoption of sound management techniques in order to prevent destruction of parks. The majority of visitors to National Parks in the DCs are domestic, in the LDCs foreign, eager to view wildlife, especially big game. In the LDCs the local population lacks financial means and interests to visit the parks. Nevertheless the locals benefit economically. Unfortunately not all are aware of this and threaten the very existence of the parks by expanding cultivation and poaching.

Poaching is a problem in parks around the world but it acquires most destructive dimensions in some LDCs. As mentioned above, in parts of Africa the words for game and meat are identical and some tribal customs (Masai) expect a young warrior to kill a lion to prove his manhood. Be it tradition or need of food, the locals pay little respect to hunting bans in national parks. Use of modern weapons and demand for souvenirs, trophies and even aphrodisiacs (rhino horns in the Far East) create conditions where the very survival of wildlife is threatened.

Animal species which suffered from poaching may recover if proper conditions are restored. However, loss of habitat for the animals is still more dangerous. And this is increasingly the case in the developing world because of escalating population pressure on forested areas including national parks.

It seems that there are various steps to be taken to change the situation. First is the strict enforcement of nature protection, second the environmental education of locals especially children. If necessary, game-proof fences should be constructed. Also creation of buffer zones with licensed hunting may help. Game ranching also decreases poaching by supplying the souvenir shops with necessary merchandise. Many developed countries have introduced restrictions on importation of certain animal trophies.

6.5.3.2. Marine Parks

The notion of National Parks is clearly associated with the terrestrial environment. Indeed, terrestrial ecosystems have been correctly perceived as endangered by modern civilization. Therefore, the land areas above the high water marks until very recently have focussed the attention and efforts for resource preservation. Maritime resources have been for long time regarded as inexhaustible. Moreover, because of the biological self cleaning ability of oceans, the seas of the world have been used as dumping areas for all sorts of refuse. Only a number of catastrophic oil spills in 1960s and 1970s convinced the public and more importantly the governments, that there is time for action because the marine environments are equally fragile and can deteriorate at an appalling pace. A wide array of measures have been undertaken in order to improve the situation, among them establishment of marine parks for protection of marine ecosystems.

The definition of a marine park is as follows: "It is an area of land in contact with the sea, whether submerged or emerged: in the first case it is a sub-marine parkland, in the second it may be totally separated from the continental mass, as an island, or it may maintain contact with the continent, taking on several geographic forms such as a peninsula, a cape or any other of the various land formations which have contact with the sea" (Marsh, J., 1970:119). In Canada "where marine parks are established adjacent to existing coastal national parks the boundary between the two will be the mean high water mark" (National Marine Parks Policy, 1986:5).

The marine parks are represented at almost any level of government responsible for them. However, the National Park level is leading in every respect. Also historically the idea of establishing national marine parks began with the 1962 First World Conference on National Parks. Today there are many marine parks all over the world. Australia, US and Canada are leading: a good example is the Virgin Islands National Park. Canada is considering a number of proposals, Australia is protecting its Great Barrier reef, Kenya's marine parks at Watamu and Malindi seem to be in good shape, West Germany went through the difficult process of establishing a Wattenmeer National Park, Japan established the Inamurasaki National Park. The Soviet Union has established its first sea preserve of 630 sq. km in the Bay of Peter the Great near the Isles of Rimskiy Korsakov in the Sea of Japan (Land 1988, l:3).

The main problems encountered by establishing marine parks: local interests and the oil and fishing industry. The management problems are numerous: control of shell and coral collection , protection of reefs against destruction, protection of endangered species and of unique or representative ecosystems, control of pollution and other detrimental impacts, preservation of archaeological and historical sites, especially shipwrecks. A very difficult problem is fishing. Recreational fishing (including spear fishing) can be regulated relatively easily. More difficult is commercial fishing by locals who may have relied on it for centuries. It seems that in many cases these traditional activities may coexist with the conservation principle of the National Park. For this reason it may be advisable that the administration of some marine parks is shared with government agencies responsible for fishing or even maritime transportation. Another problem is that many aquatic species are migrating and hence their entire habitat cannot be protected.

6.5.3.3. The US National Park System

The US National Park System (NPS) is the oldest (Yellowstone 1872) and most developed system in the world. The visitation figure of 300 million annually has also no matach worldwide. It encompasses 337 units (1985) with a total area of 321,734 square kilometers. There are at present 19 designations within the system, preceded by the word "national": park, monument, preserve, lakeshore, seashore, river, wild and scenic riverway, historic site, historic park, military park, battlefield park, battlefield site, battlefield, memorial, memorial grove, recreation area, parkway, center for performing arts. These designations may be grouped according to their purpose into three categories: natural, historical and recreational areas. The main mandate of natural areas (national parks, monuments and preserves) is preservation of nature with subordinated recreational function. Historic areas constitute historic and archaeological museums (mostly open air). The function of recreation areas is to provide extraurban (with emphasis on nearurban) recreation in pleasing environments. The US National Park System is administered by the National Park Service which was established by the 1916 National Parks Act, the first such agency in the world. The US National Parks Service (USNPS) is part of the Department of Interior. In 1985 there were 48 national parks with a total area of 194,285 square km. The largest and best known unit in the 48 contiguous states is Yellowstone with about 9,000 km^2, the most visited the Great Smoky Mountains N.P. (North Carolina, Tennessee) with over 12 million annual visits. Other famous national parks are Yosemite, California, Grand Canyon, Arizona, Everglades, Florida, Hawaii Volcanoes and Olympic, Washington. Only one national park functions as an archaeological preserve. It is Mesa Verde National Park which would appear to fit better to the category of national monument.

The majority of national monuments are historical and archaeological. However, the preservation of geological, paleontological and biological features is the purpose of over 40% of units. The total area of 77 national monuments was 19,134 km^2 (1985).

The spatial distribution of units within the National Park System is uneven: the western part of the country concentrates disproportionately more units devoted to preservation of nature than the much more populated East. However, the East contains more historical units with high visitation. For more than 15 years the overall annual visitation figure of the system surpasses that of the US population. At present the annual number of recreational visits excedes 300 million.and is growing constantly. An interesting trend in park visitation is worthwhile to note: the visitation increases substatially in parks close to metropolitan centers because of better accessibility and stronger accommodation basis. The visitation in the remote parks has been flat for a number of years because of distance from the major markets and also as a result of limited development. Thus the spatial patterns of visitation is also uneven: 90% of visistation occurs in one third of the parks (Baker 1986:67)

The US occupies a special place in the history of national parks because here the modern idea of nature preservation was first born and implemented in the form of a national park system which today has only few parallels in the world. Nash in his "Wilderness and the American Mind" writes that the wilderness idea in the 19th century United States constituted an attempt to establish identity vis-a-vis Europe. The Old World was proud of its architecture and other monuments of the past, the US had the grandeur of unspoiled nature: the wilderness. Preservation of these "people's cathedrals and churches" (John Muir) became initially an almost religious romantic idea gradually evolving into a scientific concept, especially under the influence of Darwin and his followers. In the 1860s the publications of G. Marsh appeared of which the most important was "Man and Nature: Or, Physical Geography as modified by Human Action". Marsh certainly influenced the American concepts of nature conservation. Unfortunately, these concepts did not play a very important practical role for decades to come. Instead profits from recreation and not conservation was the focus. The 1872 establishment of Yellowstone National Park served commercial interests rather than protection of nature. The park was established as "a pleasure ground for the benefit and enjoyment of the people." (McBoyle, G., Sommerville, E.:,1976). For many years lumbering, hunting, trapping, dynamite fishing, deliberate burning of forests etc. was allowed. Although the presidency of Theodore Roosevelt (1901-1909) constituted an important stage in the development of the "Conservation Movement" (When Ted Roosevelt saw the Grand Canyon for the first time in 1903 he said: Leave it as it is. You cannot improve on it.") only the 1916 National Parks Act initiated the phasing out of activities incompatible with the idea of national parks. The Act stated that the aim of national parks was:

> "to conserve the scenery and the natural and historic objects and wildlife therein and provide for the enjoyment of the same in such a manner and by such means as will leave them unimpaired for the enjoyment of future generations."

Thus the two major functions of national parks - preservation and recreation - had been recognized as the sole purpose of national parks and multiple use has been gradually eliminated. For many years after the passing of the 1916 National Parks Act, recreation frequently supported by commercial interests, dominated the parks. The problem became more acute in 1950s and 1960s when parks visitation skyrocketed thus increasing the pressure on ecosystems. The studies of the ORRRC commission in the early 1960s indicated that the trend would continue. The response of the US National Park Service was to institute a program called Mission 66. In the framework of Mission 66 almost one billion dollars were spent over ten years to increase the carrying capacity of park facilities by constructing new roads, campgrounds and other services. Also the National Park System was expanded through provision of funds to acquire more lands. As an outgrowth of the ORRRC commission, a federal agency, the Bureau of Outdoor Recreation was created to coordinate the work of government at all levels in respect of meeting anticipated recreational needs of the population.

All these endeavors increased the recreational opportunities for American people but did little to preserve nature. An important step in this direction was the 1964 Wilderness Act which enabled various public agencies to designate part of their holdings as protected wilderness.

The 1960s marked an important turn in National Park policy. At this time the parks stopped to be treated as playgrounds for recreation and the preservation function has gradually acquired the priority status. It does not mean that the struggle between developers and conservationists has stopped. On the contrary, it has intensified, especially in th 1980s during both tenures of the Reagan administration which has been characterized by its pro-development stance. Indeed, the National Park System is under pressure from economic development interests looking for park land as a source of energy, timber and other economic uses. Especially strong has been the pressure on National Wildlife Refuges (e.g. Arctic National Wildlife Refuge in Alaska) which are managed under the multiple use principle. The mounting tendencies to increase lumbering, grazing, farming, oil and gas extraction, and also consumptive recreational uses like hunting and fishing dismay the conservationists who fear the destruction of wildlife habitats.

Agencies in the US whose mandate is the preservation of nature are also under pressure from economic development and environmental deterioration occurring outside their boundaries: acid rain, air and water pollution, construction of hydropower plants, increase of logging in national forest buffer zones, water control and drainage projects, resorts and other settlements built nearby etc.

Another area of concern for the National Park Service have been the budgetary cuts during the Reagan administration. The available funds have been inadequate for both expansion of the park system and adequate maintenance of existing parks.

All these park policy issues require immediate remedial action in order to enable the parks to fulfil their mandate.

6.5.3.4. The Canadian National Parks System

The appreciation and understanding of wilderness and the development of the concept of nature preservation developed in Canada later than in the United States. The reason was that Canada has more space and natural resources. The feeling of unlimited bountiness of nature prevailed for many years until relatively recently. Therefore, there was no sense of urgency to protect the natural heritage, there was no strong tradition in this respect like the US conservation movement with famous writers and activists of the stature of Thoreau, John Muir or Stephen Mather.

The history of national parks in Canada started in 1885 when the federal government set aside a small reserve in the Rocky Mountains which was later to become the nucleus of Banff National Park. Like in the United States the initial idea was not the desire to preserve unspoiled nature but rather a development connected with commercial exploitation of natural resources. The value of Banff consisted in its potential to draw tourists to the resort located in a beautiful mountain landscape and containing mineral hot springs. Therefore, Banff was created as "a public park and pleasure ground for the benefit, advantage and enjoyment of the people" (McBoyle, G., Sommervile, E., 1976: 195) a very similar formulation to that with respect to Yellowstone N.P. The newly built Canadian Pacific Railway was bringing tourists to the quickly expanding townsite of Banff, and a large CPR hotel was built. Other developments were not forgotten: lumbering, mining, grazing and hunting within the Banff National Park boundaries were not only allowed but encouraged for many years to come, as economically profitable activities.

Limitations to these activities were imposed during the first World War but only the 1930 National Parks Act constituted a beginning of gradual abandonment of the multiple use principle not only in theory but also in practice.

There were already 14 national parks in Canada when the National Parks Act was passed in 1930. The act by its "dedication clause" provided the basic mandate for nature conservation:

> "The parks are hereby dedicated to the people of Canada for their benefit, education and enjoyment, subject to the provisions of this act and the regulations, and such parks shall be maintained and made use of so as to leave them unimpaired for the enjoyment of future generations."

This formulation was interpreted as an encouragement that Canada's nature heritage must be made available for people and that the primary importance should be accorded to the recreation function of national parks. Thus recreation dominated Canadian National Parks for the next 34 years (1930-1964): townsite and summer resort development, provision of services and recreation opportunities incompatible with the spirit of nature preservation spread all over the system. The 1964 "National Parks Policy" statement was a turning point establishing a clear priority of conservation over recreation. A similar paper further detailing the Parks Canada policy was tabled in the House of Commons in 1979. The present philosophy is that the basic obligation of Parks Canada is to preserve unique and representative areas and provide only such recreation opportunities which are fully compatible with this basic obligation. Other tasks of Parks Canada with respect to the public is nature interpretation and education for the tourists, and provision of research opportunities for scientists. A good example of such important research activity is the acid rain research conducted in Kouchibuguac N.P., New Brunswick. The increased public interest in nature conservation in the parks was reflected in the establishment of the National and Provincial Parks Association (today Canadian Parks and Wilderness Socity), a non-profit, non-governmental organization aimed at stimulating public participation and provision of leadership on environmental matters in parks.

The Canadian National Parks Service - CNPS (before 1988 Parks Canada) is the federal government organization responsible for national parks and equivalents. It has been part of the Department of Environment since 1979. The Canadian Parks Service has in addition to the central Ottawa Office also five regional offices

(Calgary, Winnipeg, Cornwall, Quebec City and Halifax) each responsible for a certain number of regionally grouped parks.

There are 35 national parks in Canada with an area of about 180,000 km^2 which is the largest in the world but constitutes less than 2% of Canada's territory. Canadian national parks are devoted to preservation of natural heritage. In fact, according to the "Parks Canada Policy," approved by the federal government in 1979 "national parks are intended to protect representative examples of the diversity of Canada's landscape and marine areas for the benefit of present and future generations" (Parks Canada Policy 1985:28). At present the existing national parks represent only 20 out of 39 terrestrial regions and three out of twenty-eight marine regions identified by CNPS. Similarly like in the US the spatial distribution of national parks represents a reverse to the settlement pattern. It is highly skewed in favor of the West (the four Western provinces contain 55% of the national park area) and increasingly the North (40% of the area). The East with its concentrations of population has relatively few and smaller parks (about 5% of the national park area). It is doubtful if this regional imbalance will be ever corrected because the establishment of new national parks in populated areas is increasingly more difficult. In fact, nowadays the provincial governments are more and more reluctant to cooperate with the federal government in transferring land to national administration and control and foregoing development in perpetuity. In this connection the creation of a new national park in British Columbia in 1988, the South Moresby Island N.P., should be regarded as a significant success. It was the first national park established in the provinces since 1971 (Pukaskwa N.P., Ontario) if one disregards the small (815 km^2) Grassland N.P. Saskatchewan, established in 1981 and the tiny Bruce Peninsula N.P., Ontario in 1986. The South Moresby is an especially difficult example for the creation of a new National Park because it involves large payment by the federal government of opportunity costs (recompensation for lost revenue) to logging and mining companies, and in the more distant future the settlement of the native claims to the area (since the addition of the word "reserve" to the name of the park)

The establishment of new national parks is theoretically easier in territories than in provinces. The Yukon and Northwest Territories are still directly dependent from the federal government and the majority of land is still under federal ownership. However, there are complications with native land claims. Therefore, pending the resolution of these claims, most of the new northern national parks carry the designation "National Park Reserve." In the 1980s two large northern national parks were established, in 1984 the Northern Yukon N.P. (10,000 km^2) and in 1986 the Ellesmere Island N.P. (38,000 km^2). Unfortunately, these new additions are unlikely to take the user pressure off the southern parks because of limited access. Also, time is running out in all of Canada for the establishment of new National Parks because of the advancing destruction of natural ecosystems by human action. Therefore, it is imperative, according to the 1987 Report of the Task Force on Park Establishment, to quicken the pace of land designation as national and provincial parks aimed at preservation of what is left in terms of outstanding natural environments.

The CNPS is also responsible (in addition to National Parks) for National Historic Parks and Sites and Heritage Canals whose mandate is to preserve monuments and structures of the past. Their number is growing quickly. Total attendance at Canadian national parks and historic parks and sites was more than 25 million in 1987. (Tourism Intelligence Bull. 1988, Feb.)

According to the 1985 Parks Canada Policy document three new types of protected lands have been proposed: 1) Canadian Landmarks protecting exceptional natural sites 2) Canadian Heritage Rivers representing outstanding examples of the major river environments 3) Heritage Buildings, representing Canadian architectural and cultural heritage.

Turning to specific units in the Canadian national park system one finds a number of internationally most remarkable examples. As far as size is concerned the Wood Buffalo National Park in Alberta and Northwest Territories is world's largest (not counting the Greenland and Botswana National Parks which are little more than mere demarcations of boundaries on the map). Wood Buffalo National Park has an area of 45,000 km^2 which surpasses the territory of Denmark or Netherlands and is 5 times larger than Yellowstone National Park. Other famous Canadian National Parks include Banff and Jasper in the Rockies. Also unique is Kluane N.P. Reserve in Yukon, 22,000 km^2. At present, there is no province in Canada which does not have at least one national park.

6.5.3.5. Parks planning and management in North America

The necessity for planning and management

The idea prevailing in the initial period of North American national park development was that nature protection was automatic within the parks and one should focus on providing recreational opportunities for tourists. In later years, when the necessity for preservation became more and more apparent, it was clear that simple protection would not suffice that use of the parks natural resources would inevitably lead to their deterioration. Also contemporary global and regional interdependence may lead to undesirable changes in the parks. Hence the necessity of reaching beyond simple passive protection towards more active conservation policy involving even occasional human interference into parks ecosystems. These measures to better fulfill the mandate of environmental conservation amount to planning and management of parks. Moreover, proper planning and management not only guarantees protection of nature but may prove that there is no dichotomy between preservation and use although the reconciliation between these two is by no means an easy task.

Following this line of reasoning there has been a growing recognition that planning and management are absolutely indispensable in national parks environments which are neither resource pools for use nor museums of static preservation but dynamic life support systems vulnerable to improper use, even if thc use is only recreational.

Planning and management in parks has also to reconcile various visitor categories: people who look for relaxation and exercise, people who look for environmental education, instruction and inspiration and finally scientists who want to conduct research. Clearly, parks cannot provide all things to all people, nevertheless it is possible by means of planning and management to fit these categories of people into three types of zones within the park without interference assuming, of course, a certain minimal size of the park. These zones are:

1. zones devoted to relaxation
2. intermediate zones providing wilderness experience
3. restricted zones of wilderness, which provide laboratories for scientific research.

Planning for national parks

There are three types of planning for parks: (only the first two will be discussed here)
1) System planning
2) Management (or master) planning
3) Site planning

1. System planning

System planning is park planning on the scale of the whole country or a major region. It relates to the creation of a park system which would contain all the outstanding, unique and representative natural areas of the country, a system which would meet the national and regional demand for park land, a system which would try to minimize regional imbalances in the provision of parks and finally a system which would eliminate inter-agency imbalances.

It is, of course, impossible to develop a park system which would be perfect from all points of view, however, proper system planning can achieve a situation which is as close to the ideal as feasible. There are many obstacles which prevent creation of such a park system, some of them worthwhile mentioning here. In the US, with its considerable federal land holdings, the problem is in designation of land for national parks which is often opposed by resource development interests. In Canada the acquisition of new park land requires not only adequate funding for purchase of such land but also frequently difficult negotiations with the provinces which may have jurisdiction over the area in question. The provinces have their own park systems competing in many ways with the national parks. Besides, there is a question of resources, both existing and potential, which are jealously guarded for future economic development. Therefore, any owner, be it a private person or the lower levels of government, is reluctant to transfer land to the federal government's jurisdiction without any incumbrance i.e. forgoing in perpetuity its rights. The CNPS may try to entice the provinces by providing special funds for tax and other losses, and by contributing to the development of facilities in adjacent areas which constitute economic stimuli in addition to jobs directly associated with the new park. This is the give-and-take tactic connected with negotiations about the creation of new parks between the two levels of government.

Another problem of park systems planning are the inter-agency imbalances. It seems that the situation in both US and Canada requires some rethinking and reevaluation. Jurisdictional and policy conflicts between all three levels of government have inhibited the development of national park systems. It is anomalous that Sangus Iron Works in Massachussets, Ford's Theater in Washington, the former federal prison on Alcatraz (San Francisco) and even the White House are managed by the United States National Park Service whose main function is and should be to protect wilderness (thus natural rather than cultural and historical resources). Similar trends of expanding the mandate may be observed in the case of the CNPS which, in an effort to widen the range of park experiences, moves into the area of providing intensive forms of recreation previously reserved for other levels of government. At the same time, provincial and even local governments are transgressing their traditionally established mandates in parklands designation which lie, respectively, at the intermediate and at the user oriented (or facility oriented) level of environmental modification continuum. It seems that, especially in the period of budgetary restraints, limited manpower and enormous increases in public demands for all forms of recreation, the transgression of their mandates by the three levels of government in the United States and Canada leads to harmful imbalances. The United States National Park Service and CNPS as federal agencies should concentrate in the field of resource-oriented, extensive forms of recreation where the conservation function is paramount as opposed to recreation, the local government level should be responsible for user-oriented (opportunity centered) intensive urban and suburban recreation and the state and provincial governments for intermediate (multiple centered) recreation. Only such clear-cut limits in parkland designation, planning and management may alleviate the imbalances.

In conformity with these views the parts of Redwood N.P. California owned and managed as state parks should be transferred totally to the federal jurisdication and such units like Gateway National Recreation Area or Golden Gate National Recreation area should be operated by local governments and not the USNPS. For historic and cultural units of national significance, like Gettysburg, separate specialized federal agencies have to be created. Similar is this author's opinion with respect to Canadian historical and cultural resources. At present the CNPS is concerned with not only natural but also historical and cultural heritage.

2. Management (master) planning

Once a new national park is created within well established boundaries the first step is to work on its management or master plan. A management plan includes a zoning plan designating various land uses with respect to present and future needs. An ideal arrangement of zones within a park should start in the middle of the park with a wilderness preserve. The access to this area is limited or restricted and the size rather small. The main function of such a zone is to provide undisturbed ecosystems for scientific research. This zone should be surrounded by a wilderness zone with only trail access for visitors and nature conservation as the main function. This is the large core area of the park usually covering the major part (over 90%) of the park. Again this area is surrounded by a narrow recreation zone with road access. At the outer periphery of this zone which constitutes the park boundary lie the nodes of intensive use including service areas. The territorial extent of this zone should be kept at an indispensable minimum. This arrangment reflects the spatial concentration of people on a relatively small percentage of the entire area of the park and the gradual transition from the predominance of the conservation function to use.

Such an "ideal" concentric arrangement of resources and consequently of the zones within the park boundaries never occurs in real space. Nevertheless, the principle of gradual transition from the extremes of preservation to use should be adhered to, to create sufficient buffers for the areas where the preservation function is paramount. The zoning systems in the world's national parks and equivalent areas differs from country to country. As an example of such system one could use the Canadian zoning system valid at present (Parks Canada Policy 1985:30). This system is made-up of following five classes (zones).

1. Special Preservation. Access strictly controlled or may be prohibited altogether. No motorized access or man-made facilities.
2. Wilderness. Extensive dispersed use with minimal impact. Limited primitive visitor facilities. No motorized access.
3. Natural Environment. Low-density recreation with minimum of facilities. Limited motorized access permitted but non-motorized preferred.
4. Outdoor Recreation. Limited areas for recreation and education. Motorized access permitted.
5. Park services. Small areas of intensive use. Towns, visitor centers, visitor services.

Zoning systems constitute important parts of the management plans which are formulated separately for each park. In Canada the development of a management plan for a national park is a long process taking usually 2 to 3 years in several stages. Without going into details the general outline of this process is that the public is usually consulted about issues to be addressed and has an opportunity to review plan alternatives and eventually a draft plan before the final park management plan is submitted to the Minister of Environment for appoval. Public participation in management planning reflects the democratic concern of the population for the natural heritage of the nation (Hoole A. 1978).

Important aspects of management planning include the relationship of national parks with the surrounding areas and the problems of regional integration of the parks. Indeed, the natural ecosystems of the parks extend far beyond the formal boundaries of the parks which, despite their sometimes considerable size, constitute integral parts of larger regions with common issues and both natural and human interactions. Therefore, what happens outside the parks impacts heavily on park's ecosystems. Examples of such interactions abound. The water regimes within the parks are parts of larger regional regimes. Any changes in use outside the parks (e.g. hydro-development) impacts on park ecosystems. Thus the water regime of the Everglades N.P., second largest in the US after Yellowstone, are threatened by diminishing water supply and water polution due to outside industrial and residential development. However, in wet years the situation is reversed and flooding threatens necessitating human engineering intervention. The Wood Buffalo N.P. is threatened by Peace River hydrodevelopment. The projected geothermal energy development nearby would endanger the geysers of Yellowstone and Lassen N.P., California. The exploitation (mainly logging) of National Forests, increased in 1980s, has reduced the range of grizzly bears, bison and elk. Near Glacier N.P., Montana, explosives used in seismic exploration for oil and gas are disturbing wildlife and tourists and a proposed coal mine would pollute waters flowing into the park.

These examples of human encroachment on the areas surrounding the parks illustrate the need to create of buffer zones outside the park boundaries to avoid sharp lines of division between natural and cultural landscapes like in the case of Riding Mountain N.P., Manitoba, Canada. Such buffers are usually formed by national or state forests in the U.S. and provincial parks or forests in Canada. If public property is not available to buffer the parks then the government, private conservation organizations or environmentally-minded individuals may consider land purchase or, at least, purchase of protective conservation easements around the park periphery. As an example of private initiative in this respect, one could give the purchases of farmlands at the boundary of Kruger N.P., in South Africa by nature lovers.

These buffer zones should be subject to scientific management and careful monitoring of developments which could harm the natural ecosystems of the parks. Also the problems of poaching should be dealt with not only directly by law enforcement but also by means of environmental education of local people. In order to reduce the recreation pressure on park resources it is important to develop alternative recreation opportunities in near-urban and park-adjacent areas.

These common problems between parks and their surroundings imply an integration of parks management planning with joint regional planning, aimed at developing parks and their regions as a unit. The integration of parks as parts of regional systems also facilitates federal - intermediate government level negotiations pertaining to creation of new parks, diminishing the reluctance of state and provincial governments to grant park lands to the federal governemnt. Regional integration implies cooperation between the US National Parks Service or the CNPS and those responsible for park periphery (buffer zones, visitor service areas and other lands adjacent to the parks). The problems in concluding bilateral and multilateral agreements consist not only of reluctance of the partners to enter such agreements but also is compounded by the complexity of jurisdictional make-up over lands outside the parks involving frequently all levels of government and private lands, private industry, functioning communities etc. all with widely differing objectives frequently not compatible with the objective of national parks: preservation of nationally significant landscapes.

Despite various problems associated with the establishment of cooperative agreements between national parks and those responsible for lands outside park this important issue should be followed energetically in order to achieve not only mutual consensus but also mutual satisfaction.

In order to achieve their objective the planners in the US and Canada are using various tools and techniques which deserve mentioning, e.g. the US Forest Service uses a decision framework called Limits of Acceptable Change (LAC) and Canadian Park Service the Environmental Assessment and Review Process (EARP). Both methods are instrumental in achieving planning goals which harmonize conservation of ecosystems with use including tourism/recreation.

Problems of park management

The existence of the best management plan does not constitute the guarantee that the park mandate will be fulfilled. Many governments, especially in LDCs lack the will and/or resources to enforce the plan and thus protect nature.

Of course, the parks should be provided with access roads, water, sanitation facilities, garbage collection, fuel for campfires and other facilities and supplies subject to budget constraints. This comes first to one's mind when thinking about park management. However, a closer look at this task leads to the conclusion that park management is much more than that. Take for example both North American National Park systems, the oldest and finest in the world. Despite the long traditions and experience, the people responsible for park management face formidable problems caused mainly by increasing visitor pressure especially in the world famous parks like Yellowstone or Grand Canyon. After the surge of 1960's and early 1970's the number of visitors are increasing at about 3% anually in 1980s. The parks fill with people and traffic. The ecosystems suffer from the impact of visitors and other internal and external impacts. In response to the deteriorating situation, management measures should be taken and techniques used to cope with the situation. These response strategies aimed at improved nature preservation and visitor satisfaction could be divided into two groups: management of people and management of the park environment. The management of people could be presented in the form of an open-ended list of management techniques which may be applied if a concrete situation warrants the introduction of such measures.

1) Affect visition

a) in time: encourage off-season visitation.

b) in area: encourage visitation of less known parks.

2) Educate tourists and the public at large in the wilderness ethic with respect to environment-conscious behavior: elimination of noise (incl. radio), alcohol, careless smoking, littering, poaching.

3) Restrict or limit visitation, especially in ecologically fragile areas of parks by introducing quota admission systems of reservations, advance booking, hike permits, boat traffic restrictionas, entrance or user fees, maximum number of days.

4) Minimize the construction of roads.

5) Design of roads for low speed and scenic viewing rather than catering to high speed motorists. Best roads: one-way narrow winding lanes out of sight of the oncoming lane.
6) Ban motor traffic and substitute it by public shuttle bus service.
7) Promote non-facility oriented wilderness recreation.
8) Minimize construction of facilities within NP.
9) Leave only absolutely essential facilities.
10) Locate essential facilities in ecologically resistant areas.
11) Concentrate the essential facilities in nodes rather than dispersed throughout the park. This minimizes the disruption of nature.
12) Limit the number of campgrounds within parks.
13) Rotate the use of campgrounds in order to ensure their ecological recovery.
14) Introduce stricter control of townsites existing within some parks.
15) Enforce law and order against vandalism, drug abuse, crime and poaching.
16) Phase out and/or locate outside the park boundaries all non-essential facilities incompatible with nature preservation like commercial services, stores, service stations, hotels, and motels. Staff housing, visitor centers and park administratinos should be also preferably located outside parks.
17) Phase out incompatible recreational activities like attending movies, dancing in a disco, drinking beer, roller-skating, waterskiing, motorboating, golf, tennis, bowling (bowling alleys), snowmobiling, swimming in a pool, ATV. In some cases horse riding maybe destructive to environment. The use of sightseeing aircraft may also be limited in some parks (e.g. Grand Canyon N.P.). Recreational activities appropriate for NP should be permitted: hiking, canoeing, cross-country skiing, snowshoeing, camping, mountaineering, nature study and interpretation, nature photography, painting, swimming, man-powered boating.

To manage the amount of park visitors is an action fully compatible with park objectives but to manage park environment may cause objections. Only " a hands-off policy" would, fully conform with the objectives of nature conservation. However, after agreeing that theoretically "a hands-off policy" would probably be the best, one has to state that such a policy is not feasible in the national parks surrounded by human landscapes and townscapes. Indeed, such a policy was actually never implemented during the more than 100 years of history of North American national parks. There has been always a measure of human intervention into the dynamics of natural park ecosystems, be it control of fires or control of predators. Indeed, throughout the years the park environments were subject to changes which proved not always advantageous. Frequently, the climax conifers have substituted for broad leave trees as is the case in some western parts of the continent thus decreasing the grazing areas for herbivorous animals. Also the abundance of overmature and dead trees has increased the danger of disastrous megafires and beetle infestation. Thus the long standing fire control policy in parks has been revised acknowledging that in certain cases natural fires should not be interfered with because they play an important role in the evolution of natural ecosystem and protection of some plant species like the giant sequoias in Yosemite and Sequoia National Park. Moreover, a policy of controlled fires or prescribed burnings i.e. fires deliberately set by the park service has been adopted, if deemed necessary. Another example of habitat manipulation is spraying to eliminate pests or diseases or destroying exotic vegetation (Florida).

In terms of wildlife management the national parks certainly created the ecological niches indispensable for the survival of many species close to extinction, e.g. the survival of the whooping crane can be, at least partially attributed to such niches in the Wood Buffalo National Park. The preservation of such animals as the North American bison, European visent, rhinoceros, elephant, lion, orangutan, etc. have been made possible because of the existence of national parks. One could certainly quote many other examples of such preservation within the parks. However, preservation of animals in their natural environments brings also problems which have to be dealt with by proper management techniques. Too many herbivorous animals, especially ungulates lead to overgrazing, soil erosion, starvation and diseases among the animals. The management techniques to cope with these problems are controlled hunting and/or reintroduction of predators which were formerly exterminated "to save" ungulates. Another technique is the relocation of animals to other areas which is preferable to killing them: e.g. it has been used for a number of years in Elk Island N.P., Alberta. In Kejimkujik N.P., Nova Scotia a racoon control program was implemented because the racoons, proliferating especially near campgrounds, destroyed turtle eggs.

In some parks the "exotic species" introduced by humans in the past constitute or constituted problems e.g. goats and cats in the Galapagos National Park, Equador or burros in Grand Canyon N.P., Arizona. These last animals were removed from the park in 1980-81. The elimination of "exotic species" from the Galapagos has met with only partial success so far.

Another management problem is the interaction between wildlife and the park visitors. A superficial observer may state that the animals have accepted the tourists as a tolerable, if not necessarily compatible, part of park ecosystems. The animals to a large extent have lost their natural fear of humans who hunt with their cameras rather than guns. However, here start the problems which should be dealt with expeditiously in order to protect the people from animals and the animals from people. The animals who at least partially have lost their fear are more accessible to people and, being unpredictable, can harm them. Here are some examples: Rodents mistake a finger for a peanut and bite the visitor. Such a bite may transmit disease like bubonic plague. Tourists approach animals too closely in order to observe them or to take pictures. There are cases of bisons or wild boars goring people. Even deer were reported to attack visitors when they or their offspring were perceived as threatened.

In some parks like Glacier, Yellowstone, Banff there have been a number of dangerous encounters between tourists and bears, especially grizzlies. Some of these encounters have resulted in injuries or deaths of tourists which have been increasing with increased visitation. The management response to these unfortunate events is to remove the nuisance bears to remote parts of parks or even to shoot them. Other management tools are the introduction of bear-proof garbage containers, efficient removal of all garbage from campgrounds to dumps outside the park, forbidding feeding the bears.

It is obvious that animal behaviour cannot be changed to prevent harm to visitors. Therefore, tourists have to learn and adjust to animals in the park. Thus education of the public becomes an important part of park management. Tourists should not feed animals, not approach them too closely, especially bears with cubs, not to hike alone and when hiking to watch for poisonous snakes and make some noise in order not to surprise the bears. These simple precautions will increase the security of park visitors. As for the security of animals the policy of not feeding animals with inappropriate diet, or still better not feeding them, at all, should be the park rule.

6.6. CONSERVATION OF ECOSYSTEMS

The need to preserve natural ecosystems reaches well beyond the limits of tourism/recreation. The need to preserve the genetic pool for future generations is one of such examples. The scope of tourism/recreation reveals various dimensions of this issue. Economic considerations head the list. However, one should never forget also that there are some important non-economic reasons like the necessity to preserve human mental and spiritual welfare. Yet, the task to preserve the natural ecosystems is not easy. The humankind confronted with fast deteriorating natural ecosystems has to recognize that natural resources are limited and increasingly scarce. Therefore, a reactive attitude to environmental emergencies has to be substituted by an proactive one with respect to conservation. The list of conservation measures presently being undertaken all over the world is open-ended. The most important task is to set aside representative samples of all types of ecosystems in a country. To achieve this the governments need funds to purchase private land because only public ownership can guarantee conservation. Thus the prime task of society is to maximize funding for land acquisition. However, acquisition of land may be too expensive. Therefore, other techniques are used. Normal land ownership is in "fee simple." This means no incumberance within the legal bounds. To prevent environmental deterioration and to preserve scenic beauty the government may purchase from the private owners the rights to develop their land in an area where the natural ecosystems are considered worthwhile to be preserved. Examples of such rights purchase are scenic easements or zonings of various degrees of development limitations.

The land and its resources saved from human impact, both interior and exterior, will be preserved. If the resources had suffered some past damage, they will finally, allowing time, regenerate naturally. However, this regeneration process may last too long or be even impossible in certain circumstances. In such cases an active conservation strategy may be more suitable than simple preservation. A new form of such a strategy is worth mentioning: restoration ecology. Restoration of prairies, forests, wetlands etc. represents an important extension of environmentalism beyond pollution control and preservation of flora and fauna. Actually it is not entirely new: there are many early examples of reclamation of quarries and open-pit mines which today are major tourist attractions. But contemporary reclamation ecology is much more intrusive and wider in scope. Whole ecosystems, destroyed by human impact, are being restored. Hundreds of projects are in progress. Specialized firms are at work in both DCs and LDCs. Most interesting examples in US are described by J. Berger (1987). Especially important is the restoration of wetlands. Indeed, there is a wide scope for action in all ecosystems which have been impacted by humans. One of the best known of such large-scale restoration projects is the restoration of about 1075 sq. km of tropical dry forest in the north western part of Costa Rica (Guanacaste N.P.)

which is now largely in pasture. If left alone this land would take 500 years to recover to its original state. The application of active restorative measures will shorten the recovery time to about 100 years.

The most crucial problem to overcome is funding. The situation in this respect is difficult in the DCs and hopeless in the LDCs. In the DCs the governments have to allocate funds for preservation from a shrinking economic pie in competition with innumerable other options. Private financing is miniscule. The LDCs are development-oriented. They are using their natural resources without regard for the future. They do not care and cannot afford funds as for the "luxury" of preservation, especially if suffering under the burden of international debt. This is certainly a tragic situation. Increasingly, one has to find outside funding for conservation projects in the DCs and this seems to be the only workable solution. Some recent developments indicate interesting trends in this respect. In July 1987 a U.S. organization Conservation International bought $650,000 of deeply discounted value of Bolivia's foreign debt for $100,000. In exchange the Bolivian government made a commitment to protect about 15,000 km^2 of Amazon lowlands in the Beni river basin. Similar "debt-for-nature" swaps are on the drawing boards in other parts of the world. Not only private conservation agencies like World Wildlife Fund are actively engaged in "debt-for-nature" swaps but also supranational intergovernmental organizations like World Bank. Especially targeted are the heavily indebted Latin American countries like Brazil, Peru and Mexico. Of course, these projects will not solve these countries' debt problems but they certainly will make important contributions to the conservation of biological diversity of ecosystems like tropical rain forests, vital to the survival of life on our planet. The "debt-for-nature" swaps will also increase the hard-currency earnings from tourism. This will be an important additional reward for conservation of natural heritage.

6.7. HUMAN RESOURCES (HUMAN ELEMENTS OF SUPPLY)

6.7.1 Classification

The human element of tourism supply (human resources) could be subdivided into two categories: human (cultural) attractions and infrastructure. There are site attractions and event attractions. Site attractions are always there, accessible throughout the days like museums, churches, temples etc. Event attractions are time-limited e.g. ethnic music, cinema or theatre festivals. Outstanding examples of site attractions, called mega-attractions, are the Egyptian pyramids, Louvre museum in Paris, London Tower, Vatican City. They are more difficult to determine as to their scope (e.g. Louvre could be extended to the City of Paris) than the event attractions. Outstanding examples of event attractions called mega-events are: Olympic Games, Calgary Stampede, Winnipeg Folklorama, Quebec Winter Festival, Munich Octoberfest, Oberammergau Passion Plays, New Orleans Mardi Gras, Rio's carnival, etc.. At the first glance the site attractions could be evaluated as better economic performers than event attractions because of the permanancy of the former and short duration of the latter. Such an observation is certainly correct. However, an appropriate marketing strategy may blurr the distinction between the two in order to maximize the economic effects, e.g. the venues of former Olympic games may become site attractions and the anniversary of construction of London or Pisa Tower may become an event attraction.

The second element of human supply is the infrastructure (facilities). The infrastructure, cannot be regarded as attraction although certainly there are elements of attraction in some of its components. The tourists are using two types of infrastructure:

1) General (basic, socio-economic) infrastructure

- transportation systems
- communication systems (telephone, telex, telegram, radio, TV, postal services, newspapers, etc.)
- utilities (gas, electricity, water supply, sewage and garbage disposal)
- banking systems
- social elements of infrastructure, sometimes distinguished as a separate social infrastructure (public safety, health services, educational system)

2) Tourism infrastructure (facilities) sometimes called superstructure. It could be subdivided into two categories: primary and secondary. Primary tourist enterprises work exclusively or mainly for tourism. Hence their collective designation - tourism industry:

- Travel agents selling tourism services in retail directly to the public.
- Tour operators, wholesalers sell package tours using travel agents as intermediaries.

- Passenger transportation carriers (airlines, railroad companies, bus companies, car rentals etc.)
- Accommodation establishments (called hospitality industry).
- Restaurants and other food catering establishments working for tourists.
- Entertainment companies working only for tourists.
- Tourist guides.
- Souvenir shops.
- Tourism and recreation equipment enterprises.
- Specialized catering companies e.g. outfitters.
- Government and private organizations planning, regulating and marketing tourism.

Secondary tourism enterprises rely only partially on tourism business. Otherwise they cater for local population e.g. banks, restaurants, theatres, zoos, museums, etc.. Their dual role does not allow to designate them generally as part of tourism industry although they, indeed, partly play such a role. Therefore, one could suggest that they may be recognized as such if they do business mainly with tourists.

6.7.2. Characteristic features of human resources

1) They are more spatially concentrated, to a large extent in cities, than natural resources.
2) They are associated with more capital, labor and management inputs than natural resources. Because of these inputs their carrying capacity is greater. In fact, the carrying capacity grows proportionately to inputs of these factors.
3) Because of the spatial concentration they need less space and provide opportunities for high density recreation and tourism.
4) Being spatially concentrated and less vulnerable to demand pressure than natural attractions human attractions divert part of the demand from natural supply which is in many places endangered by use and overuse. This circumstance should be understood and appreciated.

6.7.3 Human attractions

Natural attractions i.e. nature in all its beauty and lifegiving qualities, even if it is to a degree human-manipulated, constitutes the most important tourism attraction.

But it is followed closely by another set of attractions connected with people and their works. Frequently these attractions are called socio-cultural to underline the fact that not individuals are meant but human groups which through social interaction create what is known as culture. Culture finds its geographical imprint in cultural landscapes. First of all, these are people of the host area and their attitudes towards tourists which are to a large extent influenced by their cultural traditions. Further, such characteristics of host population like religion, language, education, handicrafts, art, music and dance, national costumes, work, leisure and eating habits (gastronomy) and their technical scientific and economic achievements should be regarded as tourist attractions. Also their historical past with its monuments, their traditions and customs should not be forgotten. Finally the most visible and relatively stable form of cultural expression - architecture - should be mentioned.

6.7.4 Geographical distribution of human attractions

Geographically the human attractions can be subdivided into following subgroups: urban attractions, rural attractions and "footloose" attractions (may be located anywhere).

1) Urban attractions

Of course, the cities with their historical monuments and traditions remain focal points of cultural tourism: Paris, London, Rome, Athens, Venice, Amsterdam and Florence with their museums, cathedrals, famous buildings, landmarks, monuments etc. are some European examples. Big cities lure the tourists also increasingly to their economic and scientific attractions: ports, airports, industrial enterprizes (e.g. brewery tours!!!). Also ther takes place the majority of event attractions (theatre performances, festivals, fairs, sport events, etc.).

One of the important attractions, mainly located in urban areas, is shopping. Indeed, the majority of tourists are interested in the purchase of various objects of art specific for the area. The prices may range from cheap souvenirs, frequently made in East Asian sweat shops (Hong Kong) to expensive objects of art, antiques, etc. Eskimo carvings, Flemmish embroidery, Persian carpets are examples of articles willingly purchased by tourists. However, tourists should be warmed that in some areas, especially in the Mediterranean, there is a whole industry supplying fake antiques. Some of the antiques offered for sale are genuine but acquired illegally by museum robbery or grave robbery (e.g. Mexico, Guatemala), or illegal digs (e.g. Italy, Greece). Many countries have imposed export restrictions on exportation of antiquities (e.g. The Soviet Union, Greece, Cyprus, Syria, Turkey).

Urban areas are also centers of religious tourism. Even if the object of pilgrimage was initially located in rural setting, nevertheless, invariably a more or less sizable city has developed, especially with growing numbers of pilgrims e.g. Lourdes, France gets well over 4 million tourists annually. Such places like Jerusalem, Mekka, Fatima (Portugal), Varanasi (India), may serve as other examples attracting millions of pilgrims. Most religious attractions belong to the category of site attractions. However, also event attractions such as religious festivals or papal visits play an important role. One should not ignore their significant economic impact.

2) Rural human attractions

Rural areas do not lack such types of attractions like castles (e.g. Loire Chateaux) fortresses, palaces, monasteries, abbeys, temples, pagodas, battlefields. In North America some of the old homesteads and fortresses have been converted to museums. Even World War II German concentration camps (e.g. Oswiecim - Auschwitz, Poland) receive millions of visitors. The same may be told about mega-creations of modern technology e.g. hydrodams (Itaipu, Brazil) or space centers (Cape Canaveral, Florida). In developing countries ruins of ancient civilizations: Maya (Chichen-Itza, Tikal), Inca (Machu Picchu, Peru), Roman (Baalbek, Lebanon), Egyptian (Pyramids of Gizeh, Egypt), Bhuddist (Borobodur, Java, Indonesia) Hindu (Ankor Wat, Cambodia), Nabatean (Petra, Jordan).

Some tourists not only sightsee and enjoy the culture of the countryside but also, albeit temporarily, choose to pursue olden lifestyles acting e.g. as cowboys in the West or as gold prospectors in Yukon or Alaska.

The countryside attracts tourists not only with its natural landscapes but also with unique cultural landscapes (e.g. Japan, China, rice terraces in Northern Luzon, Philippines, Valley of the Nile in Egypt, fishing villages in Portugal or Newfoundland, polderlands in the Netherlands).

One of the most outstanding features of cultural landscapes in rural areas and townscapes in urban settings is architecture which may enjoy the eye of the viewer by its beauty, uniqueness, variety and harmony with the natural landscape. On the other hand uniformly cosmopolitan forms of purely functional architecture can destroy the most attractive landscape or townscape

3) Footloose attractions

The third category are "footloose" human attractions i.e. they may be enjoyed almost anywhere. This category has been in part artificially created (artificial attractions) for the use of visitors or is a non-localized product of the past cultural development which is nowadays shown to tourists.

Artificial or contrived attractions are created "out of nothing" for the purpose of mass entertainment like amusement parks or theme parks (super-amusement parks). These complexes can be created almost anywhere, preferably at short distance from a large market. No special natural setting is necessary because capital investment and labor substitute for it in order to provide for high density recreation and entertainment. Examples of such complexes are: Disneyland, Anaheim, California, Disneyworld, Orlando, Florida, Sea World, San Diego, California, Lake Havasu, Arizona, Epcot, Florida,, Edmonton Mall etc. To the category of artificial attractions belong also computerized indoor golf courses, artificial iceskating rinks, various Skansens (open air museums of mainly traditional farm architecture reassembled in one place), artificial mountains constructed from solid wastes, rubble, construction fill, spoil earth canal construction etc. They may be used for various forms of recreation (Teufelsberg, West Berlin, Mt. Trashmore near Norfolk, Virginia, Blackstrap Mt. near Saskatoon, Saskatchewan). In the field of water-based recreation one could cite such examples as swimming pools, artificial surfs, mechanical water ski towing facilities substituting for more expensive, noisy and water-polluting motor-boats.

"Footloose" artificial attractions perform a very important role by inducing the tourists to stay longer at destination and thus increasing their total spending. For this purpose many destinations are provided with museums, exhibitions of local art, culture and history etc. in addition to the main attraction (excample - Niagara Falls). Tourists are encouraged to visit local factories, to meet with local people. There is no limit to human ingenuity in the attempts at tourists' dollar. Well known examples are the London Bridge at Lake Havasu, Arizona and the former British liner "Queen Mary" at Long Beach, California.

To the category of artificial attractions belong also the event attractions. Most of them are "footloose." They offer high density entertainment for thousands of tourists like fairs, festivals, rodeos etc. Some of the cultural attractions may have an artificial tinge because of their flimsy historical background (e.g. Dracula tours in Romania).

A variety of artificial attractions are points or lines of interest which are not distinguishable from their environment but regarded as important. Their characteristic feature is that in contrast to other artificial attractions they require very little, if any, investment, mainly for monuments. Examples of such points are the geographical centers of North America in Minot, North Dakota and of the continental US in northern Kansas near the interstate highway 281. The center of Asia lies near the town of Kizil, Tuwa, USSR. The poles as tour destinations are also gaining on popularity. As far as lines are concerned, people are attracted to the equator, the tropics and the zero meridian. Usually more or less elaborate monuments mark such spots. One of the most imposing is in Shantou, China, marking the Tropic of Cancer.

Recreation and entertainment are the major allure of artificial attractions. The artificial recreational facilities conform increasingly with the modern trend from passive to more active participation. However, new elements are appearing in some of them: people want to be not only amused and entertained but also instructed and enlightened.

Footloose cultural attractions consist to a large extent of ethnicity and all the cultural attributes connected with it (language, customs, songs, dances, old traditions and legends, etc.). The interest in folklore is widely met by providing performances of dance and song ensembles, folk orchestras, etc. In many places there are still amateur groups but one can count on them less and less. If they achieve a high level of artistry they turn professional anyway and the tourists are charged. This commercialization may be associated sometimes with loss of authenticity but in majority of cases the tourist demand for folklore contributes to revival and/or sustenance of folkloric traditions which otherwise may have been swept away by modernity.

The interest in ethnicity culminates in the so called "ethnic tourism" connected with visits of former emigrants or their descendants to the country of their origin. Some tourists may be interested in VFR, others in geneaology i.e. they may be looking for documents pertaining to their forefathers in parish offices, government registers, etc. The overseas Irish visiting Ireland are especially keen at that. The host countries are very much aware of the economic and political potential of ethnic tourism and trying to promote it (e.g. Peoples Republic of China, Poland).

The tourists's interest for folklore leads sometimes to rather unusual performances e.g. firewalking on the Fijian island of Beqa. A weird case has been reported by the press in 1973 when tribesman in Papua-New Guinea expressed their willingness to eat human flesh taken from the morgue at a folk festival. The government, of course, declined the offer (Winnipeg Free Press, 15-6-1973).

6.7.5. Tourism and the preservation of culture and traditions

Since our planet has become "a global village" and entered the modern age of conformity culture has fallen frequently victim of "modernization." The geographical cultural differentiation tends to lessen and even gives place to uniformity. Uniformity should be welcomed in technical matters all over the world (standards, measures, weights, etc.) but must be avoided in matters of culture. Tourism in its quest for variety is the best ally in preservation of cultural identity of regions and nations.

Indeed, there are many examples of tourism providing financial and organizational prerequisites for preservation of cultural treasures. Many human attractions are financially self-supporting and do not need gifts or grants. Preservation of many cathedrals, castles, fortresses, water mills, private and other old buildings which were decaying for lack of funds, has been assured by tourism. They function as museums, thriving hotels, (e.g. paradores in Spain, pusadas in Portugal) restaurants and other tourist facilities. In North America abandoned churches, banks, railroad stations, and other nostalgic places are being preserved that way. Tourism pays for preservation of old farm building in the form of Skansen museums, initiated in Skandinavia but now spreading all of the world. Some of these old buildings are being preserved as tourist settlements like Faakersee in Carinthia, Austria where ancient farm buildings have been reassembled and adapted to modern tourist use. The oldest house is from 1496. On the Yugslav Dalmatian coast a whole fishing village, located on a tombola, has been converted to a hotel (Svati Stephan). One of the contributions of environmental movements and tourism (which at least partially foots the bill) to considerable improvements in urban environments have been the recent reclamations of many historic inner city areas. Examples abound in Europe and North America: pedestrian malls with no vehicular traffic, beautiful open spaces plazas, public squares, historic waterfronts, parks, ancient fortresses, etc..

In most of these cases the preservation of cultural and historical attractions is paid mainly indirectly by tourism. Sometimes the governments initiate, plan, execute the renovations and provide funding and management of cultural and historical objects in recognition of their national importance and economic (i.e.

tourist) potential. A country open to tourism (e.g. Sri Lanka, Thailand) will as a rule take better care of its cultural monuments than a country like Burma which does much less in this respect and rejects even financial help for restoration work of ancient temples and pagodas offered by international organizations.

Among the international organizations providing the leadership in preservation of monuments of ancient culture the UNESCO has to be singled out. UNESCO can boast a distinguished record in this respect: the raising of the Egyptian temple of Abu Simbel to protect it from rising waters of Lake Nasser on the Nile in 1960's, the translocation of the temple of Philae below the Assuan Dam in Egypt in 1970's, and finally the 10 years effort to reconstruct and restore the largest Buddhist temple in the World Borobodur on Java, Indonesia, dedicated in 1983. These are only the most outstanding projects of UNESCO. Restoration of ancient monuments of culture all over the world expanded the ranks of professional restorationists or conservationists. One of the leading countries in this respect is Poland. Well over thousand Polish restorationists are working on projects all over the world.

Controversial forms of tourism, playing the role of protector of disappearing cultural traditions are the dance, song and other cultural performances for tourists. Also sale of souvenirs belongs to this category. Some praise tourism for its positive impact in this respect because tourism patronizes such manifestations of cultural heritage. Others criticize the role of tourism as the travesty of culture offered for profit for mass market. Indeed, both sides in the controversy may be right. One has to analyze the situation on case to case basis. If abuses are found, remedial action is imperative. Not only that: prevention is better than cure. This is another argument supporting the concept that education and planning on the supply side of tourism are an absolute necessity.

6.7.6 Tourism industry

An indispensable element of human tourist supply is the tourism infrastructure called tourism industry. Tourism industry is composed of various elements (enterprizes, facilities, agencies, etc.) which cater to the needs of tourists. These heterogenous elements have been traditionally highly fragmented and of rather small size. This circumstance weakened the market position of the industry. However, in recent decades a trend to integration has made the industry less fragmented and consequently more cooperative and better balanced internally. The integration proceeds by way of mergers and business take-overs either horizontally (e.g. hotel chains substituting for individually owned hotels) or vertically (mergers of travel agents, tour operators, carriers and hotels). The airline industry is recently very actively expanding into hotel and tour operating sectors. A fragmented industry is more dependent from the market (demand), an integrated one gains not only on economics of scale and efficiency but also on the market clout. Indeed the supply influences demand in a direct manner, supply "manipulates," creates demand. Big corporations integrate horizontally and/or vertically in all fields not only tourism. They do not wait for the consumers to be told what they wish. They plan, execute projects, produce goods and services and then persuade the consumer that it is just exactly what he or she wants. In this way the consumer power, consumer sovereignty gradually retreats giving way to new economic realities, realities indicating a severe curtailment, if not the total disappearance of the classical market with its free forces of demand and supply. Indeed, wants start more and more to originate with the producer than with the consumer. These tendencies are obvious also in the tourism industry: the supply side of the demand/supply equation starts to act increasingly independently from the forces of classical market. It does not mean that market research in demand is bound to decline. It is a very valid part of tourism research. Only the relationships between demand and supply have became more complicated necessitating a new approach to market research, taking into account that consumer is not an undisputed sovereign anymore. The concentration of tourism industry, we are today witnessing, is far from complete. Indeed, tourism industry is still to a large extent fragmented and subject to demand fluctuations. Nevertheless, the tourism industry is becoming more complex and costly and cannot afford to be subject to the whims of the consumer. This notoriously high risk industry wants to minimize the element of risk through integrated long-range planning based on scientific premises through total destination supply (using economies of scale) and finally through an advertising campaign shaping the consumer's wants, desires and needs.

Besides an increased tendency for vertical and horizontal integration (consolidation, concentration) there is a definite trend for spatial concentration in tourism. Spatial concentration means that one large company becomes the owner and operator of resort complexes which are fully serviced integrated resort areas where the visitors use a whole array of services, preferably within walking distance. This form of concentration of tourism facilities enables the company to develop plans both physical and economic, enables them to plan and operate the resort in the interest of environment. Indeed, better environmental impact control is another positive result of concentration in the tourism business. When a single governmental agency or private corporation controls the entire destination area, it is much easier to exercise or enforce the environmental standards that are a necessary condition for any sound development.

The tourism industry sells the tourism product i.e. attractions and services rendered by tourist facilities. The tourism product cannot be stockpiled, it is "perishable": empty airplane or bus seat, not occupied reastaurant table, vacant motel room is a wasted product which has not been used and cannot be used because of time lapse. Nevertheless the industry has to be ready with the service and in many cases actually performing the service, at least partially, even when the service has not been sold. Other industries are in a better position in this respect: production mustn`t be necessarily followed by immediate sale of the product. They may stockpile for a certain time. A car not sold today may be sold at a later time without the total loss of the product as it is in the tourism industry. This situation of high fixed costs or overhead weakens the market position of the tourism industry.

6.7.7. Accommodations

6.7.7.1. General trends

Accommodation constitutes an important maybe even the most important segment of the tourism industry, sometimes regarded independently as the hospitality industry. The tourists can travel without at least some other services offered by tourism business but certainly not without some sort of accommodation - this veritable core of the industry. Tourist accommodation in its size, forms and location has been influenced by developments in the field of transportation: every mode of transportation has its partner accommodation, the stagecoach - the tavern or inn, the railroad - the railroad hotel, the motor car - the motel and the airplane - the airport hotel. Various forms of accommodation have been spreading with the development of transportation network. Because of a relatively long lead-time between the project and the commissioning of a hotel, the local supply of accommodation frequently does not match the demand created by the expansion of transport. This locational inflexibility is the general characteristic feature of the tourism industry at large, but specifically it pertains to hotel-type accommodation, which is fixed at the location of production and consumption at the same time. The consumers (the tourists) have to travel to the goods rather than the goods to the consumers as in other sectors of economy where the goods follow the consumers. Hence the problems in adjustment of accommodation capacity to the fluctuating demand in time and space.

According to the World Tourism Organization terminology there are two categories of accommodation:

1) traditional (hotels, motor hotels, motels).

2) supplementary (youth hostels, campus accommodation, campgrounds, trailer parks, hostels, vacation homes, rented rooms, rented or swapped homes and apartments, sanatoria, condominiums). They offer accommodation but not the services of a hotel. There is less comfort but the prices are lower as compared with hotels.

The first group is also called main or primary, the second - para-hotels, secondary or complementary accommodation. Accommodation facilities open more than 10 months a year usually considered as full-year accommodations. Seasonal accommodation is open less than 10 months a year. The traditional hotels are graded by a five star or other official system according to the quantity and usually also quality of services offered. The criteria used to rate hotels are by no means universal. They vary from country to country and there are many important tourist countries with no national system of star rankings (e.g. Great Britain and West Germany). The problem with rating the hotels according to the quality of their service is, that the criteria take into account items that can be counted or weighed, but not the promptness and courtesy of service. There is just too much subjectivity on the demand side in order to establish truly "objective" criteria. Tourists have different tastes in terms of style, some do not require services (e.g. swimming pools) which are taken into account in the rating and for which they have to pay because higher ranking means higher room rates. Thus there is, perhaps, no necessity to establish an internationally valid star system.

Recent trends in the traditional accommodation could be summarized as follows:

1) Hotels have not escaped the world-wide trends to economic integration, both vertical and horizontal. The vertical integration predated the horizontal. It paralleled the development of the railroad networks in the second half of the 19th century Europe and North America when the railroad companies were interested in providing accommodation for their passengers. An example of such development

is the venture of Canadian Pacific Railways (CPR) into hotel business in Western Canada with the opening of Banff Springs Hotel in 1888. Also the shipping companies built hotels in conjunction with passenger transportation. Thus the Matson Navigation Company built the famous Moana and Royal Hawaiian Hotel near Honolulu, Hawaii. Nowadays not railroads but airlines are the primary movers in the field of hotel ownership, e.g. PANAM owns the Inter-Continental Hotels Corporation and Trans World Airlines owns Hilton International.

The horizontal integration in hotel business leads to establishment of huge transnational hotel systems (chains) controlling thousands and sometimes hundreds of thousands rooms all over the world. They are mainly US financed and controlled. Good examples in order of number of rooms: Holiday Inns, Sheraton, Ramada Inns and Hilton. Their advantages in the competitive market with the independent hotels are uniformly high standards of global service, integrated computer reservation systems, economies of scale and experience in purchases, planning hotel design, construction, management, labor pool, marketing etc. Nowadays the integration of the hotel industry appears to affect especially the middle-sized enterprises. What remains on the scene are the giant chains and small family run operations as their viable alternative (Tourism Intelligence Bull. 1988, Sept.). The Soviet block countries have also huge hotel chains only one for the whole country dealing with international tourism. They are: Balkantourist (Bulgaria), Intourist (USSR), Cedok (Czechoslovakia), Interhotel (GDR), Orbis (Poland). Frequently using Western capital and expertise they have built a number of luxury hotels in recent 10-15 years. The reasons for the integration of the hotel industry in the Soviet block countries are different than in the West: here it results not from the impact of market forces but is caused by the command economy which is based on public (government) ownership of all production of goods and services.
2) One of the most important characteristics of modern hotels is their architectural (both exterior and interior) design. It is a specialized field of architecture geared towards maximum of economy and convenience for the hotel guests. One of the numerous achievements in this field has been the standardization of design aimed at increased economic efficiency and higher standards of comfort. However, the standardization leads also to undesired results which militate against the principle that tourism is interested only in standardization in technical terms e.g. electrical outlets, reservation and payment procedures. As far as culture, traditions, landscapes, townscapes etc. tourism thrives on variety. Therefore Hilton hotels in Istanbul and Tokyo, which look identically both from outside and inside, are not

desirable. Criticism of this cosmopolitan similarity of hotel architecture has resulted in changes which are gradually taking place. There are a number of beautifully designed hotels which in their interior and exterior architecture reflect the traditions and culture of a country where they are located. Examples are hotels at Morea in French Polynesia, the Hilton International's Salt Lick Lodge in Tsavo National Park in Kenya. At the island of Djerba in Tunisia it is difficult to distinguish from a distance some of the hotels from native villages. In many places new hotels will not get the building permit if they are higher than palms.

The use of local design and local building materials gives the hotels the flavor of the country, its culture and tradition. At the same time, however, efficiency does not need to be sacrificed: the technical details should remain uniform world over.

3) Numerous new hotels of today were built hundreds of years ago and used as fortresses, castles, palaces, monasteries. These architectural treasures of the past have been saved by tourism from destruction. Their upkeep costs were simply too high to serve according to their original function. Tourism took over these structures and is using them as hotels to everybody's satisfaction: the monuments are preserved and tourists thrilled and flattered to live in such places. Good examples of such solutions are publicly owned Spanish paradors and Portugese pousadas.

4) In the period since 1945 the traditional forms of hotel business have expanded to take new forms and locations. The traditional commercial hotel (called by C. Gee, 1981, "transient hotel") located in the city center has proved inconvenient for the motorized tourists. Therefore, the important addition to the accommodation supply was the motel - an off-shoot of the hotel industry catering to motorized tourists. Motel rooms are mostly accessible from the exterior. Motor hotels resemble more traditional commercial hotels. The difference is in provision of free or low cost parking on the premises. Another form is the airport hotel saving the inconvenience of travelling to the city center. Airport hotels today cater not only to transients but also serve as places for business meetings, conventions, etc. Resort hotels are located in non-urban areas and will be discussed later. One of the most recent forms of resort hotels are boatels or floating resorts. Boatel is either located at the mooring site or on a ship of various size. Some of these ships are permanently moored like the old Queen Mary at the waterfront of Long Beach, California, some could be moved from place to place according to necessity. Boatel is a link between hotel and cruise ships. First such specially built 200 room floating hotel was opened in 1988 off the Northeast coast of Australia at the Great Barrier

Reef. The name of the hotel "Four Seasons Barrier Reef." Built at the cost of $21 million by a company specialized in construction of oil rigs and floating drydocks, this resort hotel opens a new era in recreation tourism. Among other types of hotels one could mention residential hotels representing a very small segment of the hospitality industry. Residential hotels provide accommodation for people staying for longer periods of time but requiring hotel-type services. "Baby hotels" developed in Austria specialize in accommodating families with very young children.

The richest system of hotel types have the French: hotel permanent (full time), saisonier (season), garni or meublee (furnished), de la gare (RR), a voyageurs (for transients), commercial (business), de residance (residence), hotel-appartements (apartment hotel), pavillonaire (bungalow). The resort hotels are called hotel thermal, hotel de cure or hotel des stations balneaires. Hotel "classée" or "homologue" means that it has been assigned a classification within the five-star system.

5) Quality supply in hotel services has widened considerably in the last decades. The old-fashioned small luxury hotels still cater to exquisite clientele attracted by high staff per guest ratios of 2 or 3 and personalized service. Some of these hotels like Ritz in Paris or Singapore's Raffles Hotel are institutions unlikely to fade out. However, many of such hotels have closed down unable to compete. At their place (or near-by) modern, large, sometimes gigantic luxury hotels are being constructed like the "Raffles City" complex in Singapore or "Asiaworld" in Taipei, Taiwan.

Hotels compete by offering an increased array of services. In some cases the supply of services is so wide that the guest practically does not need to leave the hotel grounds in order to satisfy his or her needs. "Total destination" hotels and motels expand to resort complexes of various sizes adjusted if necessary, to local climatic conditions. This frequently opens new markets for hotels which offer special weekend discounts. Not only tourists but also locals relax in such "total destination" hotel complexes. Large resort hotels are sometimes difficult to distinguish from what we are accustomed to regard as resorts. The largest hotel in the world is the gigantic Rossija in Moscow with 3000 rooms for 6000 guests. Normally a hotel with 400 rooms is regarded in the West as large and more profitable than a smaller hotel. However, the giants are too cumbersome apart from obvious design flaws of Rossija.

The modern demand trends for accommodation indicate that tourists increasingly are staying away from large luxury hotels. Indeed, there is an excellent market for budget accommodation with minimum service. This market are budget and price-conscious business travellers and pleasure tourists. Hotel business is labor-intensive and cutting labor costs, minimizing the service and emphasising self-service may lead to very competitive prices. Recognition of these new business opportunities prompts even some luxury chains to establish new divisions which fit into this category. Examples are the Forum Division of Inter-Continental Hotels and Hampton Inn Hotels, a subsidiary of Holiday Corporation, established to meet the growing market for moderately priced (budget) limited-service hotels. In some cases hotels and motels not only limit but completely stop room service ("To protect your privacy no room service is available"). In 1983 a 1301 room "Washington Hotel" in Tokyo was opened with automated check-in service. Four machines take care of the formalities and dispense magnetic cards which open the room doors. The items taken out of room refrigerators are automatically registered by the central computer, the check-out takes 45 seconds. If needed, also personal service is available at Tokyo's Washington.

An extreme form of budget hotel is "capsule hotel" where the guests are sleeping in coffin-size reinforced fibre modules with about 1.2 x 1.2 x 2.5 m dimensions arranged usually at several levels. Individual capsules are supplied with a mattress for sleeping, a light and individually controlled air conditioner. Some modules feature telephone, TV, radio. Washroom facilities and lockers are outside. Introduced in Japan in 1979 capsule hotels are gaining foothold in other countries. They are very convenient for airports and railroad stations, especially because they can be rented by-the-hour.

Budget hotels in North America have developed into motel chains offering very decent accommodation and limited services at modest prices. Swimming pools, restaurants, convention facilities are normally not available. Other saving is a skeleton staff only. As a result a limited-service operation is roughly twice as profitable as one providing a lot of amenities (Time 15-7-1985). Also the occupancy rates seem to be higher than in luxury hotels. Within the no-frill hotels the tourists have a wide choice of options ranging from medium priced (Day Inns) to bargains (Motel 6).

Budget and medium-priced hotels enjoy a boom throughout the world and expand their market share e.g. in the US the budget segment accounted in 1987 for 18-20% of all the lodging industry's rooms as compared to about 8% in 1980 (Waters S. 1988:128). The same trend may be observed with respect to supplementary accommodation. However, there is also expansion at the other extreme: the demand for all-suite hotel accommodation is growing (Waters S. 1988:127) "The American Hotel and Motel Association has estimated that the all-suite segment will grow at an annual rate of 25 percent to 50 percent over the next five years, increasing the segment's share of the room supply from 3.5 percent to 10 percent." (Tourism Intelligence Bull. 1988, Oct.). The consequence of these trends in hospitality industry is an increase of supply on both extremes of price continuum resulting in an incredible variety of accommdation in terms of types and rates. The customers are price sensitive, they want value for money. At the same time, however, there is a sector of the market which looks first of all for comfort, security and other upscale standards and features.

6) For LDCs concerned about leakages of foreign currency small to medium size hotels are the best solution. Such hotels are frequently owned and operated by local families and are supplied locally to much greater extent than international hotel chains.

7) The geographical distribution of tourists accommodation is pretty predicatable: most of it is located in the DCs. There were about 10.6 million rooms in the world in "hotels and similar establishments" (Waters, S., 1987:117) about 29% of them in North America, 52% in Europe. It is unlikely that this spatial concentration will change substantially in the forseeable future.

6.7.7.2. Hotel management problems

We will limit ourselves to mentioning some of these problems with which the hotel management and the guests have to struggle on the day-to-day basis.

Hotel overbooking causes friction between management and guests who frequently find that their confirmed reservations are not honored. The reason for overbooking is a high percentage of no-shows and the desire of management for protection against losses. Also hotel managements endeavour to maximize their occupancy rates (the relationship of guests to available beds, expressed in percent). Other aggravations are untrained or rude staff members, accidental or deliberate padding of bills, deficiencies in wake-up service, malfunctioning of heating or air conditioning, discriminatory treatment of single travellers and especially single women, etc. The hotel management has also to cope with increased theft by guests and problems with items left by forgetful guests in their rooms after departure. The composition of guests has an impact on hotel staffing: groups of travellers (e.g. inclusive tour groups) require relatively more staff than individual travellers because they use more fully the hotel facilities (especially food service) and are less likely to look for services outside the premises.

6.7.7.3. Development of supplementary accommodation

The post World War II period has witnessed a tremendous development of supplementary accommodation. In fact, this type of accommodation has significantly increased its market share as compared with the traditional forms. The reason is price. In some areas the supply of budget hotel-motel space may be inadequate. The tourists invariably look for alternatives. A list of these alternatives is long and they will be mentioned or discussed not necessarily in order of their importance. Increasingly popular is summer renting of vacant dormitory rooms belonging to various financially strapped educational institutions (e.g. universities), house swaps for vacation even on different continents, student exchanges, youth hostels, second home ownership, time-sharing or renting of condominiums, rented or swapped apartments or vacation homes, rented private rooms (frequently in the bed and breakfast form) farm and ranch (dude ranches) accommodation etc.

The benefits of these types of accommodation reach far beyond their relative cheapness: there are health, social and cultural advantages in addition to economic benefits for guests and hosts. e.g. farm and ranch vacations enable participants to live close to nature, participate in vigorous recreation activities and enjoy local folklore and close contacts with hosts. Two of these forms of supplementary accommodation will be discussed in some detail: private holiday accommodation and camping.

6.7.7.4. Private accommodation

One of the increasingly popular forms of accommodation is private rental or ownership of rooms, suites, condominiums, houses, etc. In most cases this form allows the tourist to escape from impersonality of big hotels, to establish contacts with local population so helpful in learning more about the area or country visited. Especially advantageous in this respect are the bed and breakfast (B & B) arrangements. The advantage is not limited to contacts with locals but extends also to pricing: private accommodation is as a rule considerably cheaper than hotels. Less popular in North America than in Europe, the B & B is especially useful in accommodation peak demand: seasonal and associated with event attractions such as EXPOs.

The quality and prices of private accommodation range from modest rooms in city and especially countryside to accommodation in palaces owned by aristocracy. The former provide the house owner with additional income, the latter constitute a tourism's contribution to the upkeep of aristocratic residences which became uneconomical in modern times. Surely, these ancient castles, palaces and fortresses represent architectural values worthwhile to preserve. Tourism income is rescuing these monuments of the past from neglect, ruin and demolition. Examples abound from all over the world e.g. Great Britain and India. In some cases contacts with the aristocratic owners constitute an additional attraction, sometimes the arrangements approach hotels.

In the post-World War II Europe and North America a type of accommodation has become very fashionable which accords maximum privacy and independence and also investment which is likely to appreciate. This accommodation is variously called summer cottage, vacation home, winter chalet, vacation condominium or apartment, etc. - a privately owned second home usually exclusive property of the owner but increasingly on the time-sharing basis. The quality and prices range from primitive and cheap to luxurious and expensive, and the owners from working class to millionaires. Generally speaking the ownership of a second home constitutes a drain on discretionary income and time of the owner and therefore is associated with less participation in other forms of tourism. The ownership of second homes is also positively correlated with families, especially families having many young children. Among the reasons one could quote the expenses connected with family travel. Second home ownership helps to reduce the vacation expenses and also provides the family with quality leisure time during weekends. Not all second homes constitute structures especially built for this purpose, some of them particularly in Europe e.g. in Switzerland are converted farmhouses which makes them in many ways more appealing. The second homes have become so popular in developed countries that in many places they have grown together into virtual cities vibrating with life in summer and empty in winter.

With respect to spatial functions there are three types of second homes:

1) Located in the near-urban zone. The family moves out during the summer from the primary residence and lives during the whole warm period of the year in the second home. Working members of the family have to commute to work longer distances than during the rest of the year.

2) The second home is located in extra-urban area too far for daily commuting. The family spends only weekends at the cottage.
3) The second home is located at considerable distance from primary residence too far to travel on weekends. Second homes of this category are frequently located in another country, e.g. Scandinavian or German-owned homes located in France, Spain or Italy, sometimes in another continent e.g. North American-owned homes located in Greece. In such a case the family moves only for vacation and the house is not used during the rest of the year or there is a time sharing arrangement.

A special category of second homes constitute mobile homes, especially popular in North America, where mobile homes with higher degree of comfort are in use. The travellers stop in specially equipped mobile-home parks, especially frequent in such vacation states like Florida. Many retirees are using these "homes on wheels" spending winters in the south and summers in the north.

The second homes are constructed in various natural environments. From this point of view one could divide them in following types:
1) Seashore environment
2) Mountain environment
3) Lakeshore

In North America second homes are real homes in nice natural setting. In Europe with its lack of space and high land prices the trend is towards purchase of condominium apartments sometimes located in huge blocks or high rise apartment buildings like e.g. near Travemünde in West Germany at the Baltic Sea or La Grande Motte in France at the Mediterranean coast. Another example is the Gold Coast, Queensland, Australia. The buyers or renters are mainly younger people who do not have the means to purchase a detached vacation home and/or want to avoid the house-and-garden keeping duties while at leisure. The environment in such resorts is almost urban. The vacationers desire to have close-by opportunities for shopping, sport and entertainment. Social contacts with neighbors in these "vacation factories" are usually minimal, like in the cities.

The economic and environmental impact of second homes on the host region constitutes a controvertical issue. The advocates of second homes argue that the often depressed host area benefits economically and environmentally as a result of second home development which broadens the tax base, injects money, provides jobs, beautifies the landscape by creating a rather dispersed, park-like settlement pattern. The detractors insist that these statements are exaggerated and even sometimes untrue. They maintain that the contributions to the local economy are rather modest. Frequently second homes constitute a burden rather than an asset for the local economy. Their owners pay too small taxes. The injection of money and creation of jobs occurs only during the initial period of construction to decrease later considerably because the owners frequently bring their own supplies and do not look for local services. The only local people who gain economically are those who sold their land, cottages, farmhouses. But the impact of these purchases inflates the real estate prices forcing many locals from their land. Also the local residents often face the costs of financing the expansion of basic infrastructure and other services, resulting from second home development, which in this way constitute a drain to local economy. Also large cottage agglomerations are not appealing environmentally. In many cases they constitute eyesores, vast slum cities despoiling the nature, polluting air and water, disturbing by noise of outboard motors. The arguments go on. The development of second homes certainly should be judged on case to case basis without the prejudice of blanket statements.

6.7.7.5. Camping

Camping is one of the most dynamic forms of supplementary accommodation which has shown significant growth since 1946 both quantitatively as well as qualitatively. The carefree world of camping attracts participation because it is a perfect escape from monotony and drudgery of everyday life, a real change of role, opportunity to "rough it" and spend wonderfully inexpensive leisure time in new natural and social environment. Families with children especially enjoy camping not only because of its economic aspects but also because it strengthens intra-family bonds by sharing experiences while removed from the pressures of urbanized society. The relative cheapness of camping tends to increase the time and perhaps to some extent the distance associated with this activity.

The camping boom since the end of World War II is mainly viewed from the quantitative point of view i.e. through the imposing statistics showing the exponentially growing participation figures. However, one frequently forgets about qualitative change which has occurred. Traditionally camping was an antithesis of urban life, a quest for isolation amid unspoiled beauty of nature. Camping grounds were located mainly in faraway places as close to nature as possible. Facilities were minimal, tenting the prevailing form.

This traditional approach to camping has changed significantly with technical progress involving both camper's and, campground equipment. As far as camper's equipment is concerned tent camping has declined relatively and new forms of "camping on wheels" have gained popularity: various car-trailer combinations ranging from tent-trailer, tent mounted on trailer to sophisticated fully self-contained house on wheels. Other forms include truck mounted camper motor homes. All these forms of Recreational Vehicles (RVs) have increased the convenience of travel and even added more than a touch of luxury to it. Also the campgrounds have changed: once small and primitive, many of them are large and highly developed with all possible equipment and amenities: electricity and water supply, garbage and waste disposal, shopping, sports and entertainment facilities, etc. Fully serviced trailer parks of today have substituted for tent campgrounds of yesterday. Whereas the traditional campers are seeking communion with nature, the majority of contemporary campers are looking for convenience and less for environmental and more for social experience i.e. for social interaction between campers. Indeed, nowadays camping attracts participation to a large extent because of companionship, community feeling, wish to meet and associate with new people ("trailer subculture"). The composition of participants has also changed. Formerly young and low income people (which meant largely the same) participated. Nowadays the income range is very wide from low to upper middle incomes. The age range of camping participants has also widened and encompasses today all age brackets. It is interesting that many retirees participate even full time in a sort of peripatetic retirement: e.g. in U.S. the winters they spent in their trailers in the south to migrate to the north in summer.

Modern camping has become a serious competition for hotels because of pricing, greater sociability among guests, in some cases even superior shopping and sporting facilities. To widen the market the camp operators are seeking to imitate hotels in the range and quality of services and amenities while beating them in terms of prices. The results are at hand: in some countries more vacationers are staying in campgrounds than in hotels. The national or regional government controlling bodies classify campgrounds according to various rating systems e.g. star system. Similarly like with hotels there are also campground chains like KOA (Kampgrounds of America). The majority of camps are private but on public lands they are owned and operated by government agency responsible.

A distinctive form of competition even with the budget hotels is, a West German invention, so called rotel, i.e. rolling cubicle hotel pulled by a bus. The rotels are normally using campgrounds for overnight stay.

According to spatial distribution, relationship to resources and degree of development the campgrounds could be divided into following categories:
1) Primitive tent camping and campgrounds in roadless scenic wilderness. This is the traditional resource based form where the emphasis is on closeness to nature and privacy.
2) Centralized resort complexes in semi-wilderness or some scenic nature setting. Intensely developed, offering a wide array of services and facilities. Length of stay as a rule more than one day because of their "terminal" function. Accessible by road.
3) Wayside transient camps differing substantially in services and amenities offered. Short stays of one day for transient tourists typical. They play a roles of a cheap motel.
4) Trailer parks in near-urban or urban areas offering sometimes natural amenities like lakes, mountains, forests. Highly developed with wide range of facilities and amenities. Trailer sites frequently on long term rental basis. Trailers used as substitutes for cottages, if insulated, even year-round.

Ideally only the first category, if located on publicly owned land, should be operated by government agencies e.g. the US National Park Service and CNPS in Canada. The public campgrounds are run on non-profit or even no self-financing basis. Other forms of camping, better serviced and maintained, should be in private hands. The northern fly-in fishing and hunting wilderness lodges in Canada, so popular with exclusive clientele, belong rather to the hotel category. If they are operated as campgrounds then they are normally privately operated, even when located on crown land.

Campgrounds, like hotels, should be run for profit. This does not mean that the social welfare function of recreation should be discontinued. Such a function should be maintained in public parks, especially wilderness parks where some measure of subsidy is, in fact, necessary. The limits of this subsidy are certainly debatable and changing in time.

There are some environmental and social problems connected with camping. First of all carrying capacity of the eco-systems, and especially vegetation, should be taken into account. Fragile environments should be avoided for campgrounds. Therefore, camp siting should be subject to careful research to optimize the conditions for the campers but at the same time to minimize the negative environmental impact. The site selection should base on soil drainage, topography, vegetation and climate analysis. The soil should be of medium texture, well drained, the terrain slightly sloped but not more than 10%, the vegetation resistant to trampling. In this respect botanists can help with resistant grasses and shrubs, the landscape architects with arrangement of trails. The vegetation at the site should open to the East in order not to obstruct the penetration of the morning sun. The camp should be shaded from other sides. However, extremely dense vegetation is also to be avoided. The distribution of comfort stations should be carefully planned. The most preferred individual sites are not too close and not too far from them.

There are a number of regulations pertaining to environmental protection of campground sites. These regulations start with the minimal distance limits of sites from a road or even trail in wilderness camping. Other regulations stipulate the size of an individual site to avoid overcrowding. Of course, the low density sites are normally associated with minimal services, the high density campgrounds as a rule are intensively developed.

The camping regulating organizations vary. The national, regional and local bodies have more practical clout than international ones. However, quality control and licencing seems to transcend national boundaries. A good example is the International Federation of Camping and Caravaning (F.I.C.C.) based in Lucerne, Switzerland. The approval of F.I.C.C. means that a campground will appear on a list which recognizes its high quality. It appears that the influence of F.I.C.C. does not extend far beyond Europe. Another organization which campaigns for quality camping is the International Office of Social Tourism (B.I.T.S.).

The social problems associated with camping are being dealt with by local authorities. This is the increased vandalism, noise, theft, alcohol and drug abuse, etc. To prevent such occurrences the interpretative, educational and policing programs are necessary.

6.7.8. Tourism settlements

6.7.8.1. Classification

One of the earliest classifications of settlements from the point of view of tourism was proposed by Grunthal in 1934 who distinguished between settlements without tourism, settlements with tourism and tourism settlements (Grunthal, 1934:12). Since that time one can find in the literature a veritable plethora of classifications. Reviewing these classifications one comes to the conclusion that tourism settlements are much more differentiated than other types of settlements which are more uniform. This results in more difficult planning problems.

The criteria used in these divisions vary. From the point of view of spatial economics one is interested in researching spatial links between tourist places and other settlements and areas. Thus one focuses on relationships between tourist settlements and settlements of tourist origin or catering and supply centers and areas for tourist industry. Also transfer (gateway) centers for tourist flows to tourist areas are of interest (e.g. Denver, Salt Lake City and Calgary for the Rocky Mountain skiing areas).

The settlement geographers have developed their own typology which takes the form and the structure of settlements as criteria of division. As far as terminology is concerned they operate with such terms as transit-gateway centers, rural tourist villages (symbiosis of agriculture with tourism), resorts, tourist cities etc. but traditionally they focus on settlement morphology.

However, from the point of view of tourism the most important general classification of settlements bases on their resources and functions. This author suggests the division into cities and resorts.

6.7.9.2. Cities

We usually associate tourism with sunny beaches, mountain peaks and recreational activities in natural setting. Cities are commonly regarded as sources of tourist flows, as places of origin and not tourist destinations. Such views disregard the fact that cities, especially large metropolitan areas constitute special attractions because of their history, culture, educational, entertainment and shopping opportunities. Such cities like Paris, London, Rome, Berlin, New York, to name only a few, have been for years powerful magnets for millions of visitors. As a result the capacity of the tourism industry, especially that of hotels, in many instances surpasses the capacity of extra-urban areas (Pearce D. 1987:179-180). In some cases (e.g. Rio de Janeiro, Capetown, San Francisco) city's attraction may also be of natural character because of particularly beautiful physical environment. Not only large but also medium-sized and small cities attract considerable numbers of tourists, e.g. Venice, Florence in Italy and Rothenburg in West Germany. Particularly high is the importance of cities for international tourism despite the fact that the average numbers of overnight stays in the cities tend to be considerably shorter than in extra-urban areas (typically 2-4 days). Other characteristic feature of city tourism is that the attractions patronized by tourists like museums, theaters, art galleries, architectural monuments, parks, restaurants, sports stadiums, rodeo arenas, entertainment and nightlife, enterprises, factories, scientific institutions, shopping facilities, etc. are simultaneously patronized by local residents, which makes the statistical evaluation of tourism impact more difficult. Many of these attractions require large numbers of visitors in order to survive economically and tourism constitutes an important factor contributing to their sustenance. In certain cases tourists may compete with locals for use of some facilities, in other instances they may help to even out the demand in off-peak periods like e.g. when the locals are on vacation or during weekends and the demand for transportation services or hotel space is lower. Weekends constitute favored time of visitation for "pure tourists," business tourism is concentrated on week days. Another function of urban tourism is that the city frequently serves as a base for excursions to the surrounding areas. The tourists leave for a day e.g. Paris to visit Versailles and other suburban palaces, they venture farther away to see the cathedral in Chartres or even the Loire Valley.

The recent growth of urban tourism promotes the counter-seasonality in tourism, helping to spread the demand more equally throughout the year because a large part of city attraction is of indoor character. Additionally, special efforts are being made in large urban centers like New York, London and Paris to attract off-season tourists to their theaters, operas and concert halls. There is another indirect advantage of the boom in urban tourism: the diversion of the demand from fragile non-urban environments to the cities contributes to the conservation of nature and decrease of negative environmental impact of tourism.

In the recent two or three decades many cities, and especially their cores, have become more attractive for tourism as a result of urban renewal programs which have successfully changed the formerly decrepit city centers into renewed clean townscapes with shopping areas, pedestrian malls, monuments, museums, theaters, open green spaces. Architecturally blending old and new, or constituting entirely new townscapes, the rejuvenated city centers and especially reclaimed historic inner city areas, appeal to millions of tourists. These favorable developments on the supply side of tourism are the results of a number of initiatives undertaken by governments and private organizations all over the developed world. In some European cities the urban renewal programs were the response for World War II destruction (e.g. Rotterdam, Warsaw), in other cities, as in North America they resulted from pressures to improve urban environments not only for esthetic but mainly for socio-economic reasons as a response to progressing decay of central cities. An example of such initiatives is the Historic Preservation Act of 1966 and the National Environmental Protection Act of 1969 in US. These and other legislations gave the green light for massive financial public and private sponsoring of urban renewal projects all over the country.

The examples of the renewal of city core areas, frequently of significant historical and artistic value, are many in almost all DCs. One could mention the British city of Bristol or Canadian cities of Vancouver, Winnipeg and Toronto. Even some core areas of some more well-off LDCs have been recently renewed (e.g. Rio de Janeiro and Curitiba in Brazil). However, the cities of the DCs are in the lead for obvious reason: considerable costs of such renewals. The decrepit part of central Paris have been reconstructed at great costs: Les Halles were "transformed" into Forum des Halles and the Pompidou Centre constructed, the Musee d'Orsay art museum was placed in the renovated old RR station etc.

The success of cities as tourist destinations brings many benefits: environmental, cultural, social and economic. The presence of tourists in the cities improves safety, makes them vibrant living communities. However, there is also another side of the coin which shows some disadvantages of the tourist boom in cities. Too much tourism, beyond the saturation point causes traffic congestion, parking problems, noise and air pollution. Therefore, management of tourism is indispensable. The techniques used include: restrictive zoning, controlled access for busses and cars, traffic-free Blue zones, encouraging city walking tours, introducing shuttle bus service etc.

A classic example of city congestion and pollution caused by local and tourism traffic was Florence. Belatedly, action was taken in 1988 by restricting the motorized traffic in inner city. In this way not only the traffic situation has been improved, but also pollution, especially harmful to marble architecture and monuments, cut significantly.

Tourism management in self-contained historic restorations in museum-cities like Pompei or Williamsburg, Virginia is easier because there no vehicular traffic is allowed. Therefore, the management techniques applied are much simpler.

Despite the problems the future of high-density urban tourism seems to be bright as the interests in cultural and socio-economic attractions grow with the increasing educational levels of visitors. Thus city tourism, although only part, normally by far not the most important part of local economy, plays and is going to play a vital role in the life of cities.

The spatial patterns of accommodation facilities in urban areas reflect not only the needs of tourists but also the historical expansion of the cities. The small and medium size hotels, located within the Central Business District (CBD) or on its fringes, cater to business travellers, shopping-, sightseeing- and entertainment-oriented visitors. However, large modern luxury hotels have to locate outside the city core, sometimes at considerable distance. Their location is dependent mainly on availability of space and the development of urban transportation systems. Recently the airport hotels have increased their capacities considerably, catering mainly to business travellers and transients on connecting flights through the air traffic hubs (gateways) e.g. Toronto, Canada.

Cities are not only centers of urban tourism. One of their multiple functions is provision of urban recreation for visitors and locals. In fact, urban recreation is the fastest growing form of out-of-home recreation. To achieve this aim one has to provide facilities like swimming pools, tennis courts etc. However, provision of urban green space, mainly in the form of various types of parks, seems to be the task of paramount importance. The economic, physiological and psychological benefits of urban open (green) space have been discussed by many authors (e.g. Driver, B. et al 1978). Without going into details one should indicate that provision of urban green space is an integral part of city planning and that it gains increased attention of local authorities in more recent time. The mistakes of the past are being corrected despite formidable obstacles. The development of the recreational services has also suffered recently as a result of increased costs of liability insurance, especially in the US.

A separate specialized form of urban tourism are pilgrimage cities which develop relentlessly from small villages, and frequently in places where there was no settlement at all. Examples of such urban centers specialized almost entirely in catering to pilgrims are Lourdes, France and Fatima, Portugal. There are, however, also important cities where the religious tourism constitutes only one of multiple city functions e.g. Rome, Italy and Czestochowa, Poland. Also Jerusalem belongs to this group. Despite the shortness of the visit, centers of pilgrimage should be regarded as one of the major contributors to urban tourism with conspicuous economic impact on tourism business. The accommodation and food catering facilities are mainly of mass budget quality. The seasonality is dependent not so much on climate but rather on event attractions scheduled at various times of the year.

6.7.8.3. Resorts

In contrast to most cities, tourist resorts are specialized settlements where there is relatively little economic activity not connected with tourism. They differ very much in size. Some are as small as one hotel with 100-200 rooms which blurs the distinction between a resort and a hotel. In such cases a term "resort hotel" is used. On the opposite side of the continuum there are "resort towns," or even "resort cities" of truly urban dimensions (Waikiki Beach, Hawaii). Most guests stay at a resort more than 4 nights ("vacation resorts") although some are patronized also on weekends. Resorts differ substantially in the quality of services provided and consequently in the rates paid by guests. They range from simple domestic resorts in India and modest "ashrams" in the DCs to utterly luxurious exquisite "club" members-only resorts in secluded islands of the Caribbean. The business is normally seasonal, although in cases where the resort performs both summer and winter functions, the off-season could be relatively short. There is a definite trend to extend the season of operation. To achieve this various marketing strategies are used e.g. organization of various festivals, conventions, meetings, courses, provision of more recreation activities (golf, tennis, swimming pools). Very useful for marketing of a resort is a prestige image and good reputation concerning the attitudes of locals towards tourists. In some resorts there is a higher percentage of visitors in older age brackets. This allegedly hurts their image (e.g. Miami, Florida). Indeed, retirees frequently choose to establish their permanent domicile in the pleasant resort environment. This reportedly discourages the patronage by visitors in younger age brackets, especially the youth.

Characteristic features of resorts

A modern resort must have a number of characteristic features in order to be successful. Of course, not all these features are a "must" for a single resort.

1) Natural features: attractive natural setting with distant panoramic views, scenic mountains, waterfalls, lakes, beaches, existence of mineral springs of preferably thermal waters, mild macro- and micro-climatic conditions which would provide relief from either excessive heat or cold. Extra-urban location is preferred for resorts.

2) Accessibility - not too large distance from the market, easy swift and comfortable means of transportation. Some resorts have special airports. (e.g. Punta Cana, Dominican Republic)

3) Good general (basic) infrastructure.

4) Good tourist infrastructure (superstructure) i.e. accommodation, food, shopping facilities, internal transportation.

5) A package of attractions, entertainment, and wide range of recreational activities which through their diversity and quality would create conditions of self-contained "total destination" or "resort complex." Most resorts are self-contained in terms of points 3-5 (e.g. White Sulphur Springs, West Virginia). The "resort complexes" with their self-contained entertainment and recreational facilities promote the feeling of unity, cohesion and social togetherness among the visitors. This applies especially to Club Mediterranée members (GMs) who are living in "villages." Such terminology conveys the sense of human scale and social togetherness.

The resorts could be classified as follows:

1) Health resorts
2) Seaside resorts
3) Mountain resorts
4) Climatic resorts
5) Retirement communities

Health resorts

Health resorts are based on balneo-climatic resources. They are also called spas and represent what could be called "classical" resorts as historically the earliest. The emphasis is on cure by means of scientific application of balneology and hydrotherapy. Spatial morphology assuming flatland location: centripetal around the core (curative facilities, concert hall, promenade, park, etc.).

The curative properties of water and mud were probably discovered in prehistoric times (Lehane, P., 1974:405-408). The Romans attached great importance to these properties and used them in a number of health resorts located all over their empire, even in distant periphery e.g. Aquae Sulis (Bath) in Britain. During the Dark Ages following the demise of the Roman Empire the health spas were forgotten and banned as centers of evil serving the loathed body and not the soul. Their gradual revival started in 16th and 17th century Europe and was associated with the development of balneology, the science of therapeutic use of baths, and later of hydrotherapy, scientific use of water for curative purposes. Health resorts acquired a new name "spa" which originally meant a mineral spring but later has been associated with a health resort in South-Eastern Belgium founded in the middle of 16th century. During the 18th and first part of the 19th century European health resorts achieved the peak of their development. Their functions and forms changed but their importance grew consistently until about the middle of 19th century.

The curative functions of mineral and thermal waters used both internally (drinking) and externally (bathing) and of mud lost gradually their primary importance. With the time social and cultural functions of spas became prevalent and the medical aspects were frequently used only as little more than excuse for socializing. As a result the morphology of spas changed. In addition to bath houses and mineral water drinking halls also lake piers, promenades, parks, gardens, concert halls, amusement parks, zoos, theaters, exhibition buildings, casinos, etc. were developed or constructed. Social and cultural life focussed the attention of visitors, "taking the waters" was mostly an excuse. The leading inland spas of Europe, most of them developed in 18th and 19th centuries were: Bade-Baden, Wiesbaden, Homburg, Ems, Nassau in Germany, Vichy, Plombieres in France, Karlsbad (now Karlovy Vary) and Marienbad (now Marianske Lazne) in Austria-Hungary, Koslovodsk, Piatigosk, Mineralye Vody in Russia. The North American resorts developed in the 19th century. Some of them date from the pre-railroad era (Saratoga Springs, N.Y., White Sulphur Springs, Virginia) but only the railroad proved to be their decisive "boost factor." Their mineral springs which brought them fame initially, were soon ignored or have played a marginal role. Their activities have concentrated on socializing. Saratoga Springs has focussed already in the second half of 19th century on horse racing, concerts, dances, etc. Alberta's

Banff was developed at the end of the 19th century in connection with the Canadian transcontinental railroad. Its initial attraction - hot mineral water - has also gradually given way to other activities.

During the 19th century the gradual decline of inland spas occurred, initially only in relative terms as compared to other resorts. Later many resorts declined also in absolute terms. The reasons were many: first of all it was the competition of seaside resorts and in the 20th century, mountain sport resorts, and also the competition from other forms of tourism like e.g. itinerant domestic and international tourism. But beside competition there have been the loss of credibility in the curative powers of mineral waters. The medical profession has relied increasingly on other therapeutical methods to combat disease. Thus many traditional health resorts, especially located in colder regions with a short season, have vanished from the map. Others had to adjust to survive in new conditions. This adjustment in many cases meant loss of identity as a spa, or radical change to a new role e.g. of a dormitory suburb for a large metropolitan center. Some Belgian spas are today inhabited mainly by commuters (car, train). Thus many health resorts have lost to a large extent their character. The same may be told about transformation of some spas into retirement communities e.g. St. Petersburg, Florida. In some cases introduction of other economic activities have contributed to the loss of health resort identity. Such incompatible developments have taken place in former German spas Wiesbaden und Satzgitter which today are just large industrial centers.

However, while some health resorts have ceased to perform their original function in the modern age, others have adjusted more or less successfully by reorganizing their priorities or tapping new resources. The less successful draw on visitors of modest means patronizing them in the framework of social tourism like sanatoria or rest homes run by social insurance agencies (Vichy, France). Some have focussed on attracting exhibitions, conventions, business meetings and cultural festivals mainly during shoulder month or (sometimes even in off-season) or have become gambling centers or developed winter sport facilities. Others have concentrated on preserving their spa functions by marketing and emphasizing their old traditions and lobbying governments and the public. The fast deteriorating Vichy recently started the construction of a $50 million health and beauty center in order to stop the decline. The health resorts are fighting for thier survival. In order to do it efficiently they have organized themselves in a number of national and international associations united under the umbrella of Federation

Internationale du Thermalisme et du Climatisme which has been created in order to represent their interests. In a statement issued in 1973 the federation declares that "thermal and climatic treatment" in spas does not constitute an alternative but rather a supplementation to treatment by conventional medicine (Kaspar, C., 1976, 2:2-6).

All this means that the traditional health resorts of Europe and North America are somehow surviving at the end of the 20th century. They were able even to expand slightly in other parts of the world (e.g. the mud and mineral baths at the Dead Sea, Israel and Jordan) However, there are other modern types of spas developing recently. They approach the health care from another angle. It is the active part of health care, not only passive as has been the tradition. This means that the guest should not only bath, drink waters and sit in fresh air but also participate in sports like golf, swimming, tennis, jogging, etc. In this connection many traditional health resorts have shifted their functions to sport, especially winter sport centers like e.g. St. Moritz, Switzerland. Many modern health resorts are focussing on weight loss by exercise and diet with fairly regimented programs. Other resorts feature general health care: exercise, nutrition and relaxation (stress management). In some not only physical but also spiritual well-being is sought (yoga and meditation). All these modern types of health resorts are expanding quickly in the DCs. They conform closely to recent trends for healthy life styles. No wonder that the corporate sector promotes its meetings and seminars in spa-oriented resorts finding that "a relaxed healthy environment improves productivity and the mind learns and retains learning better" (Tourism Itelligence Bull. 1988, Feb.).

Seaside resorts

The inland health resorts in the pre-railroad era catered as a rule to the rich classes of aristocracy and later to capitalists. Because of high costs of travel and stay in the spas, the middle and lower classes were excluded. However, in 18th century England and later on the continent a new competition for spas, which held a more democratic promise, appeared: the seaside resort. This new form of holiday resort developed as a result of "discovery" of curative properties of sea water which was used mainly externally (for bathing) but also in the initial stages internally (for drinking). Yet, similarily to inland spas, soon the social functions became predominant: promenades, piers, parks, concert and dancing halls, assembly rooms, libraries, theaters and later in 19th century, gambling casinos were indispensable fixtures of such British seaside resorts as Brighton, Margate,

Scarborough, Weymouth. The patronage of these resorts was initially limited to high society but in 19th century gradually also other classes started to participate mainly as day trippers (excursionists). With the development of railroads spreading in Europe and North America since mid-19th century, the patronage of middle and lower classes has grown consistently, and this increased visitation marked the great expansion of coastal resorts. Some of them based their existence on middle class clientelle, e.g. Blackpool in North West England. The democratization of seaside resorts in Britain and later on the continent was possible only because of rising incomes of employees and gradual introduction of regular paid vacations.

In Britain the seaside resorts eclipsed the inland spas already about 1800 and in Continental Europe in the second part of the 19th century. Their spatial diffusion extended farther and farther away from the industrial cores of Britain and the continental Europe. The rich have been the pioneers in the quest for periphery in the development of new coastal resorts. This process in some respects continues today, with the increasing geographical mobility, which is the highest for the upper social strata. The instruments of this mobility - the railroad and later the automobile and the airplane have enabled the almost perfect penetration of space in the quest for new resort periphery. With the time the earliest seaside resorts located at short distances from densely populated urbanized areas have lost their pure resort character by transforming to commuter suburbs. This has happened to a large extent to such coastal resorts like Brighton in England and Belgian seaside resorts. The seaside resorts in continental Europe, especially in more southern locations have been booming in both 19th and 20th centuries to the point when many of them surpassed their carrying capacities. The French Riviera and the Biarritz area were the early leaders in 19th century but 20th century has seen exponential development almost everywhere in the Mediterranean area and in Algarve region of Portugal. Black Sea coastal resorts like Yalta have been developed since 19th century at the Crimea. In the 20th century the North Caucasus coastal development took place between Sochi and Sukhumi. Most of the Bulgarian and Romanian sea resorts have been founded only after World War II. It is interesting that some mineral spas of the USSR combine the attributes of both seaside and mineral resorts e.g. Matsesta in North Caucaus and Yevpatoria (Eupatoria) in the Crimea. American seaside resorts developed in the second half of 19th century, first close to the industrial heartland (Atlantic City) and then in the beginning of the 20th century farther away (Sarasota, Florida).

After World War II some of the older "classical" seaside resorts have faced a number of problems. This made drastic adjustments necessary to save them from demise. One could illustrate these problems on the examples of Atlantic City and Miami Beach. Atlantic City still booming in 1920s had later suffered from pollution, beach erosion, urban blight and decay, urbanization and competition of now easily accessible more southern resorts, notably in Florida, the Caribbean and Mexico with longer seasons and more attractive natural environment. As a result the patronage has decreased catastrophically and the mere existence of the resort was threatened. The remedy was casino gambling. Introduced in 1977 it brought the expected boom to the resort. The main asset is the proximity to the Boswah Megalopolis, the largest conurbation in the world located between Boston, Mass. and Washington, D.C. Taking advantage of better accessibility to the market Atlantic City outperforms Las Vegas not only in visitation but also in gambling revenues. Atlantic City today is the most popular resort in the US (Tourism Intelligence Bull. 1987, July), mainly because of its location. Its competitor - Las Vegas had an earlier start: it developed soon after the World War II but does not enjoy similar accessibility from large markets. Gambling as a major source of revenue in European resorts has a much longer history reaching back to the second half of the 19th century (Monte Carlo, Homburg in Germany).

The main reason for Miami Beach's decline as year-round vacationland was beach erosion. During the unrestrained and unplanned resort development in the early 1920s and later in the 1950s, the hotels and condominiums were put up almost at the water's edge atop the protective dunes which were destroyed in the process of construction. As a result many buildings not only lost their beaches but also were threatened by the waves. The remedy came in the form of largest beach restoration project ever attempted. After several years of work by the US Corps of Engineers, the $64 million project was completed in 1982, restoring almost 20 km of beaches in Miami Beach and neighboring communities. Other problem of this resort is the bad publicity resulting from frequent racial unrest in the nearby Miami, rising crime, drug traffic and a poor image of the resort connected with its "galloping giatrification." It seems that at present the resort is able to cope with these problems. There are signs of modest revival such as increased investments and growing patronage by younger people and foreigners. However, Miami Beach still has to cope with problems of congestion and pollution and face the competition of more exotic and increasingly accessible destinations in the Caribbean. Also the booming central Florida (Orlando area) draws visitors from South Florida.

Another competing factor is the shift in the vacation-taking patterns towards winter ski holidays. Despite these obstacles it seems that the revival of Miami Beach is under way as a result of measures taken including skilful marketing efforts. Legalization of gambling has been also helpful.

Treating casino gambling as the panacea for the economic ills of resorts proved to be rather controvertial strategy because its introduction causes a number of severe socio-economic problems like infiltration of organized crime, prostitution, excessive urban development, land price inflation harming local inhabitants of modest financial means etc. Atlantic City certainly has its share of these problems. Nevertheless, gambling in resorts, for a variety of reasons, seems to be spreading all over the world including the LDCs (e.g. Malaysia's Genting Highlands). Governments seem to close their eyes at problems, focussing rather on tax revenues as principal incentive for sanctioning legal gambling. The legalization of gambling in a number of Spanish resorts in late 1970s was an abrupt about-face of the Spanish government which was traditionally opposed to it. The reason was mainly to keep more residents at home as a measure to improve Spain's balance of payment. The government came to the sound conclusion that Spanish gamblers should loose their money at home rather than abroad. However, in a number of countries (e.g. in the Caribbean and Paraguay) casino gambling is still restricted to tourists. Locals are barred from casinos.

At the present time the seaside resorts constitute the most popular type of resort. There are thousands of them all over the world. Among the most famous one could list Brighton and Blackpool in England, Ostend in Belgium, St. Malo, and Deauville in Northern France, Biarritz, Nice and Cannes in Southern France, Rimini in Italy, Yalta and Sochi in the USSR, Atlantic City and Miami in the U.S., Punta del Este in Uruguay, Mar del Plata in Argentina. Many of these "classical resorts" are located in mid-latitudes and most northerly of them (Northern Hemisphere)suffer from short season and unreliable weather conditions.

However, the development of seaside resorts is by no means limited to mid-latitudes and to the DCs. A number of important resorts have been created in the LDCs, mainly after 1945, and as a rule in lower latitudes like Mexico (e.g. Acapulco, Cancun), Caribbean (e.g. Montego Bay, Jamaica), Morocco (Agadir), Tunesia (Jerba), Kenya (area North of Mombassa), India (Kovalam, Goa area), Thailand (Phuket, Pataya), Indonesia (Bali), Fiji, Samoa.

Although the seaside resorts are located in various natural environments one could list some of the physical characteristics which make a successful coastal resort: sandy beaches, scenic panoramas from hills or cliffs. Marshes, river mouths, etc. are not advisable as locations. An ideal supply of all the desirable features is only sometimes available in one place. A good example is Algarve, Portugal.

Some seaside resorts have developed in relationship to pre-existing settlement (fishing port, fishing village). Such places as Torremolinos at the Spanish Costa del Sol or St. Tropez at the French Riviera were "discovered" as picturesque by nature and tradition lovers (e.g. artists). Since their tourist development, grafted on pre-existing settlement, they have lost their initial solitude character and became crowded with tourists and frequently with their boats displacing the original fishing boats from ports. Another category of seaside resorts are those developed specifically for this purpose. Many of the "classical" 19th century European resorts belong to this category. They have been planned and constructed on previously uninhabited or very thinly inhabited land. Modern seaside resorts develop mainly in such environments. A good example of such completely new resort developments is the Languedoc-Rousillon area in SW France spread on about 180 kilometers of Mediterranean Coast. This gigantic tourist development took a quarter century to build (since 1963) creating 5 large resorts and a number of smaller ones with the total capacity of almost one million tourist beds and 20 ports for pleasure boats. It's a veritable "new Florida." An example of completely new single resort development is Nueva Andalucia on the southern Spanish coast accommodating almost 50,000 tourists. The largest bullring in southern Spain is an unusual although typical amenity here. The "vacation villages" of Club Mediterranée (Club Med) are on much smaller scale (up to 1000 beds). These, mostly seaside resorts, are scattered all over the world and exceed one hundred. Despite their smaller size in comparison to mega-resorts mentioned above, the Club Med resorts also constitute the "final destinations". They are self-contained and offer a number of supervised sport activities and nightly entertainment, all on a voluntary basis. The relationship between staff (GOs) and guests (GMs) is one of equality, friendship and co-operation.

Resorts for very rich people fall into two categories: the rich can luxuriate either in seclusion (no telephones, no ads, members only) of small resorts (up to 200 rooms) or splurge in extravagant luxury of large resorts (over 1000 rooms). An example of the first category is Caneel Bay, St. John, US Virgin Islands (170 rooms). The second category is represented by Hyatt Regency, Waicoloa at the Big

Island of Hawaii. At $360 million it is the most expensive single resort development in the world (Time 27-2-1989). With 1,241 rooms it is, of course, self-contained, and offers luxury to super-luxury accommodation (presidential suite at $2,500 a night). A whole range of activities are offered, among them inland hunting safari, various helicopter trips and frolicking with eight trained dolphins in a protected salt water lagoon. All the activities carry amazingly high price tags.

Spatial morphology and functions of seaside resorts

The traditional (19th century) spatial arrangement of a European seaside resort is linear paralleling the coast with greatest inland extension in the center. The ocean or seafront orientation coincides with the linear alignment of resources: the ocean and the beach where the recreational activities of the visitors are taking place. The marine zone and the beach have been discussed in section 6.4.1. The resort proper may be identified with two zones running parallel to the coast: shoreland and hinterland. The shoreland is the most important functional part of the resort illustrating the close relationship between function and form.

The shoreland (Fig. 6.2) starts with the dike. The dike was built initially by waves which at certain line accumulate more material than take back away to the sea. This natural dike is usually reinforced or built completely artificially for protection against sea erosion. It is normally a cement structure often reinforced by boulders, piles etc. At the dike a major part of resort's life is taking place: pedestrians enjoy promenades, and parks. Also here at the seafront, the casino and the tourist information office are usually located. Farther inland runs a major thoroughfare for vehicular traffic. The shore boulevard constitutes a major roadway parallel to the coast. Inland from it the first block of buildings consists of top accommodation facilities with the premium seafront rooms in close contact with the beach and the sea. Here also are some top seafront cafes, restaurants and sometimes luxury shops located. The quality of the hotels deteriorates inland in the transition zone towards the Central Business District (CBD) which is usually spatially separated from the seafront accommodations. This change of functions reflects not only the preferences of the visitors but also the decrease in land values towards inland.

Farther inland from the CBD the intensity of the land use continues to decrease within the last zone of the resort, the hinterland. Similarily to the shoreland it extends parallel to the coast and performs all possible services for the resort. Here are the residences of local population, warehouses and business indirectly catering to tourism.

Figure 6.2

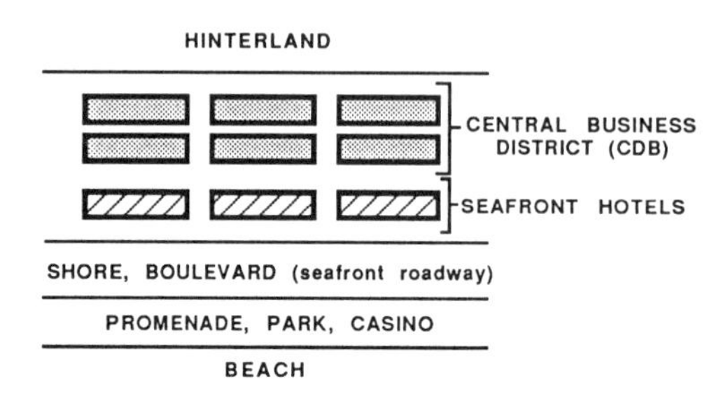

Seaside resorts developed on offshore sandbars (lidos), built mainly by sea currents parallel to the coast, developed a somewhat different spatial morphology than the traditional resorts because they have a second seafront, that faces the lagoon. Good examples for this type of morphology are the American Atlantic resorts, especially in New Jersey. They have been analysed by C. Stansfield (1969). The arrangement of the resorts is linear, similarily like in the European resorts. However, there are some differences, only partially connected with environmental factors. First difference characteristic for most American coastal resorts, at least in New Jersey, is the Recreational Business District (RBD) catering directly to the neds of tourists. The RBD is sited directly inland from the seashore promenade. According to Stansfield, the development of the RBD is mainly due to "the absence of a seafront roadway separating the promenade from the nearest possible frontal trading development sites" (Stansfield 1969: 132). Such a roadway usually exists in 19th century European seaside resorts separating the strolling pedestrians from shopping facilities. The American RBDs in New Jersey are located at the seashore pedestrian promenade called locally "boardwalk." The first boardwalk was constructed in 1870 in Atlantic City, New Jersey. It is a "wooden pedestrian promenade, slightly elevated on pilings" (Stansfield, 1969:131). The American RBD consists of various types of food and beverage catering businesses, amusements facilities, gift shops and shops catering to tourist needs. The RBD is closed in off-season. The CBD is located farther inland. If the lagoon is too narrow to allow the development of all zones of a resort, like in Cancun, Mexico or Miami

Beach, Florida then the vehicular access road bisects the hotel zone and the CBD and hinterland have to develop on the mainland.

A characteristic feature of resorts located on sandbars which are interrupted by sea access channels is the development of marinas on the inland side of the lido facing the lagoon. The environmental drawback for tourist development of this area are marshes and tidal flats with the associated mosquito population. If these obstacles are removed then the inland locations of marinas on lidos become very desirable, especially for the owners who need an easy access to their yachts and pleasure boats.

The traditional spatial development of seaside resorts has resulted in a linear or ribbon zonal extention of the resort parallel to the beach (fig.6.2). It did not present any major environmental or access problems in the pre-automobile era. However, with the exponential growth of coastal tourism after the World War II, significant disadvantages of the traditional parallel structure soon became apparent. First of all, the desire to maximize the amount of hotel rooms at the seafront leads to construction of high-rise hotels in the first block from the beach. This creates a concrete barrier both visual and real separating the inland part of the shoreland from the sea. There have been also access problems resulting from traffic congestion and noise pollution near the sea front. Thus the traditional development blocks both views and access to the seafront of the resort. It is also frequently connected with environmental degradation. These negative phenomena are especially acute near the resort center diminishing towards its periphery.

Clair Gunn (1971) suggests a number of solutions to cope with problems created by traditional spatial morphology of seaside resorts. His ideas are reflected in differing variations in the design of many modern seaside resorts built starting with 1960s. Gunn introduces a concept of "concentrated dispersal", the clustering of major coastal tourism/recreation complexes across all resource zones of the coast (Fig. 6.3). Mass circulation must be placed outside the tourism/recreation coastal zones. The tourism development should proceed in complexes or segments in order to retain integrity. Short access links tie the clusters with the main circulation system.

Figure 6.3

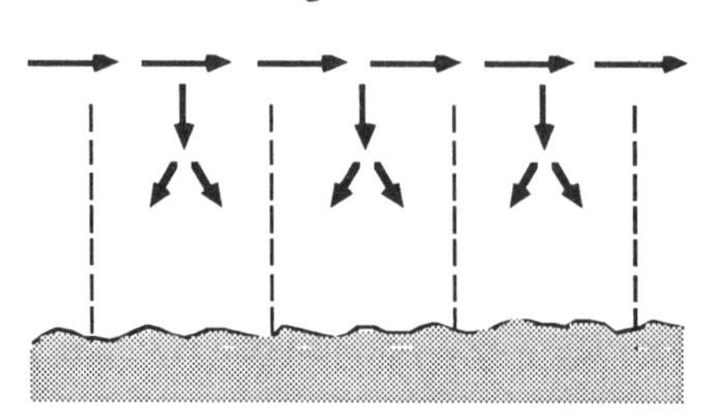

Such a segmented development has many advantages: a single stop allows the tourist to see as many things as possible, have a meal, shop, see the night life. Clustering is better adapted to modern travel, because, according to Gunn, it provides a superior mix of services and better land environment than "a polarized and fragmented approach." It is also advantageous for business by concentrating mass markets.

Gunn gives an example of cluster development promoted in Oahu, Hawaii which he finds superior to the traditional strip development. This concept provides for grouping of high-rise hotels into "visual volumes or envelopes" which enhances views and access to the beach. The high density clusters are separated by open space. The major traffic road parallel to the coast is located inland with branch access roads leading to each segment (Fig. 6.4).

Figure 6.4

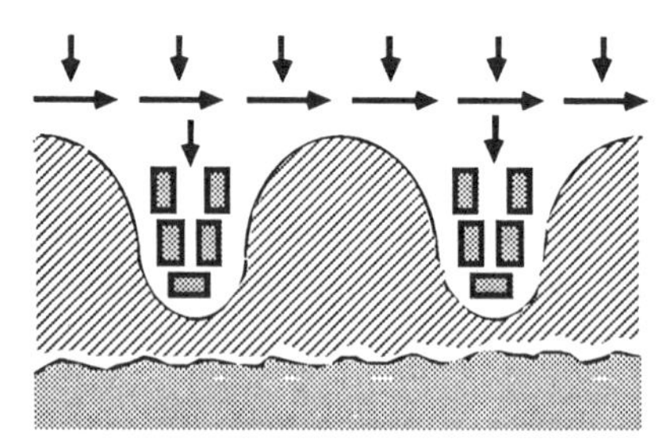

An interesting solution of the problems resulting from linear structure of seaside resorts offer the new French Mediterranean resorts developed since the mid-1960s. They have been analysed by Douglas Pearce (1978a). The French seaside resorts, described by Pearce, reflect the tendency of modern holiday resort in almost all locations to close spatial association between tourist's accommodation and recreational activities. Thus, as the modern skier likes to step out of the hotel almost directly into the valley skilift station, similarily the seaside resort visitor is interested in having the recreational facilities at the doorstep. Pearce discusses the concept of "cité lacustre" implemented in several French Mediterranean resorts. These resorts constitute replicas of Venice in miniature. The tourists leave their cars at the entrance. Within the resort only pedestrian traffic is allowed. The low-rise accommodations are located at interconnected islands facing the canals with berths for the boats. This arrangement provides for maximum mooring space and enables the vacationers to step out of the door to their boats. The situation is similar in case of marinas where, as Pearce put it, "the juxtaposition of accommodation and berths is less immediate than in the case of cités lacustres."

In the second half of 1960s the French government in conjunction with private enterprise started the development of a number of large coastal resorts in the Languedoc-Rousillon area of the Mediterranean. Here the recreational focus is not so much on boating as is the case of cité lacustre, but encompasses a wide array of water-related recreational activities. Pearce illustrates this type of spatial resort morphology on the example of La Grande Motte, one of the most prominent resorts in the area with about 43,000 beds. The ribbon development here is avoided by placing the main road at about 400-700 meters inland parallel to the beach and the major access road at about 200 meters and also roughly parallel to the shore. Secondary roads, terminating in car parks about 50-100 meters from the sea, run perpendicular to this major access road. This system effectively bars the automobile from the beach area (a scourge of Daytona Beach, Florida) and encourages pedestrian traffic. Indeed, no part of the resort is farther away from the beach than 10-15 minutes walk. In order to avoid over-crowding the number of beds in La Grande Motte has been carefully matched with the capacity of the beach at 600-800 beach users per one hectare of beach. Accommodations in La Grand Motte are high density apartments near the beach and farther inland individual houses and a campground. The multiple-unit dwellings are beautifully designed and they do not constitute a visual or access barrier to the beach. The accommodation density decreases inland from the beach.

Mountain resorts.

They developed initially in Europe as climatic health resorts with summer as the main season. After the World War II with the spectacular development of winter sports, especially down hill skiing they became multifunctional, mostly with two seasons: winter and summer. Examples of such resorts: Banff, Jasper in Canada, Vail, Snowmass in U.S., Chamonix in France, Davos, Zermatt, Gstaad in Switzerland, St. Anton, Lech, Zurs in Austria.

Many of the mountain resorts have developed on the shores of glacial lakes like Interlaken and Lugano, Switzerland or Bariloche, Argentina. This additional scenic feature enhances their attraction during the summer season enabling them to compete to a certain extent with seaside resorts by providing water-oriented recreational activities. Mountain resorts without lakes have normally large swimming pools. The year-round operation is the most desirable feature of mountain resorts. The French call them polyvalant. Nevertheless also these resorts experience peaks and valleys of demand, the peaks occur during the Christmas-New Year period, at Easter and in July-August. Some of the mountain resort may have only one season. Only the summer season oriented towards the lake may be possible in cases of lower altitude when commuting to skiing areas is too time consuming and/or costly. A reverse situation occurs if the resort is located at very high altitudes and the season is only in winter, summer being too cool for general public, especially for families. Stadel (1982) observes the trend of new Alpine ski resorts being located at higher altitudes than previously. The examples are Megeve developed at 1100 meters after the World War I and Courchevel after the World War II at 1850 meters.

The reason for this "up-hill" movement of modern ski resorts is to bring the skiers as close as possible to skiing areas which have been recently developed at higher elevations in order to extend the season. To achieve perfect solution for the skier i.e. to enable him or her to step out of the hotel and walk only few meters to the lift, the resort is arranged linearly in a valley with up-hill transportation facilities built on both slopes of the valley.

Stadel (1982) criticizes the negative environmental impact of high altitude modern ski resort concept, developed in the French Alps (Swiss and Austrian resorts are mainly grafted on existing settlements). These resorts are usually developed by large companies responsible for planning, design, construction and operation. A resort complex of high-rise buildings among Alpine meadows is an obvious eyesore in addition to other negative impacts like the destruction of fragile

Alpine vegetation and soil erosion. Certainly these para-urban "total destination resorts" exceed the physical, ecological and socio-psychological carrying capacity of the environment. Stadel advocates the limits of growth of Alpine ski resorts as an indespensible planning tool. He also gives examples of new construction in some Alpine resorts in Switzerland and Austria being subject to ridgid standards (size, building materials, architecture) "to preserve an image of a traditional mountain village in a rural cultural landscape" (Stadel 1982:9). Another widely criticized aspect of tourist development in the mountains is the "consumption of space," especially by second homes scattered all over the landscape. Here also "limits of growth" principle should be applied.

The Rocky Mountains are in better shape environmentally than the Alps. Nevertheless, also here the negative impact of overdevelopment and congestion is visible in some areas. Cole and Dietz (1984) criticize the overdevelopment of Vail, Colorado where there is an additional source of pollution, as compared with Europe: the use of fireplaces.

Climatic resorts.

This type of resort provides relaxation in better climatic conditions than the place of tourist domicile. The emphasis is mainly on thermal aspects, e.g. in tropical and subtropical areas tourists want to escape the oppressive heat and humidity of low altitudes in order to "cool-off" in upland conditions. Until the end of WW II the climatic cure in the mid-latitude mountains was used in treatment of some diseases, notably the tuberculosis. Today curative application of climate largely lost its appeal, although its importance for general health is widely recognized. The recreation activities enjoyed in climatic resorts range from passive to most active forms.

First climatic resorts were developed in 19th century Switzerland. The classical example was Davos - center for tuberculosis sanatoria. Similar development took place in Zakopane (in Polish Tatra). During the 20th century, these resorts have gradually changed their "raison d'etre" to active recreational activities like skiing, hiking, mountaineering. Another type of climatic resorts which lost this function almost entirely were the resorts of French Riviera like Nice and Cannes which started in 18th and 19th century as climatic winter resorts for the British and later continental high society and only in 20th century have been transformed in summer playgrounds with winter off-season.

The climatic resorts in the LDCs are associated with colonialism. These are the so-called "hill stations" pioneered in 19th century by colonial powers in tropical and subtropical areas. There are good examples of them in Malaysia (Genting, Cameron Highlands), Indonesia (Bandung), Cypres (Platres), India (Simla, Darjeeling). Here the colonial masters could relax in cool mountain air at relatively short distances from centers of power and economic activity. There are also hill resorts in DCs performing a similar function e.g. Blue Mountains for Sydney or Catskills for New York City. The southern resorts which provide an escape for winter-weary tourists from the DCs of Europe and North America overlap with the next type (retirement communities). A classical example are seaside resorts in Florida.

Retirement communities,

These are leisure-oriented settlements comprised of a variety of housing types: single-family homes (mobile or conventional), rental apartments, condominium apartments or institutionalized homes. Frequently operated on a condominium-management basis. Location should be attractive from the point of view of landscape and climate. Otherwise they are pretty ubiquitous. They should be always carefully planned in order to minimize adverse social and environmental impacts. The planning should assure provision of educational, entertainment, recreational and health facilities oriented to special needs of seniors. To qualify into this category the population of the community should be composed in at least 50% of retirees. Again, Florida and California are typical areas where retirement communities are especially common.

Resort cycles

The decline of resorts at the eastern seaboard of the US and the functional changes of some European resorts indicate that these recreational communities are operating in a dynamic and not static socio-economic environment. Indeed, the fact that they pass through distinct stages in their life-cycle has been observed by Walter Christaller (1964).

"The typical course of development has the following pattern. Painters search out untouched unusual places to paint. Step by step the place develops as a so-called artist colony. Soon a cluster of poets follows, kindred to the painters; then cinema people, gourmets, and the *jeunesse doree*. The place becomes fashionable and the entrepreneur takes note. The fisherman's cottage, the shelter-huts become converted into boarding houses and hotels come on the scene. Meanwhile the painters have fled and sought out another periphery -- periphery as related to space, and metaphorically, as "forgotten" places and landscapes. Only the painters with a commercial inclination who like to do well in business remain; they capitalize on the good name of this former painter's corner and on the gullability of tourists. More and more urbanites choose this place, now *en vogue* and advertised in the newspapers. Subsequently the gourmets, and all those who seek real recreation, stay away. At last the tourist agencies come with their package rate travelling parties; now, the indulged public avoids such places. At the same time, in other places the same cycle occurs again; more and more places come into fashion, change their type, turn into everybody's tourist haunt." (Quoted after Butler 1980:5)

Gustav Zedek et al (1967:39 ff) was the first who distinguished four stages of the resort development:

1) Discovery - the news about the attractions of the place are spready by "mouth-to-mouth propaganda." 2) Quick development of the resort. 3) Congestion 4) Shifts in the composition of tourists resulting from unrestrained quantitative growth. Similar ideas were expressed by other researchers in the subsequent years (Butler, 1980). Butler suggests a general pattern of tourist area evolution which may be called - using Stansfields terminology - "the resort cycle" (Stansfield 1978).

1) Exploration stage (small number of tourists, no specific tourist facilities, high degree of contacts between tourists and locals)

2) Involvement stage (locals increasingly provide services and facilities for tourists, some marketing efforts to attract tourists)

3) Development stage (heavy marketing, local involvlement decline, external organizations engaged in providing larger and more elaborate tourist infrastructure, particularly for accommodation, number of tourists at peak periods equal or exceeding the local residents, utilization of outside labor).

4) Consolidation stage (the rate of visitation increase declines as compared with stage 3, the total number of tourists exceeds that of the locals, economic dependence on tourism, some opposition against tourism on the part of locals, particularly those not involved in the tourism industry).

5) Stagnation stage (capacity level reached or exceeded with attendant environmental, social and economic problems, the area is no longer in fashion).

6) Decline stage (inability to compete with newer attractions and resulting declining market, functional shifts: loss of tourist function, increase of other functions, particularily residential, including retirement homes).

At this stage Butler envisages five possible scenarios:

A. rejuvenation by either adding a new man-made attraction (e.g. gambling in Atlantic City) or by reorientation to previously untapped natural resources (e.g. winter sports)
B. rejuvenation at reduced rate when the modifications are minor allowing only modest growth
C. readjustment adequate to achieve more stable visitation levels
D. marked decline if no measures taken
E. catastrophic events ruin tourism completely

In connection with the question of resort cycle following remarks seem appropriate:

1) The "classical" six stages of Butler represent the best generalization of reality. However, this model differs substantially from specific areas or even certain types of resorts. Thus "instant resorts" - total destination areas planned and developed by large corporation on unpopulated land - clearly lack the two initial stages.

2) Resorts may reach stages 5 and 6 without surpassing or even reaching their carrying capacity. Their decline may be the result of technological innovation, particularily in the field of transportation, change of fashion, natural or political events, etc.

3) The shifts in the composition of tourists may occur in various ways in specific cases, especially in the initial stages of resort development. Not necessarily the gregarious allocentrics are always the resort pioneers but also frequently the shy self-centered psychocentrics who move on when the resort becomes too crowded leaving the place for more gregarious types who want to be "where the action is" i.e. in the crowd. Sometimes the pioneers are artists, meditators, freaks of various sorts etc. Pioneers are mainly people of modest means who later become disenchanted with the development and/or have to move because of rising prices in an increasingly "trendy" resort. In Torremolinos on Spain's Costa del Sol first villas and small hotels belonged to the rich (move stars, millionaire heirs, various rich eccentrics) who had to leave under the onslaught of mass tourism characteristic for stages 3-5 (Moynahan, B. 1983:164-66).

4) The problem of carrying capacity is a controversial issue. It is impossible to determine exactly because such an threshold, such a number does not exist, especially for psychological carrying capacity which is very subjective and depends from the individual psychological make-up and from the cultural background. The threshold for ecological carrying capacity is easier to determine but even here the nuture's ability for self-renewal may be underestimated or overestimated. The

same may be told about human ability to "manage nature." Therefore, it seems to be more advisable to treat the carrying capacity in stages 4-6 only as a rather wide zone substantially differing according to the type of resort area and types of tourists but not as a fixed dogmatic number which absolutly cannot be exceeded. One has to agree with G. Wall (1983:269) that the concept of carrying capcity in tourism serves a useful purpose if it "encourages tourism planners and managers to give greater consideration to environmental matters, to the qualities of the experiences available to both hosts and guests, and to specify their goals and objectives." If for practical reasons any numbers are used they should be regarded only as very general guidance subject to adjustments.

With all these caveats Butler's model constitutes a valuable guide for resort research. It shows that successful resorts should never be taken for granted. Indeed, managers and planners should be always kept on guard to improve accessibility and to diversify the attractions in order to successfully militate against the cycle progression and assure the relative stability of demand.

6.7.9. Meeting and convention tourism

6.7.9.1 Trends for expansion

Settlements of all sizes, particularily large cities and resorts serve increasingly as venues of a relatively new type of tourism: meeting and convention tourism which falls between pure recreational tourism and business tourism. Although international and national meetings have a long history, reaching back at least several hundreds of years, convention (congress) tourism is a relatively new phenomenon steadily increasing its market share in overall tourism. It showed a healthy growth even during world recessions (i.g. 1979-82) when tourism as a whole stagnated. Although the convention busines seems to fluctuate from year to year in terms of growth rate, nevertheless, it is a multi-billion dollar item in the GNP. It generated $30.3 billion in US alone in 1987, with a healthy 9.4% growth over 1986 (Tourism Intelligence Bull. 1987, July).

One could speculate about the reasons for the world boom of meeting and convention tourism. Certainly there is no one reason but a number of them reinforcing each other in a circular causation pattern. The trend towards relative decrease of transportation costs, resulting from technological charge and discount fares, creates good incentives for convention tourism. However, one could argue that tourism as a whole is benefitting from this development and that there must be some other reasons why convention tourism is so outstanding as compared with most other types of tourism. It seems that the most important reason is the need for personal contacts, face-to-face meetings of peers aimed at exchange of views, person-to-person negotiations on various issues including lucrative contracts and jobs. Indeed, some conventions are called "floating job markets." Even the impressive development of electronic communication and resulting easy access to all sorts of data and world-wide viewing of any events has not made conventions obsolete. It seems that the reverse happened: people feel isolated in front of their video display terminals. People want the human touch which enables them to function better and achieve better results in todays complicated world. Also the need for pure social interaction plays an important role. Especially in US cities, the meeting participants constitute a permanent subculture when the people break the usual rules of conventional behaviours and act like clowns or jackasses. Some conventions are just plain fun and little more. However, the motives of exchange of ideas, views, experiences and the quest for knowledge are undoubtedly prevalent and increasing their share. Indeed, one could regard convention tourism as an important tool for promotion of understanding among people.

Among other reasons for the immense success of the convention business is the fact that at least part of the costs are paid by the company and/or are tax-deductible. There is a trend on the part of employers to reward and motivate employees by financing their participation in all sorts of meetings, seminars, symposiums, training courses in various exotic spots providing ample opportunities for relaxation. In many of these instances of this "incentive travel" the afternoons are free to enjoy leisure time in attractive natural or cultural settings. Another reason for the growth of conventions is the increasing interdependence of people on local, national and international levels (globalization), requiring action by various interest groups in promotion of their goals. Indeed, conventions serve as important instruments of groups of people aiming at achievement of their purposes ranging from consumerism and government regulation to religion.

6.7.9.2. Terminology and classification

Peter Müller (1983:20) quite correctly regrets that there is no uniform definition of convention tourism. He writes that there is only a general agreement with respect to international congresses which are defined as meetings of participants from at least three countries with a minimum of one over-night stay. However, there is no agreement on many other points of importance, especially with respect to number of participants and terminology. A number of terms are used, in part synonymously: congress, convention, symposium, seminar, meeting, conference, workshop, session. One could suggest that "congress" has rather serious connotation whereas "convention" means mostly pleasurable social event, sometimes associated with socially negative behaviour like e.g. abuse of alcoholic beverages. But this distinction has not been generally accepted and the terms "congress" and "convention" are used interchangeably.

Lack of uniform definition with respect to convention tourism is followed by inadequacy of statistical and research materials. The cause and effect relationship is pretty obvious: most countries do not provide in their statistics for separate data on congress tourism which has to be investigated in special sampling research. As a result the statistical data are unreliable. There are grounds for suspicion that much of the congress business remains unreported and that much of the data is not comparable.

From the point of view of the characteristics of associations convening, one could divide conventions and meetings into following categories:

1) Business - labor - government.

2) Educational - professional. Educational meeting are frequently using campus facilities.

3) Social - fraternal. Large part of these meetings is devoted almost exclusively to amusement and fun, like Bachelor Days in New Salem, South Dakota drawing almost 2,000 participants.

4) Religious

One could also divide the meeting and convention market into two groups:

1) Business (corporate, company) meetings, constituting approximately one third of the market in terms of expenditures (Abbey, J. 1987:267)

2) Association meetings and conventions spending the remaining two thirds.

6.7.9.3. Size of the meetings

The size of the meetings varies from small corporate gathering of less than 50 persons to huge jumbo-congress of several thousand participants.

There is a tendency away from large general full-membership sessions where the key speakers hold long speeches full of generalities and platitudes. If such sessions are held at all, they tend to be rather broad and formal to allow a subsequent break-up into smaller sessions for sub-specialist groups which give the opportunity to exchange views in more personal and informal manner. In many cases, the general formal meetings are not held at all or only at long time intervals. The trend is to smaller, separate, highly specialized and/or regional meetings (Tourism Intelligence Bull. 1989, March) which may or may not be held in conjunction with general gatherings of total association membership. Frequently various specialized symposia or field trips take place before and after general meetings but even these are rarely limited to plenary sessions without sectional meetings. Whatever the specific arrangement, one could generalize that the large umbrella conventions (congresses) are organized less frequently than the smaller sub-group and regional meetings.

These trends point the future size of conventions. Large jumbo congresses of over 2,000 participants tend to decrease their market share. Intermediate size (200-2,000) is doing well, especially in the 200-500 range. Small meetings with below 200 participants (to a large extent training seminars and symposia) are quickly expanding. This indicates that location of meetings in smaller localities has an assured future.

6.7.9.4. Convention facilities

Convention tourism is a competitive business with long lead times for organization. Indeed, bookings for large conventions run 3-5 years and more in advance. In order to compete successfully one has to offer high quality product. Let's start with facilities.

In cities there are two types of facilities where conventions take place: convention hotels and convention centers. Convention hotels offer accommodation, food services and meeting facilites under one roof. Large modern convention hotels are especially popular in North America. Besides one larger room for plenary sessions also several smaller rooms for section meetings or smaller gatherings must be provided. The lobby or foyer must have ample sitting space for encounters of small groups of people.

Congress or convention centers are special multi-purpose buildings, separate from the hotels but in their immediate vicinity. Sometimes an all-weather connection between them is facilitated by an underground tunnel or an overpass over the street, which is especially important for year-round operation. Modern congress centers constitute flexible combinations of convention and exhibition facilities for trade shows. These facilities should be adjustable at short notice according to needs. There is usually a large hall accommodating several thousand participants for plenary meetings, concerts, exhibitions, and a number of smaller rooms for section meetings, seminars, etc. It is advisable that the size of these rooms is adjustable by soundproof removable (retractable) partitition walls.

There is some competition between convention hotels and convention centers but they are rather complimentary: convention centers cater to larger congresses and provide more exhibition space than the convention hotels.

6.7.9.5. Convention venues

The convention hotels and centers are as a rule located in large cities. The third type of venue is located in the countryside. It is usually a self-contained resort equipped, in addition to formerly described facilities, also with outdoor amenities catering to recreational needs of participants. Such an environment enables them to spend free time by the pool, playing golf, strolling in the park in the spirit of "one big family." This type of meeting venue contributes most to establishment of personal contacts among participants and enjoys incrasing popularity. Indeed, "meetings and conventions account for 22 percent of total resort business in the US and that figure is expected to grow" (Tourism Intelligence Bull. 1988, April).

In most cases the sites or venues of conventions should comply with certain standards:

- Good accessibility within national and international transportation networks i.e. availability of an international airport with trunk airline connections. Also major highway connection is necessary. Sometimes interesting en route stopovers constitute an asset for a congress venue.
- Attractive natural and/or cultural environment.

- Good general and tourist infrastructure.
- Political and social stability guaranteeing personal safety of the participants.
- Location of convention facilities on downtown's periphery.
- Location of convention facilities in the immediate vicinity or under the same roof with accommodation and food services.
- High standards of convention facilities.

A successful convention site should offer a flexible attractive product, to the participants, a mix creating optimal conditions not only for work and mutual contacts but also for relaxation, entertainment and fun. Convention site also should offer to participants "value for money" - high cost effectiveness. Only such a site may be competitive on the convention market.

The choice of convention venue poses for the organizers a number of diffiicult questions to solve. The business is competitive and various sites woo for conventions with letters, folders, submissions and even travelling salespeople in order to boost the image of the host locality as the most attractive and best equipped for accommodating conventions and their participants. The image of the venue must, if possible, be tailored to the character of the convention. At any rate there should be no conflict with the association's self-image and the image of the venue. Venues with strong stereotypic images like Las Vegas or Atlantic City with their gambling casinos certainly are not suitable for religious meetings.

Another point for convention organizers to be taken into account while considering the choice of venue are the costs. The participants as well as organizers want value for their money. That is obvious. However, this value should be adjusted to the level the participants can afford. New York City or Paris are certainly very prestigious venues but their high price levels causes many associations to avoid such expensive places and look for venues with lower prices like e.g. Winnipeg, Manitoba, one of the cheapest convention venues in North America, which in addition to low costs offers one of the largest convention centers in Canada for 6,000 participants. Winnipeg also provides for excellent personal safety of its guests which constitutes an additional asset. Of course, the level of price consciousness varies among associations looking for most appropriate venues for their convention. The venue should be also tailored to the character and duration of the meeting. Airport hotels are the best choice for one day conferences because they save time and costs of transfers to downtown. Self-contained resorts with good recreation facilities may be the best choice if distractions of big cities are to be avoided and an atmosphere of togetherness among participants, also after formal meetings is important.

The choice of venue, planning and organization of congresses contains so many variables and presents so many problems that sometimes professional convention planners (e.g. Meeting Planners International - MPI) are necessary to advise and arrange medium size conventions for a 5-10% commission or retainer fee according to the degree of involvement. Large conventions (over 2,000 participants) are as a rule organized by own permanent staff of the association. There are some umbrella organizations which in various degrees supervise or at least register congresses. In the US it is the International Association of Convention and Visitor Bureaus, in Europe the Union of International Associations with the seat in Brussels. In the US there is a special Meetings and Conventions magazine.

6.7.9.6. Spatial distribution

Europe is the world leader as convention venue. Especially the cities like Paris, Geneva, London, Brussels, Strassbourg, Hamburg, Berlin, Vienna, Rome excel in attracting congresses. All of them have huge congress centers equipped with meeting and exhibition halls. The largest facility is in West Berlin (completed in 1976). It has a meeting hall for 7,500.

The second region competing for the congress business is North America. Similarily like in Europe every large city has a congress (convention) center as a matter of prestige. The number of convention centers in US is now 60 and growing fast (Tourism Intelligence Bull. 1987, Feb.). North America leads the world in the number of convention hotels able to host medium size congresses. The most important congress venues in North America are: New York City, Chicago, Detroit, Atlantic City, Houston (Astrodome is the largest facility).

The third region which is rather new at the scene of convention business is East Asia: Japan (Tokyo, Kyoto), Hong-Kong, The Philippines (Manila) and Singapore. The facilities are modern, most of them recently completed. The attraction is in exotism, interesting natural and cultural setting, relatively low costs (excluding Japan).

The newly emerging congress venue is the Australia-New Zealand region.

With the emergence of the Third World as political power and the progress in means of air transportation the developing countries started to increase their share in congress business since 1960s. The attraction of far-away exotic places has been irresistable. However, this trend is meeting some significant obstacles as the political and economic conditions in many of these countries fail to guarantee personal safety of participants and adequate general and tourism infrastructures. Political unrest and criminality constitute effective deterrence for any tourists but congresses particularily require an atmosphere of calm, stability and security. Also, in times of oil crisis venues located at longer distances from markets are loosing clientelle. Therefore, it is difficult to predict the future role the LDC are going to play in congress business. Certainly, their increased participation as host countries should be welcomed and promoted as an important tool to their economic and political advancement. The most important congress venues in the LDCs with excellent facilities are Rio de Janeiro, Brazil, Nairobi, Kenya and Manila (Philippines).

6.7.9.7. Economic impact

The economic impact of conventions results from the length of stay, average daily expenditures, the size of the party, the pre-and post-congress arrangements, etc. The average duration of conventions has a slight tendency to diminish. At present it is about 3-4 days (Müller, 1983:30) but not long ago it was 5 days. (Der Spiegel, 1973, 15:170) This conforms with general tendency to shorter stays in tourism industry as a whole. However, the length of stay of convention participants remains above that of other tourists.

With respect to expenditures it is a well known fact that daily spending of meeting and convention participants surpasses the other tourists. Müller (1983:3) estimates that congress participants spend from one third to one half more per day. The structure of expenditures varies according to different sources (Müller, P. 1983; Abbey, J., 1987; Waters, S., 1987). One could generalize that about 40-50% is spent on accommodation, 25% for food and 11% for shopping.

The economic impact of conventions is also correlated to the spending and behaviour patterns of various groups of participants. This is the reason why certain conventions are sought after more by the venue representatives than others. Wealthy bankers, medical doctors, auto dealers constitute the prime clientele known as free spenders. On the other end of the financial spectrum are educational, scientific, social service and religious associations whose members are more price conscious thrifty spenders. They frequently look for modest accommodations, meals and facilities e.g. at University campuses. Bikers meetings are regarded as outright disadvantageous for the host community because of their association with problems like litter, property damage, noise and crime.

The larger the size of a party arriving for a convention the greater its economic impact. Therefore the organizers apply all possible marketing strategies in order to induce the participants to bring along members of family or friends. The figures given in various publications range from 0.3 to 0.6 accompanying person per one participant. The largest party size is in US amounting to 2.4 persons (Abbey, J., 1987:268). It is reasonable to expect that these figures have the tendency to decrease in the long range as the size of parties in tourism diminishes. Besides the accompanying persons, congresses attract also some additional paying clientelle: journalists, exhibition and security people, salespeople. All these convention visitors constitute an excellent market for future repeat business provided they like the attractions and services.

The economic impact of conventions also increases if the members participate in pre- and/or post-convention arrangements (seminars, symposia, field trips, etc.) This is frequently the case, especially if the venue and its region offer special attractions or interest for the participants. The pre- and post-convention events and tours are either offered by the organizers or in form of special packages by tour operators. Conventions constitute an excellent opportunity for extended family vacation in an attractive setting. As the travel costs of the participant are at least partially paid, many families and friends of members decide to go along and extend their stay sometimes well beyond the duration of the convention. It is certainly an economical and interesting option.

Another economic asset of conventions and meetings is that they tend to diminish seasonality - a real scourge of tourism business. Conventions as a rule do not take place in high season but rather in the shoulder periods. Sometimes they take place even in the off-season or "value season." The most frequent month on the northern hemisphere are April-June and September-October. The sequence in terms of number of participants in US is October, May, June, April, September. As for the days of week the most popular are Monday-Thursday with about two thirds of all conventions (Müller, 1983:32).

Most convention centers operate normally in the red and are subsidised by governments in recognition of the fact that the business spinoff from congresses brings considerable benefits to the host community and increases the tax base which pays for the operating losses of the convention centers (Waters 1988:16). To improve their finances the convention centers are used in off-season for various consumer shows (e.g. auto shows) attracting mainly local visitors.

The future of the meeting and convention market looks optimistic and one could expect the expansion trend to continue. However, there are few clouds on the horizon. The most important threat is teleconferencing, an advancement in telecommunication which enables the executives to hold a meeting without travelling (Abbey, J., 1987:271). It cuts travel costs and saves the executive time. The impact of teleconferencing on the participation in meetings is not clear at this moment and has to be researched. It constitutes one of the variables of change constantly affecting the tourism business. However, it is reasonable to expect that the impact of teleconferencing will be much stronger on business travel than conventions because conventions contain a more pronounced pleasure element and personal contacts seem to be more important for the participants.

CONCLUSIONS

This book contains the discussion of world trends in tourism and recreation within the parameters of our contemporary civilization. The conclusion is inevitable: in the framework of our civilization, tourism and recreation have moved from the relatively unimportant margins to a very salient position. Indeed, tourism and recreation are spreading all over the world at an unprecedented scale. As a result their impacts, both positive and negative, are increasingly evident. Everyone, even in the most remote corner of our planet, is participating or at least in some way affected. The evaluation of the economic, socio-cultural and environmental impacts receives an expanding attention of the research, media and even politicians. This author deals with these issues only tangentially, because, as stated in the preface, such an evaluation was beyond the scope of this book. The impression of the reader, that the author represents an essentially pro-tourism position, is certainly correct. However, this does not mean that the clearly observable evidence of the negative impacts of tourism should be ignored. This criticism is well documented in the research literature and is only to a certain extent reflected in the contents of the book because its focus is on trends and not on impacts. As of impacts, the opinion of this author is that, like the other branches of economy, also tourism exerts frequently harmful influences. Nevertheless, he thinks that some criticism of tourism is one-sided, almost entirely concentrated on the negative impacts without taking the realities of life and the undeniable economic, social and cultural benefits of tourism into account. And the benefits, in this author's opinion, on balance outweigh the disadvantages. Nobody is suggesting the liquidation of industry or agriculture because they may be occasionally harmful. What is to be done is to endeavor, as much as possible, to minimize the negative effects and to maximize the advantages. There is proof enough that this is entirely feasible. Therefore, this author opposes extreme radicals who suggest to "abolish" tourism completely, as an allegedly totally negative socio-economic phenomenon, at least in certain countries (mainly LDCs) or their parts. These radicals resemble the extremists in the animal rights movement who oppose any killing of animals. Occasionally the animal rights extremists resort to violence in support of their views. As for tourism, there were already some violent anti-tourist incidents, however, this author has registered only one threat in professional literature: "Someday, there is going to be a bombing of a hotel. I hope that is very soon" (Ecumenical Coalition 1986: 38). The author is far from accusing all the detractors of tourism of violence but he encourages them to take a more balanced view. One has to cope with tourism instead of destroying it. Tourism is here and now and it will be here, probably as long as the humanity exists. Like other branches of economy, it is an inseparable component in the fabric of our civilization. Certainly, it is not desirable that tourism develops uncontrolled and unregulated everywhere, anytime and on an unlimited scale. This is the reason why the opinion of researchers and the interested parties, especially the local population, should be always taken into account while considering any tourism development.

REFERENCES

Abbey, J., 1979 Does Life-Style Profiling Work? Journal of Travel Research, Vol. 18, 1:8-14

Abbey, J.R., 1987 The Convention and Meetings Sector-Its Operation and Research Needs. In: Ritchie, J.R.B., Goeldner, C.R., eds. Travel, Tourism and Hospitality Reseach. J. Wiley & Sons, New York, pp. 265-302.

An analysis of demand Trends for Tourist accommodation in Canada. 1969 Acres Ltd., Toronto, Vol. I and Vol. II. Vol. I., p 71.

Azar, V.I., 1975. Vvedienie V Ekonomiku Turizma Izdatelstvo Ekonomika Moscow

Baker, P., 1986. Tourism and the National Parks. Parks and Recreation Vol. 21(10):50-53,67)

Baretje, R., Defert, P., 1972. Aspects economiques du Tourisme Editions Berger-Levrault, Paris

Barnet, E.M., 1971. Tomorrow's Travelers and how to attract them. Revue de Tourisme, No. 3, 106-109; No. 4, 145-148.

Bartos, R., 1982. Women and Travel. Journal of Travel Research, Vol. 20, 4:3-9.

Bella L. 1987. Parks for Profit Harvest House: Montreal

Bernay, E., 1971. Movers and Shakers. Harper's/The Atlantic, New York.

Berger, J.J. 1987 "Restoring the earth: how Americans are working to renew our damaged environment." Doubleday, 264 p.

Bernecker, P., 1962. Probleme der Fremdenverkehrs-organisation. Revue de Tourisme, No. 4, 181-187.

Blazey, M., 1987. The difference between participants and non-participants in a Senior Travel Program. Journal of Travel Research, Vol. 26, 1:7-12.

Bleile, G., 1976. Der Einfluss der Konjunktur auf den Tourismus in der Bundesrepublik Deutschland. Tourist Review, 1:2-7.

Boerjan, B., 1974. Measuring attitudes for tourism marketing strategies. Tourism Review No. 3:86-93.

Brachon, P., 1968. Naissance du Tourisme, Thomas Cook. Connaissance du Monde, No. 120:52-64.

Brockman, C.F., 1959. Recreational use of wild lands. N.Y. McGraw-Hill, 346 p.

Bronowski J., 1973. The Ascent of Man, Little Brown, Boston.

Brunsden, D., 1971. Ever Moving Hillsides. Geographical Magazine August:759-764.

Bryant, B., Morrison, A., 1980. Travel Market Segmentation and the Implementation of Market Strategies.

Burck, G., 19709. There'll be less leisure than you think. Fortune, March 1970, 86-89; 162-166.

Burkart, A., Medlik, S., 1975. The management of Tourism. London.

Burton, T., 1971. Experiments in Recreation Research. Totowa, N.J., Rowman and Littlefield.

Burton, T.L. (ed.) 1970 Recreation Research and Planning. A Symposium. University of Birmingham, Urban and Regional Studies, No. 1, London, Allen & Unwin Ltd., 276 p.

Butler, R.W., 1980. The concept of a tourist area cycle of evolution: implications for management of resources. Canadian Geographer, 1:5-12.

Cabelli, V., 1979. The relationship of illness among swimmers to the quality of marine bathing waters. Proceedings, Pacific Science Congress, Committee N. Khabarovski, USSR.

Cask, M., 1981. Segmenting the Vacationer Market: Identifying Vacation Preferences, Demographics, and Magazine Readership for each Group. Journal of Travel Research, Vol. 20, 2:29-34.

Casson, L., 1974. Travel in the ancient world. London, Allen & Union,.384 p.

Chadie, D., Mieczkowski A. Changes in Demographic Structures: Implications for Canadian International Tourism. Touriscope 1988 Statistics Canada, Ottawa 1988.

Chadwick, R.A., 1987. Concepts, Definitions and Measures Used in Travel and Tourism Research. In: Travel, Tourism, and Hospitality Research. Ritchie, J.R.B., Goeldner, Ch.R., (eds.), J. Wiley and Sons, N.Y., 47-61.

Chib, S., 1968. Assessment of the Types and Volume of Domestic and Foreign Tourism Demand. In: United Nations Interregional Seminar on Tourism Development. Berne, 21-Oct.-2. Nov. 1968,.

Choy, D., 1984. Tourism Development. The case of American Samoa. Annals of Tourism Research, Vol. 11, 4:473-590.

Christaller, W., 1964. Some considerations of Tourism location in Europe: the peripheral regions - underdeveloped countries - recreation areas. Papers of the Regional Science Association European Congress, Lund, Vol. 12, p. 95-105.

Chubb, M., Chubb, H., 1981. One Third of Our Time. John Wiley & Sons, N.Y.

Clawson, M., 1972. America's Land and its Uses. John Hopkins Press, Baltimore.

Clawson, M., 1960. The Dynamics of Park Demand. N.Y. p.39.
Clawson, M., 1964. How much Leisure, Now and in the Future? in: Leisure in America: blessing or curse? Charlesworth, J.C. (ed.) American Academy of Political and Social Science, p.96.
Clawson, M., Knetsch, J.L., 1966. Economics of Outdoor Recreation. John Hopkins Press, p.328.
Cole, D., Dietz, J. 1984. The changing Rocky Mountain Region. Focus, 34, 4:1-11.
Collingwood, R.G., 1938. The Principles of Art. Oxford.
Cosgrove, I., Jackson, R., 1972. The Geography of Recreation and leisure Hutchinson University Library, London.
Council of Europe, 1988. European Conference on Mountain Regions. Strassbourg.
Dacharry, M., 1981. Geographie du transport aerien. Litec Paris.
Datzer, R., 1988. Die Senioren Werden Aktiver. Review de Tourisme, No. 1, 21-23.
Davy, T., Rowe, L.A., 1964. in: Leisure in America: blessing or curse? Charlesworth, J.C. (ed.) American Academy of Political and Social Science, p.96
Defert, P., 1966. La localisation touristique. Problems theoriques et pratiques. Berne, Editions Gurten, 143 p.
De Grazia, S. 1962. Of Time, Work and Leisure. New York, 559 p.
Dictionary of Sociology 1944. Philosophical Library. New York.
Dower, M., 1970. Leisure - Its impact on Man and the Land. Geography, Vol. 55, Part 3, No. 248, 253-260.
Drakatos, C., 1987. "Seasonal Concentration of Tourism in Greece" Annals of Tourism Research Vol. 14:582-586.
Driver, B.L. (ed.), 1970. Elements of Outdoor Recreation Planning. University of Michigan Press, Ann Arbor, 317 p.
Duchet, R., 1949. Le Tourisme A Travers Les Ages sa place dans la vie moderne. Vigot, Paris, 235 p.
Dulles, F.R., 1965. A history of Recreation. America learns to play. N.Y., Appleton-Century Drafts Comp., 2nd ed., 446 p.
Durden, D., 1966. Use of empty areas. in: Darling F., Milton, J. (eds.): Future Environments of North America. The Natural History Press, Garden City, N.Y., 767 p.

Ecumenical Coalition on Third World Tourism 1986. "Third World Pople and Tourism: Approaches to a Dialogue". Bangkok 145 p.

Eliot, E., 1974. Travel. CMA Journal, February 2.

Environment Canada, 1981. Federal Policy on Land Use. Ottawa.

Enzensberger, H.M., 1967. Eine Theorie Des Tourismus. F.R.G. Frankfurt, in Einzelheiten I, Suhrkamp Verlag, 212 p.

Escourrov, P., 1977. La duree de la saison Touristique en Europe. Travel Research Journal, 31-40.

Everyman's United Nations. 1968. A complete handbook of the Activities and Evolution of the United Nations during its First Twenty Years, 1945-1965. 8th ed., March 1968, United Nations, N.Y., 589 p.

Farina, J., 1961. The Social and Cultural Aspects of Recreation. Resources for Tomorrow. Conference Background Papers Vol. 2. Ottawa, Queen's Printer, 941-950.

Feige, M., 1988. Hypothesen Zur Quantitaven Entwicklung Des Fremdenverkehrs In Der Bundesrepublik Deutschland in Den Nachsten 25 Jahren. Review de Tourisme, No. 1, 16-21.

Field, E., 1964. Some Aspects of traveling in Stuart England. Traffic Quarterly, Vol. XVIII, July, 406-420.

Fitzgibbon, J., 1979. Segmentation analysis underused, overused, abused. TTRA Annual Conference Proceedings: 43-47.

Fourastié, J., 1960. The Causes of Wealth. Translated-edited by T. Caplow. The Free Press of Glencoe, Ill., 246 p.

Fourastié, J., 1965. Les 40000 heurs. Paris, Laffont, 246 p.

Galbraith, J.K., 1967. The new industrial state. Boston, Houghton Mifflin, 427 p.

Galton, F., ed. 1864. Vacation Tourists and Notes of Travel in 1862-3. London, MacMillan, 524 p.

Gee, C., 1981. Resort Development and Management The Educational Institute of the AHMA East Lansing, Michigan.

Gerasimov, I.P., et al. 1970. Current geographical problems in recreational planning. Soviet Geography: Review and Translation, Vol. 11, 3:189-198.

Ginier, J., 1969. Les Touristes e trangers en France pendant l'ete. Ed. Genin, M.T., Paris, 643 p.

Green, F., 1978. Recreation Vehicles. A perspective. Annals of Tourism Research Vol. 5, 4:429-439.

Graff, J., 1987. Parks and attraction: thrive or survive. TTRA Conference: 9-16.

Griffiths, J., 1976 Climate and the environment. the atmospheric impact on man. London: P. Elek.148 p.
Grunthal, A., 1934. Probleme der Fremdenverkehrs - geographie. Schriftreihe des Forschungsinstituts für den Fremdenverkehr, F.R.G. Berlin, Heft 9.
Gunn, C. 1971. A New Approach to Coastal Tourism Development 12 p. Mimeographed TTRA Conference.
Hahn, H., 1971. Psychologische Aspekte des Erholungsverkehrs. in: Ruppert, K., Der Tourismus und seine Perspektiven für Südossteuropa. 6:15-31 Munchen.
Hart, J., 1966. A Systems Approach to Park Planning. Norges, Switzerland, 118 p.
Hawkins, D.E. et al. eds. 1980. Tourism Planning and Development Issues.. Washington, D.C., George Washington University (Department of Human Kinetics and Leisure Studies, 817), 473 p.
Heisler, G., 1977. Trees modify metropolitan climate and noise. Journal of Arboculture, 11:201-207.
Hendee, J.C. 1974. "A scientist's views on some current wilderness management issues.." Western Wildlands, Spring.
Hoole, A., 1978. Public Participation in Park Planning: The Riding Mountain Case. Canadian Geographer Vol. 22, 1:41-50.
Huberman, L., 1961. Man's Worldly Goods. 3rd ed., 349 p.
Hudman, L.E., 1980. Tourism - A Shrinking World. Ohio: Columbus, Grid Publishing Inc., 290 p.
Hunziker, W. 1959. Betriebswirtschaftslehre des Fremdenverkehrs. Gurtenverlag, Bern, 430 p.
Hunziker, W., Krapf, K., 1942. Grundriss der allgemeinen Fremdenverkehrslehre. Zürich.
Huth, H., 1957. Nature and the American Three Centuries of Changing Attitudes. Berkeley, University of California Press, 250 p.
Kahn H., 1967. "The Year 2000". Macmillan N.Y., 431 p.
Kahn, H. et al. 1976. The next 200 years: A scenario for America and the World. Hudson Institute, New York, Morrow, 241 p.
Kaplan, M., 1975. Leisure - Theory and Policy. John Wiley, N.Y., 444 p.
Kaplan, M., 1960. Leisure in America: A social inquiry. John Wiley, N.Y., 350 p.
Kaplan, M., 1963. Reflections on Issues of Leisure. Dynamic Aspects of Consumer Behavior VI. Foundation for Research on Human Behavior, Ann Arbor, Michigan.

Karp, G., 1981. Saving the Alps: Development Madness Threatens a Delicate Ecology. World Press Review, 28 (11).

Kaspar, C., 1976. Reflexions economique sur les stations thermales et climatique. Revue de Tourisme, 2:2-6.

Kiemstedt, H., 1967. Zur Bewertung der Langschaft für die Erholung. Beiträge zur Landespflege, Sonderheft 1, Stuttgart, 151 p.

Klimm, L., 1954. Empty areas of the Northeastern United States. Geographical Review, 3:324-345.

Kostrowicki, A., 1970. Zastosowanie metod geobotanicznych w ocenie przydatnosci terenu dla potrzeb rekreacji i wypoczynku Przeglad geograficzny XLII, 4:631-645.

Knebel, H.J., 1960. Soziologische Strukturwandlungen im modernen Tourismus. F.Enke Verlag, Stuttgart, 178 p.

Larrabee, E., Meyersohn, R., eds. 1958. Mass Leisure. Glencoe Illinois, Free Press, 429 p.

Lehane, P. 1974. Life and Death of Spas. Built-Environment, Aug. p. 406-408.

Leisure Systems Inc., 1976. Tourism in the United States-potentials and problems.

Lewis, R., Beggs, T., 1982. The Interface between National Tourism and the Hotel Industry in Promoting a Destination Area in Off-Season: The Bermuda Case. Journal of Travel Research, Vol. 20, 4:35-38.

Linder, S. 1970. The Harried Leisure Class. Columbia University Press, N.Y. and London, 182 p.

Little, A.D., 1978. National Tourism Policy Study. U.S. Government Printing Office, Washington, D.C.

Little, A.D., (ed.) 1967. Tourism and Recreation. U.S. Department of Commerce, 301 p.

Lundberg, D., 1972. The Tourist Business. Calmers Publ. Company, Chicago, 276 p.

Lundgren, J., 1974. On access to recreational lands in dynamic metropolitan hinterlands. Geographical Survey, 3.

McBoyle, G., Sommerville E. (eds) 1976 Canada's Natural Environment. Toronto, Methuen p. 264. Ch. 6, 109 p.

Magary, S., 1987. A perspective on US. airline industry, TTRA Conference.

Marsh, J., 1070. Marine Parks. In: Foster H. and Sewell (eds) Resources, Recreation and Research. University of Victoria, B.C.

Mazanec, J. 1981. The Tourism Leisure Ratio: Anticipating the Limits to Growth AIEST meeting. Institut für Fremdenverkehr der Wirtschaftsuniversität Wien.

Mazur, D., 1975. A Method of Land Analysis and Classification for the Canadian Shield Portion of Manitoba. M.A. Thesis, Dept. of Geography, University of Manitoba.

Medlik, S., 1969. Economic Importance of Tourism. Revue de Tourism. Numen Special: 38-41.

Meinke, H., 1968. Tourimus and Wirtschaftliche Entwicklung. Gottingen, Vanolenhoeck & Ruprecht, p. 104.

Mercer, D., 1973. The concept of recreational need. Journal of Leisure Research 5:37-50.

Mieczkowski, S., 1983. "The feasibility and necessity of climatic classiciation for purposes of tourism" International Geographical Union, Commission for Tourism and Leisure Symposium, Lodz:315-331.

Mieczkowski, Z. 1985. "The Trouism Climatic Index: A method of evaluating climates for tourism" Canadian Geographer, 3:220-233.

Mieczkowski, Z., 1981. Some Notes on the Geography of Tourism: A Comment. Canadian Geographer (15), 2:186-191.

Miller, N., Robinson, D., 1963. The Leisure Age. Its Challenge to Recreation. Wordsworth Publ. Comp., Belmont Calif., p. 497.

Moynaman, B., 1983. Fool's Paradise. Pan Books, London.

Müller, P., 1983. Nenue Entwicklungstendenzen im Kongress-Tourismus-Chancen für die Schweiz? If, Forschungs Institut für Fremdenverkehr, Univ. Bern: 12-48.

Nash, R., 1967. Wilderness and the American Mind. Yale Univ. Press, p. 256.

National Academy of Sciences, 1969. A Program for Outdoor Recreation Research. U.S Dept. of Interior, Bureau of Outdoor Recreation, Washington, p. 90

National Marine Parks Policy 1986. Environment Canada.

Nelson, J.G., Slale, R.C., (eds) 1968. The Canadian National Parks: today and tomorrow ..Proceedings of a conference organized by National and Provincial Park Association of Canada and the Univ. of Clagary, Oct. 9-15, 1968. Calgary, National and Provincial Parks Association of Canada

Neumeyer, M.H., Martin H., 1958. Leisure and Recreation. Ronald Press, N.Y., p. 473.

Ogilvie, F.W., 1933 The tourist movement. An economic study. London, P.S. King & Son Ltd., 228 p.

Ontario Dept. of Education, 1970. Report of the Study Committee on Recreation Services in Ontario.

Organization for Economic Co-operation and Development, 1973. Tourism in O.E.C.D. Member Countries. Paris, OECD annual publication.

Outdoor Recreation Resources Review Commission, 1962. The Future of Outdoor Recreation in Metropolitan Regions in the United States. U.S. Printing Office, Washington D.C.

Outdoor Recreation Resources Review Commission 1962a. Shoreline Recreation Resources of the United States. U.S. Printing Office, Washington, D.C., Report No. 4, p. 156.

Outdoor Recreation Resources Review Commission, 1962b. Water for Recreation - Values and Opportunities. U.S. Printing Office, Washington, D.C., Report No. 10, p.73.

Patmore, J.A., 1970. Land and Leisure. Newton Abbot, Devon, England, 320 p.

Patmore, J.A., 1973. Recreation. In Evaluating the Human Environment. Essays in Applied Geography. Dawson J.A. Doornkamp J.C. (eds). Edward Arnold Publishers Ltd., 288 p.

Pearce, D.G. 1978. Demographic Variations in International Travel. Review de Tourisme, No. 1:4-9.

Pearce, D.G., 1978a. Form and Function in French Resorts. Annals of Tourism Research, No. 1:142-156.

Perloff, H.S., Wingo, L., 1962. Urban Growth and the Planning of Outdoor Recreation. In: Trends in American Living and Outdoor Recreation. ORRRC Study Report #22:81-100.

Pfister, B., 1974. Die strukturellen Wandlungen des Fermdenverkehrs und der Fremdenverkehrswirtschaft. Revue de Tourisme, 3, 108-114.

Plog, S.C., 1987. Understanding Psychographics in Tourism Research. In: Ritchie J.R.B., Goeldner C.R., (eds), Travel, Tourism, and Hospitality Research. John Wiley & Sons, New York:203-21

Resources for Tomorrow 1961 Conference Background Papers Vol. 3., Ottawa, Queen's Printer.

Rocznick Statystyczny, Annual publication GUS: Warszawa

Rodgers, B., 1971. The sociological trends towards greater leisure, car-ownership and car-use. Ekistics 184, March: 216-219.

Rutazibwa, G., 1974. Les transports et le tourism international. Revue de Tourisme No. 3:93-99, No. 4:135-139.

Schmidhauser, H.P., 1971. Diskussionsbeitrag zur neuen Fremdenverkehrsdefinition. Revue de Tourisme, S:51-54.

Sellin, T., (ed) 1957. Recreation in the Age of Automation, The Annals of the American Academy of Political and Social Science, Vol. 313, Philadelphia, 208 p.

Sessa, A., 1971. Pour une nouvelle notion de tourism. Revue de Tourisme, No. 1:5-15.

Sheldon, P., Mak, J., 1987. The Demand for Package Tours: A Mode Choice Model. Journal of Travel Research, Vol. 25, 3, 13-17.

Sigaux,G., 1965. Histoire du Tourisme, Klausfelder, Vevey, Switzerland, 105 p.

Stadel Ch., 1982 "The Alps: Mountains in Transformation. Focus 32, 3:1-16.

Stansfield, C., 1978. Atlantic City and the resort cycle. Annals of Tourism Research, Vol. 5, 2:238-251.

Stansfield, C., 1969. Recreatioal Land Use Patterns within an American Seaside Resort. The Tourist Review 4:128-136.

Statistics Canada, 1978. Service Bulletin. Culture Statistics. Vol. 1.

Tralaeva, U., et al, 1979. Present State of and perspectives of research in development and improvement of criteria for epidemic safety of recreational marine workers. Proceedings, Pacific Science Congress, Committee M. Khabarovsk, U.S.S.R.

Taylor, G., et al, 1969. Predicting Recreation Demand Technical Report No. 7, Recreation Research and Planning Unit, Dept. of Park and Recreation Resources College of Agriculture and Natural Resources, Michigan State University 50 p.

The Cruise Business, 1982. International Tourism Quarterly, Special Report No. 43, 3:68-102.

Tourism Canada 1988. Market Assessment: British Pleasure Travellers. Ottawa.

Tourism Canada 1989A. Market Assessment: French Pleasure Travellers. Ottawa.

Tourism Canada 1989B. Market Assessment: German Pleasure Travellers. Ottawa.

Tourism Canda. Tourism Intelligence Bulletin, monthly.

Tourism Policy and International Tourism in OECD Member Countries OECD Paris, published annually.

Tourismus 1980, 1968 Fremdenverkehr zwischen Gestern und Morgen. Studienkreis für Tourismus E.V., Starnberg, 76 p.

Tverdokhlebov, I., Mironenko,N. 1979. Sistematizatsya Omovnykh Ponyatii Rekreatsyonnoy Geografii. Ekonomicheskaya geografia (Kiev) Vol 27:41-48.

Ungarisches Fremdenverkehrsamt: 1967. Fremdenverkehrs-kolloquium, Budapest, 22-24 Nov. 1966, 258 p.

The United Nations Conference of International Travel and Tourism, 1963. Resolution and Recommendations,60 p.

United Nations Inter-regional Seminar on Tourism Development, 1968. Berne, 21 Oct. - 2 Nov., 477 p.

Urlaubsreisen, 1968. Divo Institute, Frankfurt.

Vaske, J., et al 1982. Differences in Reported Satisfaction Ratings by Consumptive and Non-consumptive Recreations. Journal of Leisure Research, 14,3:195-206.

Wagner, F., 1970. Die Urlaubswelt von Morgen, Eugen Diederichs Verhag, Duesseldorf, 222 p.

Wall, G., 1983. Cycles of Capacity: a contradiction in terms? Annals of Tourism Research, Vol 10, 2:268-270.

Ward, S., Robertson, T., 1973. Consumer Behavior: Theoretical Sources. Prentice-Hall, Englewood Cliffs, N.J.

Waters, S. Travel Industry World Yearbook. The Big Picture published annually, Child and Waters Inc., N.Y.

Webster's New Collegiate Dictionary 1976. G. & C. Meriam, Springfield, Mass.

Weiss, C., 1974. Tourist Statistics Re-examined: the older traveler as a case in point. Journal of Travel Research, Vol. 13, 1:1-4.

Western Council for Travel Research, 1969. Proceedings Eleventh Annual Conference. Salt Lake City.

Williams, W.R., 1958. Recreation Places. Reinhold Publ. Corp., N.Y., 302 p.

Wolfe, R.I., 1952. Leisure: The Element of Choice. Journal of Human Ecology. Vol. 2, No. 6.

Woodside, A., 1987. Profiting the Heavy Traveler Segment. Journal of Travel Research, Vol. 25, 4:9-14.

World Centennial National Parks 1872-1872 1974. Dept. of Lands and Survey, Wellington, N.Z.

Young, G., 1973. Tourism: Blessing or Blight? Harmondsworth, Penquin Books.

Zedek, G., et al 1967. Konjunkturelle und strukturelle Aspekte des österreichischen Fremdenverkehrs. Wien.

Zedek, G., 1968. Die Soziologie im Fremdenverkehr und ihre Nutzanwendung. Revue de Tourisme, No. 3, 86-93.

Zimmermanm, E.S., 1951. World Resources and Industries New York, Harper.

INDEX

A
Abbey J., 168, 339, 344, 345, 346
Abu Simbel, 298
Acapulco, 146, 325
adiabatic fog, 253
airport hotels ,303
Algarve, 323
Allocentric, 169
All Terrain Vehicles (ATV), 116-117
Amtrak, 105-106
Ankor Wat, 294
anthropocentric approach, 261
Apex, 129
Arctic National Wildlife Refuge, 274
Aristotle, 8, 45
artificial (contrived attractions), 295-296
Asian Highway, 111
aspect (orientation) 224, 263
Association Internationale d'Experts Scientifique du Tourisme (AIEST), 25-26
asymetrical conflicts, 216-217
Atlantic City, 65, 324, 325, 328
Autobahnen, 111
autostrada del sole, 111
Auto-Train, 103, 105
average working week, 82, 84
Avicenna, 49
Azar V., 14

B
Baalbek, 194
"baby hotels", 159, 304
Baden-Baden, 56, 60, 320
Baedecker K., 62
Baiae, 47
Baker P., 273
Babi, 325
balneology, 56, 228, 319
Balzac H., 55
BAM railroad, 106
Banding, 334
Banff N.P., 66, 67, 259, 278, 320-321
Bank of Montreal, Business Review, 78
Baretje R. Defert P., 187
Barnet E., 78, 177
Bartos R., 162
Basedow J., 56
Bath, 47, 56, 320
Bermuda, 197
Bell D., 87
Bella, L., 268
Berger J., 289
berm, 244
Bernay, E., 170-171
Bernecker P., 58
Bialowieza N.P., 269
Biarritz, 61, 325
biocentric approach, 261
bioclimatology, 232
biometeorology, 232
"black market", 166
Blackpool, 61, 322, 325
Blazey M., 162
Bleile G., 185
Blue Train, 106
Bois de Boulogne, 61
bluff shoreline (falaise), 246
boatels, 303-304
Boerijan P., 179
Borobodur, 194, 298
Boswah Megalopolis, 324
Boston Commons, 263
Brachon P., 60
"bridge" (il ponte), 98
Brighton, 56, 61, 322, 325
Brockman C., 15
Bronowski J., 176
Brooks L. 151
Brunsden D., 255
Bryant B. Morrison A., 177

Bryant W., 55, 67
Brzezinski Z., 87
Burck G., 85-86
Burton T., 15, 151-153
Burkart A., Medlik S., 46
business travel, 163, 194, 199-200
Butler R., 335-337
Byron, 55

C

Cabeli V., 226
Calgary Winter Olympics, 222
Cameron Highlands, 334
Canadian National Parks Service, 276-277
Canadian Pacific Railway (CPR), 266, 302
Canadian Parks and Wilderness Society 276
Canadian Tourist Association, 31
Canal du Midi, 122
Cancun, 140, 146, 325, 328
Caneel Bay resort, 326
Capetown, 315
Capri, 61
capsule (cubicle) hotels 222, 305, 312
Cask M., 177
Casson L., 47
catabatic winds, 253
Catskills, 334
Central Business District (DBD), 317, 327, 328
Central Park (NYC), 66, 263
Chadee D., Mieczkowski Z., 157
Chadwick, R., 24-25, 29
Chamonix, 332
Chartres, 315
Chib S., 179
Chichen-Itza, 294
Christaller W., 334-335
Chubb M., Chubb H., 12, 14, 19, 264
Church F., 67
circular causation, 337
cité lacustre, 331
CITES treaty, 237
city parks, 263
Clawson M., 11, 16, 40, 64, 77, 100, 101, 107, 211
Clawson M., Knetch, J., 15, 16, 17, 18, 100, 146-147, 206, 108-210, 214-215
climatic chamber (climatron), 231
climatotherapy, 232
Club Mediteranée (Club Med), 319, 326
Cole D., Dietz J., 333
Collingwood, R., 10
Colonial Williamsburg, 264
Colva Beach, Goa, 245
comparative needs, 150
compressed work week, 92, 94-96
Computer Reservation Systems (CRS), 136
concentrated recreation, 18
Concorde, 127-128
consumptive use of wildlife, 236-240
continuum of environmental modification, 258, 263, 265, 281
convention centers, 341
Costa Brava, 245, 248
convention hotels, 340-341
Convention on International Trade in Endangered Species (CITES), 237
Cook T., 59-60
Cosgrove I., Jackson R., 13
Council of Europe, 256
countryside recreation, 18
Courchevel, 332
Cropsey J., 67
cruises, 118-122
Cunard S., 59
Czestochowa, 318

D

Dacharry M., 124
Datzer R., 160

Davos, 61, 332, 333
Davy T., Rowe L., 10
Dead Sea, 322
Darjeeling, 334
debt-for-nature swaps, 234, 290
deferred (pent up) demand, 153
Defert P., 242, 247-248
DeGrazia S., 9
Delphi, 46
demand (classification), 151-153
deregulation of airlines, 129-133
destination (definition), 145
Developed Countries (DCs), xii
Dictionary of Sociology, 16
Didot, 62
Disneyland, 264, 295
Disneyworld, 264, 295
dispersed recreation, 18, 19
distance decay function, 142
diverted demand, 152, 217
Djerba, 303
Dodone, 46
Dopolavoro, 69
Dower M., 89, 165
Drakatos C., 198
Driver B., 153-154, 180, 317
Duchet, R., 52, 53
Dulles F., 64, 65, 68, 78, 83, 118
Durden D., 216

E

Economist, 89, 104, 137, 251
Ecumenical Coalition, 347
"edge-effect", 233, 243
Edmonton mall, 295
Eliot, E., 31
Ellesmere Island, N.P., 278
emissiveness (travel intensity), 145, 166, 172, 215
Ems, 320
Engadine N.P., 269
Environment Canada, 206
Enzensberger H., 46, 47, 59, 86, 97
Epcot Center, 264
Epicurean ideas, 48
Epidauros, 46
Erasmus of Rotterdam, 52
Escourrou P., 193
Estienne C., 51
ethnic tourism, 166, 178, 296
Etna, 222
Europe highways, 110
event attractions (Hallmark events), 195, 291, 295
Everglades N.P., 272, 283
extra-urban (non-urban) recreation, 18-19
excursionists, 22-23, 38
existing (effective) demand, 151
exposure, 224
externalities, 36

F

Faakersee, 297
Farina J., 15
Fatima, 318
Federal Aviation Administration (F.A.A.), 131
Federation International du Thermalisme et du Climatisme, 321-322
Fernweh, 55
Feige M., 160
Field E., 31
Fitzgibbon J., 181
flextime, 99
Florence, 315
footloose attractions, 295-297
Fourastié J., 86, 88, 172, 211
Franklin B., 65

G

Galbraith J., 88, 148, 165, 203
Galton F., 54
gambling in resorts, 325

Gander, 135
gateway (transfer, transit) centers, 314
Gee C., 303
Genting, 325, 334
geographical approach to supply, 206
Gerasimov I., 35
giatrification, 324
Ginier J., 20
Gizeh pyramids, 294
Goa, 325
Goethe, 55
Glacier N.P., 268, 283
Golden Gate Park, 66
Graff J., 92
Grand Canyon N.P., 272, 286, 287
Grand Tour, 52, 55, 57-58
gravity models, 166
Great Barrier Reef N.P., 271
Great Smokey Mountains N.P., 272
Green F., 116
Griffiths J., 229
Grunthal A., 314
Guanacaste N.P., 289
Gunn C., 243, 329-330

H

Hahn H., 177
Hailey A., 166, 174
Harris L, 92
Harrison W., 52
Hart J. 30
Hawaii Volcanoes N.P., 272
Hawkins D., 58, 61, 68
Heissler G., 235
Hendee J., 261
Heritage USA, 264
Herodotos, 46
High Speed Train (TGV) 103-104
Hillary E., 71
hill stations, 334
hinterland, 327-328
Homburg, 320
Hoole A., 283
horizontal integration, 298-299
Hot Springs, 65
Huberman L., 63
Hudman L., 29, 30
Hugo V., 55
Humbolt A., 54
Hunziger W., Kraft K., 25-27Huth H., 54, 63, 65, 66
Huxley A., 9
Hyde Park, 49, 61
Hydrotherapy, 56, 228, 320

I

IATA, 129
incentive travel, 200, 338
Inclusive Tour Charters (ITC), 129, 195
indoor recreation, 18
induced demand, 152
Industrial Revolution, i, 58
infrastructure (basic, tourism), 204, 291-292
interlining, 130
intermodality, 112, 113, 115, 122, 128, 139
International Air Transport Association (IATA), 129, 195,
International Association of Convention and Visitor Bureaus, 343
International Association of Criptozoology, 242
International Civil Aviation Administration (ICAA), 131
International Commission on National Parks (NCNP), 267
International Federation of Camping and Caravaning (F.I.C.C.), 313
International Herald Tribune, 46
International Office of Social Tourism (B.I.T.S.), 80, 313

International Union for the Conservation of Nature and Natural Resources (IUCN), 237, 267
International Union of Official Travel Organization, 21, 23
intervening opportunity, 108, 142, 146
intervening variables, 180-181
institutionalized seasonality, 193-194
isochrones, 101
Itaipu, 294
ITC (Inclusive Group Charters), 129
itinerant tourism, 321

J

Jahn F., 56
Jasper N.P., 278, 332
Jerba, 325
Jeune Afrique, 250

K

Kahn H., 1, 87, 88, 211
Kampgrounds of America (KOA), 311
Kaplan M., 10, 14, 65, 83
Karlsbad (Karlovy Vary), 56, 320
Karp G., 217
Kaspar C., 322
Kiemstedt H., 233
Kislovodsk, 320
Kizil, 296
Klimm L., 216
Kluane N.P., 259, 278
Knebel H., 83
Kostrowicki A., 234
Kovalam, 325
Kraft Durch Freude, 69
Kultura, Paris, 69

L

LaGrande Motte, 331
Lake Havasu, 295
Lake Louise, 259
Lake Sidney Lanier, 228
Land, 269, 271
"land ethic" philosophy, 220
Languedoc-Rousillon project, 326, 331
LaPlagne, 256
Larrabee E., Meyerson R., 65, 83
Las Vegas, 324
Latent (potential) demand (needs), 150, 151-152
Lech-Zürs area, 254, 332
Lehane P., 320
Leisure Systems, 30
leisure time-money trade-offs, 87
Le Monde, 119
Leontief W., 91
Less Developed Countries (LDC) or Developing Countries, xiii
Lewis R., Beggs T., 197
lido, 249
Linder S., 88-89
Lipsius J., 52
Little D., 17, 30, 31
load factor, 124
Loch Ness, 241
Lourdes, 166, 318
Lundberg D., 32
Lundgren J., 215

M

Machu Pichu, 294
Magary S., 137
Maglev train, 104
management (master) planning (N.P.), 282-284
Manila Declaration of WTO, 23, 24, 158
Marcus Aurelius, 5
Margate, 322
Marienbad (Marianske Lazne), 56, 320

marine zone, 243
Marsh G., 67, 273
Marsh J., 271
Mazanec J., 152
Matsesta, 323
Matterhorn, 222
Mazur D., 235
McBoyle G., Sommerville E., 267, 273, 275
Megeve, 332
Meinke H., 185
Medeo skating rink, 225
Medlik S., 26
Mercer D., 149-150
Mesa Verde N.P., 272
Metroliner, 104
Mezzogiorno, 111
Miami Beach FLA, 318, 324-325
Mickiewicz A., 55
Mieczkowski Z., 37, 230, 231
Miller N., Robinson D.,8, 9, 17, 45, 64
Mineralnye Vody, 320
miniparks, 263
Minot N.D., 296
Montaigne M., 52
Monte Carlo, 56-57
Montes Albani, 47
Morea, 303
motels, 303
motor hotels, 303
Mount Everest, 222
Mount Royal, 66
Mont Saint Helen's, 222
Moynahan B., 336
Muir J., 67, 273, 259
Müller P., 338-339, 344, 345
multi-stop (itinerant) tourism (touring), 146
multiple use principle, 216-217, 264
Murray J., 62
Mürren, 253

N

Nahanni N.P., 259
Nakhodka, 106
Nash, R., 273
Nassau, 535
National Academy of Sciences, 19
National Forests (US), 264
National Marine Parks Policy, 271
national park definition, 267
National Parks, 265-288
National Parks Service (US), 272
National Wilderness Preservation System, 260
natural seasonality, 192-194
necrogeography, 221
neighborhood parks, 263
Nelson J., Scaree R., 34
Neumeyer M., 17
Newly Industralized Countries (NICs), 92
Newsweek, 92, 137, 160
New York Times, 79, 84, 102, 114, 115, 116, 119, 127, 132, 195, 224, 226, 239
Niagara, 66, 295
Nice, 56, 61, 325, 333
non-consumptive use of wildlife, 239-241
non-governmental organizations (NGOs), 93
normals, 230
normative needs, 150
North American Waterfowl Managmeent Program, 235
Northern Yukon N.P., 278
Nueva Andalucia, 326

O

occupancy rates, 190
OECD, 21, 22
Ogilvie F., 21
Olmsted F., 66

Olympia, 46
Olympic Games, 46
Olympic N.P., 272
one-stop (destination) tourism, 145-146
opportunity costs, 268, 277
Ontario Dept. of Education, 15
Orient Express, 102, 106
origin (definition), 145
Orlando area, 324
Ostende, 61, 325
Ostia, 47
outdoor recreation, 18-19
Outdoor Recreation Resources Review Commission (ORRRC), 18, 225-226, 246, 274

P

package tours, 128
Pan American Highway, 111
paradores, 297
participation, 150
Pataya, 325
Patmore J., 15, 77, 148
Pax Romana, 47
Pearce D., 162, 315, 330-331
Pegge S., 20
Perloff H., Wings L., 214
Pestalozzi H., 56
Petra, 294
Pfister B., 77
Philai (temple), 298
Phuket, 325
Piatigorsk, 56, 320
Pieniny N.P., 268
Pitsunda, 249
Platres, 334
Plog S., 169-170
Plombiers, 320
polyvalant 196, 332
Portillo, 254
POSH, 118
positioning cruises, 121
"post-industrial" society (age), 2, 87
potential demand, 151
pousadas, 297
Prevention, 91
price elasticity of demand, 128, 131, 149, 186-189
primary elasticity, 184
primary (inherent) resources, 205
primary response strategies to seasonality, 194-195
proactive attitude, 189
Protagoras, 51
Provincial Forests (Canada), 264
Provincial Parks (Canada), 264
psychocentric, 169
psychographics, 168-170
pull factors, 145
Punta del Este, 325
"pure tourists", 23
Puritanism, 63
push factors, 146, 167
Pushkin A., 55
Puteoli, 47
Pyatigorsk, 56

Q

"Queen Elizabeth 2", 118

R

radiation fog, 253
Raffles Hotel, 304
reactive attitude, 289
Recreational Busines District (RBD), 328
Recreational Vehicle (RV), 115-117, 311
regional airlines, 128, 132
regional parks, 264
religious tourism, 294
residential hotels, 304
resort cities (towns), 318
resort complex, 319

resort hotel, 318
resource-based recreation, 18, 143, 209
Resources for Tomorrow, 151
restoration ecology, 289
retiremnet communities, 334
Rhodes, 47
Rideau Canal, 122
Riding Mountain N.P., 259, 283
Rigibahn, 252
Rimini, 247
Rio de Janeiro, 315, 316
Ritchie B., 182
Ritz hotels, 60, 304
Rocznik Statystyczny, 125
Rodgers B., 109-110
Rome, 47, 50, 318
Rostow W., 87
Roosevelt T., 273
Rossija Hotel, 304
Rostow W., 87
rotel, 312
Rothenburg, 315
Rousseau J., 54, 56, 71
Rutazibwa G., 102

S

safari hunting, 238
Saint Moritz, 322
San Diego zoo, 241
San Francisco, 315
Santiago de Compostela, 50, 166
Saratoga Springs, 65, 320
Scarborough, 323
scheduled airlines, 128
Schelling F., 54
Schmidhauser H., 28-29
Scott W., 55
secondary elasticity, 184
secondary (derived) resources, 205
Sellin T., 8, 75, 77
Shakespeare, 53
Shannon, 135
Sheldon P., Mak J., 160
Shelley, 55
Shinkansen, 104
Sessa A., 26
shoulder period, 98, 143
Sierra Club, 259
Sigeaux G., 20, 47, 52, 53, 59
silviculture, 264
Simla, 334
single use principle, 265
site attractions, 290-291
Skansen, 104, 264, 295, 297
Snowmass, 332
Sochi, 323
social carrying capacity, 214
social seasonality, 193-194
social tourism, 39, 80, 158, 321
Sorrento, 61
South Moresby Island N.P., 277
space monopoly, 213-214
Spa (Belgium), 56, 320
spas (health resorts), 56, 319-322
Spiegel, 78, 97, 161, 233, 344
Stadel Ch., 332-333
standard work week, 82, 91
Stansfield, 328, 335
Statistics Canada, 85
State Parks (US), 264
Stendhal H., 20, 62
Sterne L., 54
STOL aircraft, 134
strand, 244
substitutability in recreation, 152
substitute demand, 152
Sukhumi, 323
Sustainable Development, 256
sustained yield principle, 237, 264
Svati Stephan, 297
system planning (N.P.), 280-281

T

Talaeva V., 226
Taormina, 247

Tatra N.P., 268
Taylor G., 150
technotronic era, 87
temperature inversion, 253, 256
theme parks, 263
Thoreau D., 67
tidal range, 24
Time, 89, 105, 117, 119, 121, 122, 130, 156, 305, 327
time-budget, 13
time (classification), 82
Toffler A., 5
total destination (resorts), 256, 304, 319, 333
tourism (types), 37-41
Tourism Canada, 97, 177
tourism/leisure ratio, 152
Tourism Policy and International Tourism, 121
Tourism Intelligence Bulletin, 79, 80, 86, 89, 96, 97, 105, 115, 126, 132, 156, 160, 195, 278, 306, 322, 324, 337, 340, 341, 343
Tourismus 1980, 77, 78
tourist experience (stages), 146-147
tourist product, 145
Toynbee A., 10
Train à Grande Vitess (TGV), 103-104
Trans-Canada Highway, 111
Trans-Europe Express train system, 103
Trans-Siberian railroad, 106
transit area, 145-146
Trans-Sahara road, 111
travel intensity, see emissivness
Travel Price Index, 186, 187
trunk carriers, 128
Travo N.P., 303
Turner W., 55
Tverdokhlebov I., Mironenko N., 35
"twinning" of national parks, 268

U

UNESCO, 298
Ungarisches Verkehrsamt, 8
United Nations Conference on International Travel and Tourism, 22
UN Interregional Seminar on Tourism Development, 23, 29
United Nations List of National Parks and Equivalent Reserves, 267-268
United Nations Statistical Commission, 22
Universal Declaration of Human Rights, 83
Urlaubsreisen, 222
user-oriented recreation, 18, 143, 209

V

Vail, 332
Valpariso, 248
Vaske J., 239
Venice, 315
Verblen T., 66
Verne J., 60, 62
Versailles, 315
vertical drop (denivelation), 254
vertical integration, 134, 298-299
Via Rail, 105
Vichy, 395, 320, 321
Virgin Islands N.P., 271
visitor (definition), 21-22
Visits to Friends and Relatives (VFR), 199, 296
Volcanoes National Park, 240
VTOL aircraft, 134, 252

W

Wagner F. 47
Waikiki Beach, 318
Waicoloa resort, 326-327
Wall G., 337

Ward S., Robertson T., 181
Waters S., 1, 30, 22, 112, 114, 116, 119, 120, 124, 126, 130, 163, 305, 306, 344, 345
Waterton Lakes-Glacier International Peace Park, 268
Watamu N.P., 271
Wattenmeer N.P., 271
week actually worked, 82
Weiss C., 160, 161
Western Council of Travel Research, 30
wetlands, 235
Weymouth, 323
White Sulphur Springs, 65, 320
Wiesbaden 320, 321
Williams W., 64
wind shear, 134
Winnipeg Free Press, 79, 105, 106, 180, 239, 256, 268, 297
Wolfe R., 65, 84
Wood Buffalo N.P., 278, 283, 287
Woodside A., 165
woopies, 160
work ethic, xii, 175
World Centennial National Parks, 267
World Tourism Organization (WTO), 21, 23, 158
World Travel, 23, 158, 203
World Wildlife Fund, 237

Y

Yalta, 61, 323
Yellowstone N.P., 67, 272, 273, 283
Yosemite N.P., 272
Young G., 26
yuppies, 158

Z

Zakopane, 333
Zedek G., 77, 335
Zermatt, 332
Zimmerman E., 206
zoos, 241